Einstein's Apple

Homogeneous Einstein Fields

Einstein's Apple
Homogeneous Einstein Fields

Engelbert L. Schucking • Eugene J. Surowitz

World Scientific

NEW JERSEY • LONDON • SINGAPORE • BEIJING • SHANGHAI • HONG KONG • TAIPEI • CHENNAI

Published by

World Scientific Publishing Co. Pte. Ltd.
5 Toh Tuck Link, Singapore 596224
USA office: 27 Warren Street, Suite 401-402, Hackensack, NJ 07601
UK office: 57 Shelton Street, Covent Garden, London WC2H 9HE

Library of Congress Cataloging-in-Publication Data
Schucking, E. L. (Engelbert L.), author.
Einstein's apple : homogeneous Einstein fields / by Engelbert L. Schucking (New York University, USA), Eugene J. Surowitz (New York University, USA).
pages cm
Includes bibliographical references and index.
ISBN 978-9814630078 (hardcover : alk. paper) -- ISBN 9814630071 (hardcover : alk. paper)
1. Gravitational fields. 2. Relativity (Physics) 3. Equivalence principle (Physics) I. Surowitz, Eugene J., author. II. Title.
QC178.S35 2015
530.154'23--dc23

2014040803

British Library Cataloguing-in-Publication Data
A catalogue record for this book is available from the British Library.

In-house Editor: Christopher Teo

Typeset by Stallion Press
Email: enquiries@stallionpress.com

Printed in Singapore

This work is dedicated to

Alex Harvey

who inspired our effort to put all the pieces together.

PREFACE

The role of homogeneous gravitational fields in the formulation of the equivalence principle and in the foundation of Einstein's theory of gravitation is well known. However, the original treatment of these concepts was done in terms of Newtonian gravity and for small velocities. We believe, therefore, that it is necessary to treat homogeneous fields relativistically in Einstein's theory of gravitation. In this book we discuss how this can be done for manifolds with simply transitive isometry groups and we mention possible applications.

Since the results presented here are far from complete, we are aware of the preliminary character of our investigation. What we have tried to do is to study the concept of homogeneous fields in Riemannian manifolds from different points of view in an exploratory spirit. This leads to a certain amount of repetition in the different chapters which we hope the reader will excuse. Included are some straightforward calculations that we would expect to readily find in the literature but appear to be confined to papers in obscure forms or are not in the common bibliographies and texts.

We would like to acknowledge the assistance and support that we have have received from various individuals during the gestation of this work. We would especially like to thank Alex Harvey, Friedrich Hehl, Malcolm MacCallum, István Ozsváth, and Andrzej Trautman. Jie Zhao's contributions, published and unpublished, have been gratefully appreciated. Peter Bergmann should be mentioned for his lifelong efforts to clarify the content of Einstein's Theory of Gravitation. New York University provided the facilities where most of the work was carried out.

CONTENTS

FIGURES

CHAPTER 0

"THE HAPPIEST THOUGHT OF MY LIFE"

A. The Principle of Equivalence

In a speech given in Kyoto on December 14, 1922, Albert Einstein remembered:

> "I was sitting on a chair in my patent office in Bern. Suddenly a thought struck me: If a man falls freely, he would not feel his weight. I was taken aback. The simple thought experiment made a deep impression on me. It was what led me to the theory of gravity."

This epiphany, that he once termed *"der glücklichste Gedanke meines Lebens"* [the happiest thought of my life], was an unusual vision in 1907. In the history of science it is referred to as Einstein's first principle of equivalence and we call it Einstein's Apple.

Einstein's apple was not envisioned after watching the antics of orbiting astronauts on television, sky-diving clubs did not yet exist, and platform diving was not yet a sports category of the freshly revived Olympic Games. How could this thought have struck him? Had he just been dealing with patent applications covering the safety of elevators?

Three years earlier, in 1904, the Otis Elevator Company installed in Chicago, Illinois, the first gearless traction electric elevator apparatus, that was of the direct drive type, known as the "1:1 elevator". This first modern electric elevator made its way to Europe where, on Zürich's Bahnhofstrasse and elsewhere in Switzerland, buildings went up that needed elevators. It would have been natural for Director Friedrich Haller at the Swiss Patent Office in Bern to put applications involving electro-mechanical machinery on the desk of Einstein, his expert (second class) with expertise in electromagnetism. It would not be surprising to find Einstein's signature approving (or disapproving) a patent application for elevator's in 1907.

However, only one patent application with Einstein's comments from his years as a Swiss patent examiner has survived. It does not concern elevators [Flückinger, 1974]. The comments of the patent examiners in Haller's files have not been preserved by the Swiss bureaucracy. Einstein's expert opinions on patents were destroyed eighteen years after the files were closed. We'll probably never know how Einstein got his inspiration.

In his review "The Relativity Principle and the Conclusions drawn from it" [Einstein, 1907], Einstein formulated his principle of equivalence for the first time. He wrote:

> "We consider two systems Σ_1 and Σ_2 in motion. Let Σ_1 be accelerated in the direction of its X-axis, and let γ be the (temporally constant) magnitude of that acceleration. Σ_2 shall be at rest, but it shall be located in a homogeneous gravitational field that imparts to all objects an acceleration $-\gamma$ in the direction of the X-axis."

The next sentence contains the principle of equivalence:

> "As far as we know, the physical laws with respect to Σ_1 do not differ from those with respect to Σ_2; this is based on the fact that all bodies are equally accelerated in the gravitational field."

It was this last fact that had inspired Einstein and prompted Sir Hermann Bondi, Master of Churchill College in Cambridge, England, to the observation:

> "If a bird watching physicist falls off a cliff, he doesn't worry about his binoculars, they fall with him."

Although nowhere stated in the *"Principia"*, one may assume that Isaac Newton was already familiar with the principle of equivalence. In Proposition 6 of Book 3 of his Principia [Newton, 1687], Newton describes his precise pendulum experiments with "gold, silver, lead, glass, sand, common salt, wood, water and wheat" testing equality—as we would say now—of inertial and passive gravitational mass. His treatment in Proposition 26 of Book 3 in the Principia dealing with the perturbation by the Sun of the lunar orbit around the Earth, discovered by Tycho Brahe and called the "Variation", leaves no doubt that Newton knew how to transform away a homogeneous gravitational field.

The apparent enigmatic equality of inertial and passive gravitational mass was also still a prize question at the beginning of the twentieth century. The Academy of Sciences in Göttingen, Germany, had offered the Beneke Prize in 1906 for proving this equality by experiment and theory. The Baron Roland Eötvös won three-fourths of this prize (3,400 of 4,500 Marks); only three-fourths, because he had only done the experiments and had not attempted a theoretical explanation [Runge, 1909].

The principle of equivalence was not new in Newton's theory of gravitation. New was Einstein's extension to all of physics. He wrote:

> "At our present state of experience we have thus no reason to assume that the systems Σ_1 and Σ_2 differ from each other in any respect, and in the discussion that follows, we shall therefore assume the complete physical equivalence of a gravitational field and a corresponding acceleration of the reference system."

B. Where the Principle of Equivalence Leads

This basic observation guided Einstein in the formulation of his theory of gravitation. The principle of equivalence says, roughly, that all bodies, independent of their nature, experience the same acceleration in a gravitational field of given strength. Newton checked this principle by experiments with pendulums which were later refined by F. W. Bessel.

In his famous elevator *gedanken* experiment Einstein saw that this experimental fact of Newtonian gravitation could also be described as the equivalence between motion in a homogeneous gravitational field and motion in an accelerated frame of reference. In his 1908 paper [Einstein, 1908] he used this equivalence to show that special relativity then demands the existence of a gravitational "redshift" for clocks in a gravitational field. His derivation assumes velocities small compared to c (the speed of light) and accelerations g such that $gL/c^2 << 1$, where L is a characteristic length of the reference system [Bergmann, 1976].

Einstein never gave an exact definition of a homogeneous gravitational field nor a relativistically invariant definition of an extended accelerated reference system. The formulation of the principle of equivalence remained, therefore, somewhat vague. John Synge wrote

> "The Principle of Equivalence performed the essential office of midwife at the birth of general relativity... I suggest that the midwife be now buried with appropriate honours and the facts of absolute spacetime be faced." [Synge, 1960]

After Einstein had described gravitational fields as curvature of spacetime, the equivalence principle was thought to have been incorporated into the theory by the feature that non-spinning particles, independent of their nature, describe geodesics of spacetime when subject to the forces of gravity and inertia only. This formulation, however, was far removed from the original one since the notion of the Newtonian gravitational field had now disappeared from Einstein's theory. Only the gradient of the gradient of the gravitational potential survived in Einstein's theory as certain space-time components of the Riemann tensor.

Interest in homogeneous fields has a number of roots, some quite old. Even before the advent of the theory of general relativity, the concept was introduced by Einstein in his initial explorations of the principle of equivalence [Einstein, 1911]. After the introduction of general relativity it was discussed in detail by Levi-Civita [Levi-Civita, 1917] [Schücking, 1985A].

For years the homogeneous gravitational field has been central to the study of radiation by a charged particle undergoing uniform acceleration. Also, the concept has long since been implicitly utilized in the realm of Newtonian mechanics when discussing so-called *fictitious* fields. It is characteristic of these fictitious fields that they can be transformed away. If one wants to give a precise meaning to the notion that a homogeneous gravitational field can be "transformed away", one needs to know what a homogeneous field is in Einstein's theory.

When T. Levi-Civita treated this concept in terms of general relativity in [Levi-Civita, 1917], it was as a special coordinate system in Minkowski space, now widely known as Rindler coordinates [Rindler, 1966]. Physical phenomena in frames adapted to these coordinates have been widely discussed in connection with the interpretation of Hawking radiation [Hawking, 1975] [Lee, 1986] [Unruh, 1976]. One has to point out, however, that the Levi-Civita field is by no means homogeneous in the customary sense that field strength is independent of position.

Our objective is to establish a generic formalism for various specific applications. For instance, it is well-known that, in certain areas of the study of spacetimes, the tangent space provides a first order local approximation. We are interested in the utilization of *homogeneous spaces* to provide the next higher order local approximation. Secondly, homogeneous fields provide a means of understanding and treating, in a comprehensive fashion, fictitious fields encountered in Newtonian dynamics. Thirdly, in some instances, homogeneous fields may be considered as limits of physically significant gravitational fields. Fourthly, homogeneous fields may provide a useful tool in analyzing gauge fields of the Kaluza type.

The pursuit of our objective has led us to implement a programme suggested by Élie Cartan in lectures presented at the Sorbonne during 1926–27 [Cartan, 1927]. A consequence has been to expose an unobvious relationship between homogeneous fields and torsion. The *fictitious* fields are fields of acceleration induced torsion by teleparallelism. **[See Appendix H for some of Hessenberg's initial development along this line.]**

To develop his theory of gravitation, Einstein had equated the action of gravity with the effects of acceleration:

$$\text{gravitational attraction} \equiv \text{acceleration}\,. \tag{0.B.0}$$

As his ideas developed, the geometrical idea of gravity as curvature became dominant. Subsequently it has become a paradigm in its strength on the analysis of astrophysical phenomena. The right hand side of the equivalence equation has been neglected. Early on, Max Planck had pointed out that Einstein's constructions were limited to low velocities; in fact, they are limited to local events in space and time.

A new definition for the notion of a homogeneous field is proposed here which applies to flat and homogeneous spacetimes in Einstein's theory of gravitation. It incorporates a principle of equivalence which holds for extended spacetime regions. The new definition runs as follows: Let the metric of a pseudo-Riemannian spacetime be

$$ds^2 = \left[\omega^0\right]^2 - \left[\omega^1\right]^2 - \left[\omega^2\right]^2 - \left[\omega^3\right]^2 \equiv \eta_{jk}\,\omega^j\,\omega^k \tag{0.B.1}$$

with differential forms

$$\omega^j\left(x^k\right) \tag{0.B.2}$$

where Latin indices $j, k \in 0, 1, 2, 3$. These differential forms uniquely define connection forms

$$\omega^j{}_k\left(x^l\right) \tag{0.B.3}$$

by means of the equations [Flanders, 1963] [DeWitt-Morette, 1977]

$$d\omega^j - \omega^j{}_k \wedge \omega^k = 0, \qquad \omega_{jk} + \omega_{kj} = 0, \qquad \omega_{jk} \equiv \eta_{jl}\,\omega^l{}_k\,. \tag{0.B.4}$$

The symbol "$\wedge$" denotes the skew-symmetric product of differential forms. Development of the connection forms (0.B.3) in terms of the one-forms (0.B.2) gives

$$\omega^j{}_k = g^j{}_{kl}\,\omega^l\,. \tag{0.B.5}$$

The coefficients $g^j{}_{kl}$, which specify the contribution of each basis differential one-form to the decomposition, are termed the *physical components* of the gravitational field. One now defines:

> "A spacetime allows a homogeneous gravitational field if the manifold allows a frame ω^l in which the coefficients $g^j{}_{kl}$ are all constants, independent of position in spacetime."

The coefficients are 24 independent scalar functions with respect to coordinate transformations. These scalars satisfy

$$g_{jkl} = -g_{kjl}\,. \tag{0.B.6}$$

Under a constant homogeneous Lorentz transformation of the frames,

$$\boldsymbol{\omega}^j\left(x^k\right) \longrightarrow L^j{}_l\,\boldsymbol{\omega}'^l\left(x^k\right) \tag{0.B.7}$$

where the transformation satisfies

$$L^j{}_l\,\eta_{jk}\,L^k{}_m = \eta_{lm} \tag{0.B.8}$$

and in which

$$L^j{}_l = constant\,, \tag{0.B.9}$$

the coefficients $g^j{}_{kl}$ transform as components of a third rank tensor. Such homogeneous fields exist if and only if the spacetime metric is invariant under a simply transitive group of isometries. This includes Minkowski-space. Examples of such homogeneous fields in Minkowski-space will be studied.

We first discuss the general formalism and apply it to a rigidly rotating system followed by specialization to two-dimensional spaces. We then discuss three-dimensional spaces for which the standard approach is by means of the well-understood Bianchi symmetries [Ellis, 1969]. Though this approach is inadequate for the study of higher dimensional cases some of the concepts and techniques employed can be utilized.

We focus on the *connection coefficients* as the essential descriptive elements of the various possible spaces and their role as *field strengths*. The viewpoint here is that because the Riemann tensor is quadratic in the connection coefficients the latter may be considered *square roots* of the components of the former. They are, from both physical and mathematical points of view, simpler quantities and their study should lead to a better understanding of the space. One can tell more about certain aspects of the dynamical behavior of a spacetime by examining the geodesics, that is, connection coefficients, than by studying the geodesic deviation, that is, components of the Riemann tensor.

Among the more interesting areas of the investigation are flat spaces which *a fortiori* are homogeneous. The existence of such non-vanishing field strengths in Minkowski space is extremely interesting—not least because this lack of uniqueness is not yet understood. It indicates the existence of homogeneous *gauge* transformations by which source-free fictitious gravitational fields can be transformed away. One has in this way a coordinate independent description of the homogeneous gravitational fields which entered into the early formulation of Einstein's theory of gravitation.

A further study of such fields appears useful for several reasons. Firstly, an understanding of the reasons for the lack of uniqueness would facilitate the use of fictitious fields on a systematic basis. Secondly, the existence of non-vanishing field strengths in the case of Minkowski space suggests that the traditional usage of rotating frames in Newtonian mechanics for studying Coriolis and centrifugal forces might be extended to special relativistic mechanics. Thirdly, and possibly the most interesting, the techniques being developed might help interpret gauge theories of the Kaluza type. Such theories may be set in a five-dimensional Minkowski space. The expectation is that one might be able to show that constant electric and magnetic fields may be transformed away in a manner similar to homogeneous gravitational fields in the lower dimensional case. The analog of the Einstein elevator becomes here the "Kaluza elevator" that is built from matter of the same e/m ratio as that of the particles one studies in it.

It will be shown that this procedure can also give homogeneous fields defined as above in Einstein's theory [Geroch, 1969] [Schücking, 1985A]. Since gravitational fields induced by masses and inertial fields induced by accelerations are held to be physically equivalent, we call both "*Einstein fields*".

Before we go on with the discussion, we have to say more about the reference system.

C. The Reference System

Einstein's reference system was based on identical clocks and rigid bodies. Through the work of Gustav Herglotz, Max Born, and Max von Laue, it was soon realized that rigid bodies do not form suitable reference systems. The ideas of Hermann Minkowski and Henri Poincaré allow us to describe the reference systems of special relativity more clearly. The Minkowski spacetime with a metric given by the line element

$$ds^2 = \eta_{\mu\nu}\, dx^\mu dx^\nu \,, \qquad \mu, \nu \in 0, 1, 2, 3 \,, \tag{0.C.1}$$

with

$$\eta_{\mu\nu} = \begin{pmatrix} 1 & 0 & 0 & 0 \\ 0 & -1 & 0 & 0 \\ 0 & 0 & -1 & 0 \\ 0 & 0 & 0 & -1 \end{pmatrix} \tag{0.C.2}$$

has a set of distinguished coordinates x^μ. Under a Poincaré transformation

$$x'^\mu = \Lambda^\mu{}_\nu x^\nu + a^\mu \,, \qquad a^\mu = constant \tag{0.C.3}$$

with constant $\Lambda^{\mu}{}_{\nu}$ and

$$\Lambda^{\mu}{}_{\nu}\, \eta_{\mu\rho}\, \Lambda^{\rho}{}_{\sigma} \;=\; \eta_{\nu\sigma} \tag{0.C.4}$$

the x'^{μ} again form such a set of distinguished coordinates. We call them linear orthonormal coordinates.

To keep physics and mathematics clearly defined, we introduce four constant orthonormal vector fields $\mathbf{e}_{\mu}$ such that

$$\mathbf{e}_{\mu} \;\equiv\; \frac{\partial}{\partial x^{\mu}} \tag{0.C.5}$$

for a set of distinguished coordinates x^{μ}. The orthonormal vectors along the coordinate axes just introduced by the operators $\partial/\partial x^{\nu}$ are subject to

$$dx^{\mu}\left(\frac{\partial}{\partial x^{\nu}}\right) \;=\; \delta^{\mu}{}_{\nu}\,. \tag{0.C.6}$$

The invariant characterization of a vector tangent to a manifold as a directional differentiation operator on the functions living on this manifold goes back to Sophus Lie. He used that concept in his "*Theory of Transformation Groups*", where vector fields became "infinitesimal transformations". Defining differentials like dx^{μ} as the duals of vectors like $\partial/\partial x^{\mu}$ emerged from the work of Élie Cartan and its interpretation by Erich Kaehler and Nicholas Bourbaki.

The vectors in (0.C.6) are all parallel to each other and defined on all of Minkowski spacetime. The components $V^{\nu}(x)$ of any vector field $\mathbf{V}$

$$\mathbf{V} \;=\; V^{\nu}(x)\left(\frac{\partial}{\partial x^{\nu}}\right), \tag{0.C.7}$$

also known as the "physical" components, have a metrical meaning, that is to say, their numerical values are results of physical measurements in terms of meters or seconds. If one introduces new coordinates y^{α} that are non-linear and/or non-orthonormal functions of the distinguished coordinates x^{ν}

$$y^{\alpha} \;=\; f^{\alpha}(x^{\nu})\,, \qquad \det\left[\frac{\partial y^{\alpha}}{\partial x^{\nu}}\right] \neq 0\,, \tag{0.C.8}$$

the basis vector fields $\partial/\partial y^{\alpha}$ will no longer be orthonormal and the components $V^{\alpha}(y)$ of

$$\mathbf{V} \;=\; V^{\alpha}(y)\left(\frac{\partial}{\partial y^{\alpha}}\right) \tag{0.C.9}$$

will no longer be physical components. The metric tensor $g_{\alpha\beta}$ will no longer be $\eta_{\alpha\beta}$, but instead we have

$$ds^2 = g_{\alpha\beta}(y) \frac{\partial y^\alpha}{\partial x^\mu} \frac{\partial y^\beta}{\partial x^\nu} dx^\mu \, dx^\nu = \eta_{\mu\nu} \, dx^\mu \, dx^\nu \, . \tag{0.C.10}$$

If we take the mathematician's point of view that coordinates are an arbitrary means for naming events in Minkowski spacetime, we cannot assign a direct physical meaning to the components of vector fields given in those coordinates either. However, this was not a point of view of taken by Einstein originally and became the root of misunderstandings.

If we want to know the physical components $V^\mu(x)$ in arbitrary coordinates $y^\alpha(x^\mu)$, we only have to remember the chain rule to see that

$$V^\mu(x) \frac{\partial}{\partial x^\mu} = V^\alpha(y) \frac{\partial}{\partial y^\alpha} = V^\mu(x) \frac{\partial y^\alpha(x)}{\partial x^\mu} \frac{\partial}{\partial y^\alpha} \, . \tag{0.C.11}$$

This gives the relation

$$V^\alpha(y) = V^\mu(x) \frac{\partial y^\alpha(x)}{\partial x^\mu} \tag{0.C.12}$$

for the transformation of vector components from one coordinate system to another.

The use of arbitrary coordinates in Minkowski spacetime is just a matter of convenience for adapting the coordinates to the symmetry of a given situation. We should now stress that the identification of the constant vector field with the $\mathbf{e}_\mu$, as defined in (0.C.10), is invariant under the Poincaré group. The translations evidently do nothing and constant Lorentz transformations give

$$\mathbf{e}_\nu = \mathbf{e}'_\mu \, \Lambda^\mu{}_\nu \qquad \Longleftrightarrow \qquad \frac{\partial}{\partial x^\nu} = \frac{\partial}{\partial x'^\mu} \Lambda^\mu{}_\nu \, . \tag{0.C.13}$$

We call the set of four constant orthonormal vectors, $\mathbf{e}_\nu$, in each event of Minkowski spacetime a "frame".

In the nineteenth century physicists were already using more general frames in 3-dimensional space. For instance, in problems with axial or spherical symmetry it was useful to let the frame vectors $\mathbf{i}, \mathbf{j}, \mathbf{k}$ be tangent to the orthogonal lines of constant coordinate pairs. In this way one could define, for example, the radial component of an electrical field strength. With the advent of Minkowski's spacetime in 1908, such adaptable orthonormal frames also appeared in four dimensions.

We shall define our reference system abstractly as an orthonormal frame in every point of the spacetime manifold. Nowadays, one calls this a section of the frame bundle.

D. The Physical Principle of Equivalence

We now return to the question: "What is the physical meaning of Einstein's first principle of equivalence?"

In his monograph "Relativity, The General Theory" John Synge confesses in his introduction

> " . . . I have never been able to understand this Principle."

and goes on to write:

> "Does it mean that the effects of a gravitational field are indistinguishable from the effects of an observer's acceleration? If so, it is false. In Einstein's theory, either there is a gravitational field or there is none, according as the Riemann tensor does or does not vanish. This is an absolute property; it has nothing to do with any observer's worldline. Space-time is either flat or curved, and in several places of the book I have been at considerable pains to separate truly gravitational effects due to curvature of space-time from those due to curvature of the observer's worldline (in most ordinary cases the latter predominate). The Principle of Equivalence performed the essential office of midwife at the birth of general relativity, but, as Einstein remarked, the infant would never have got beyond its long-clothes had it not been for Minkowski's concept. I suggest that the midwife be now buried with appropriate honours and the facts of absolute space-time faced." [Synge, 1960]

Are we beginning a chapter of "forensic physics" if we investigate the corpse of a principle? It is easy to agree with Synge. If one admits a metric, one also buys into the Levi-Civita connection and its Riemann tensor and this connection makes itself felt. But that was not all. From the beginning of relativity, already in the special theory, there was the question of the reference body. A reference body can be defined mathematically through a section of the frame bundle of the spacetime manifold, with Fermi's construction being an approximation for physics in some cases. So, it is the mathematical reference body, the generalization of the constant parallel orthonormal 4-vectors in Minkowski space-time that now show teleparallelism and torsion. Mathematically, we can formulate our principle as

$$\boldsymbol{\Theta}^{\mu} = d\boldsymbol{\omega}^{\mu} = \boldsymbol{\omega}^{\mu}{}_{\nu} \wedge \boldsymbol{\omega}^{\nu} . \qquad \text{(0.D.1)}$$

Its meaning is that we enlarge the set of admissible frames for the description of physical phenomena in spacetime.

Einstein tried to understand gravitation through acceleration and that made it necessary to introduce accelerated frames, that is, frames with torsion.

Although Gregorio Ricci-Curbastro had already introduced frames into manifolds in 1895, their use appeared optional until the advent of Paul Dirac's equation for spin 1/2 particles. Most physicists approached the interpretation of general relativity from the particle point of view. But fields are more important than particles—at least quantum fields. Fermi was an exception. In his study of Fermi transport, he worried about the description of an electromagnetic field in an accelerated reference system. How would the energy density be affected by the acceleration? This raises the general question of how fields are affected by torsion. The prime example for this is the Unruh effect where the field is the vacuum [Unruh, 1976]. It is clear that a mathematical section of the frame bundle does not give rise to physical effects like energy densities, etcetera. But as soon as one attaches physical objects to the frames, one has reference bodies and for the Unruh effect [Wald, 1994], what one calls "particle detectors". But, we have already seen that where light is emitted by a source at rest and absorbed by a receiver also at rest, we get redshifts in accelerated systems as measured by Pound and Rebka.

E. Newton, Mach, and Einstein

Einstein's apple is, like Newton's apple, a seminal thought of great penetrating power. It has been claimed that the story of Newton's apple was apocryphal; but a book containing the recollections of William Stukeley [Stukeley, 1936] appears to confirm it. Stukeley, a doctor, from Lincolnshire like Newton, became a close friend to Newton in Sir Isaac's last years. Stukeley describes the Summer evening when Newton, then in his eighties, recalled his thoughts from sixty years before:

> "After dinner, the weather being warm, we went into the garden and drank tea, under the shade of some apple trees, only he and myself. Amidst other discourse, he told me, he was just in the same situation, as when formerly, the notion of gravitation came into his mind. It was occasion'd by the fall of an apple, as he sat in a contemplative mood. Why should that apple always descend perpendicularly to the ground, thought he to himself. Why should it not go sideways or upwards, but constantly to the earth's centre? Assuredly the reason is, that the earth draws it. There must be a drawing power in matter: and the sum of the drawing power must be in the earth's centre, not in any side

> of the earth. Therefore does the apple fall perpendicularly, or towards the centre. If matter thus draws matter, it must be in proportion of its quantity. Therefore the apple draws the earth, as well as the earth draws the apple. That there is a power, like that we here call gravity, which extends itself into the universe."

What Newton had done was to look at gravity from a new frame whose origin was in the center of the Earth.

In 1907 Einstein did not show the equivalence of acceleration and gravitation described by spacetime curvature. He did not show either the equivalence of geodesics and non-geodesics or the equivalence of rotating and non-rotating systems.

What he did, we now can see more clearly, was the introduction of accelerated reference systems exhibiting torsion through distant parallelism. There were physical consequences that needed to be checked for these systems, like the constancy of the speed of light independent of acceleration, no influence of acceleration on the rate of clocks and the length of standards. As far as these assumptions have been tested, they appear to be in order.

In 1911 Einstein formulated an equivalence principle that involved relative acceleration in an attempt to introduce ideas of Ernst Mach into his theory [Einstein, 1911]. This did not prove to be a happy idea since this notion makes mathematical sense only for bodies having the same 4-velocity and, from a physical point of view, accelerations are absolute. These ideas gave the theory its name.

However, as John Stachel [Stachel, 1980] pointed out, it was the older idea of 1907 that guided him through Ehrenfest's paradox of the rotating disc to Riemannian geometry. We can now see that going from Levi-Civita's connection in Minkowski's spacetime to teleparallelism opens the door to going back from torsion to a Levi-Civita connection with curvature.

CHAPTER 1

ACCELERATED FRAMES

A. Born Motion

Born's motion stands in analogy to an observer's motion on the circle

$$x^2 + y^2 = \rho^2 \tag{1.A.1}$$

of radius ρ in the $(x\text{-}y)$-plane. If the motion is at constant speed along the circle, an observer experiences a constant acceleration in the direction opposite from the center of motion. In Born motion, we require an observer to feel constant acceleration in Minkowski spacetime.

We introduce into the $(x\text{-}t)$-plane (with $c = 1$) "Born-polar" coordinates

$$t = \rho \sinh \tau\,, \qquad x = \rho \cosh \tau\,, \tag{1.A.2}$$

and they yield the line element in the form

$$ds^2 = dt^2 - dx^2 = \rho^2\, d\tau^2 - d\rho^2\,, \tag{1.A.3}$$

since the differentials of equations (1.A.2) are related by

$$\begin{aligned} dt &= d\rho\, \sinh \tau + \rho\, \cosh \tau\, d\tau\,, \\ dx &= d\rho\, \cosh \tau + \rho\, \sinh \tau\, d\tau\,. \end{aligned} \tag{1.A.4}$$

The vector form of the line element is

$$\begin{aligned} \frac{\partial}{\partial s} \otimes \frac{\partial}{\partial s} &= \frac{1}{\rho^2} \frac{\partial}{\partial \tau} \otimes \frac{\partial}{\partial \tau} + \frac{\partial}{\partial \rho} \otimes \frac{\partial}{\partial \rho} \\ &= \mathbf{e}_0 \otimes \mathbf{e}_0 + \mathbf{e}_1 \otimes \mathbf{e}_1 \end{aligned} \tag{1.A.5}$$

where the vector fields

$$\mathbf{e}_0 \equiv \frac{1}{\rho}\frac{\partial}{\partial \tau}, \qquad \mathbf{e}_1 \equiv \frac{\partial}{\partial \rho} \tag{1.A.6}$$

are orthonormal. The coordinate lines $\rho = constant$ are the hyperbolae

$$x^2 - t^2 = \rho^2 \tag{1.A.7}$$

of Born motion. Its time-like worldlines, with proper time s as parameter,

$$\tau = \frac{s}{\rho}, \qquad \rho = constant \tag{1.A.8}$$

have the unit tangent vector

$$\left(\frac{dt}{ds}, \frac{dx}{ds}\right) = \left(\cosh\frac{s}{\rho}, \sinh\frac{s}{\rho}\right) \equiv \mathbf{e}_0 . \tag{1.A.9}$$

The acceleration vector is given by

$$\left(\frac{d^2t}{ds^2}, \frac{d^2x}{ds^2}\right) = \frac{1}{\rho}\left(\sinh\frac{s}{\rho}, \cosh\frac{s}{\rho}\right) \equiv \frac{1}{\rho}\,\mathbf{e}_1 , \tag{1.A.10}$$

with the aid of (1.A.4). This shows that $1/\rho$ is the magnitude of the acceleration and the hyperbolae (1.A.7) are the worldlines of constant acceleration.

For Born motion, acceleration becomes the analog of the curvature of circular motion and the hyperbolic functions replace the trigonometric ones of circular motion.

It is easy to construct a picture of the vector fields $\mathbf{e}_0$ and $\mathbf{e}_1$ of (1.A.3) that show, in a region of Minkowski spacetime, the local four-velocity and four acceleration. [**See Figure 1.A.1.**]

It is clear that the frame field is invariant under motions in the new time coordinate τ but not under motions in the ρ-direction because the size of the acceleration is $1/\rho$.

Here was a problem that Einstein apparently had not noticed in his 1907 paper on the equivalence principle. In Newton's theory the acceleration, γ, could be a vector in the x-direction, constant in space and time, independent of the velocity of a body moving in the x-direction. Not so in Minkowski spacetime. There the acceleration vector has to be orthogonal to the 4-velocity; it would appear that homogeneity of the acceleration field in spacetime could no longer be achieved.

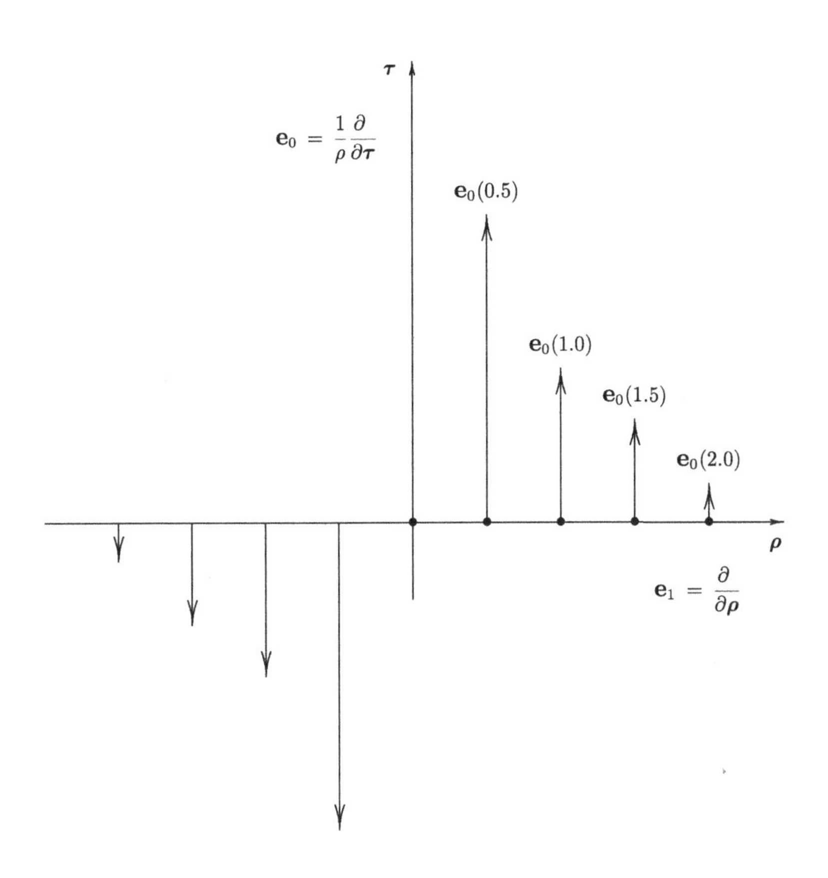

Figure 1.A.1 Born Motion Vectors

Apparently Max Planck had noticed the problem. It was four months after Einstein had mailed his paper that he sent a correction to the Jahrbuch. It began:

> "A letter by Mr. Planck induced me to add the following supplementary remark so as to prevent a misunderstanding that could arise easily: In the section 'Principle of relativity and gravitation', a reference system at rest situated in a temporally constant, homogeneous gravitational field is treated as physically equivalent to a uniformly accelerated, gravitation-free reference system. The concept 'uniformly accelerated' needs further clarification." [Einstein, 1908]

Einstein then pointed out that the equivalence was to be restricted to a body with zero velocity in the accelerated system. In a linear approximation, he concluded, this was sufficient because only linear terms had to be taken into account.

Einstein's retreat raises the question whether it is impossible to find a homogeneous uniformly accelerated reference system, or, assuming exact validity of his principle of equivalence, a homogeneous gravitational field.

B. A Homogeneous Gravitational Field

A definition of homogeneity in an n-dimensional manifold involves the existence of n linearly independent differential one-forms

$$\boldsymbol{\omega}'^{\mu}(x'^{\lambda}) = \omega'^{\mu}{}_{\alpha}(x'^{\lambda})\, dx'^{\alpha}\,, \qquad \det\left[\omega'^{\mu}{}_{\alpha}\right] \neq 0 \tag{1.B.1}$$

and n independent functions

$$x'^{\lambda} = f^{\lambda}(x^{\mu})\,, \qquad \det\left[\frac{\partial f^{\lambda}}{\partial x^{\mu}}\right] \neq 0\,. \tag{1.B.2}$$

The coordinate transformation (1.B.2) of the differential forms (1.B.1), the so-called pull-back,

$$\boldsymbol{\omega}'^{\mu}(x'^{\lambda}) = \omega'^{\mu}{}_{\beta}(x'^{\lambda})\, dx'^{\beta} = \omega'^{\mu}{}_{\beta}[\, f^{\lambda}(x^{\nu})\,]\frac{\partial f^{\beta}(x^{\nu})}{\partial x^{\alpha}}\, dx^{\alpha} \tag{1.B.3}$$

defines n new differential forms $\boldsymbol{\omega}^{\mu}(x^{\nu})$ by

$$\omega'^{\mu}{}_{\beta}[\, f^{\lambda}(x^{\nu})\,]\frac{\partial f^{\beta}(x^{\nu})}{\partial x^{\alpha}}\, dx^{\alpha} \equiv \omega^{\mu}{}_{\alpha}(x^{\nu})\, dx^{\alpha} \equiv \boldsymbol{\omega}^{\mu}(x^{\nu})\,. \tag{1.B.4}$$

If the $\omega'^{\mu}{}_{\alpha}$ and $\omega^{\mu}{}_{\alpha}$, are the same functions of their arguments

$$\omega'^{\mu}{}_{\alpha}(x^{\nu}) = \omega^{\mu}{}_{\alpha}(x^{\nu}), \tag{1.B.5}$$

we say the differential forms $\boldsymbol{\omega}^{\mu}$ are invariant.

A simple example for (1.B.5) and (1.B.2) is the translation isomorphism of $\mathbb{R}^n$ given by

$$\boldsymbol{\omega}'^{\mu}(x'^{\lambda}) = dx'^{\mu}, \qquad \boldsymbol{\omega}^{\mu}(x^{\nu}) = dx^{\mu}, \qquad x'^{\lambda} = x^{\lambda} + c^{\lambda}, \tag{1.B.6}$$

where $c^{\lambda} = constant$, which immediately yields

$$\omega'^{\mu}{}_{\alpha}(x^{\nu}) = \delta^{\mu}{}_{\alpha} = \omega^{\mu}{}_{\alpha}(x^{\nu}). \tag{1.B.7}$$

The invariance, (1.B.5), of the differential forms gives

$$\omega^{\mu}{}_{\alpha}(x'^{\lambda}) \frac{\partial f^{\alpha}(x^{\nu})}{\partial x^{\beta}} dx^{\beta} = \omega^{\mu}{}_{\beta}(x^{\nu})\, dx^{\beta}. \tag{1.B.8}$$

Introducing the inverse of $\omega^{\mu}{}_{\alpha}$ gives

$$\xi^{\gamma}{}_{\mu}(x'^{\lambda})\, \omega^{\mu}{}_{\alpha}(x'^{\lambda}) = \delta^{\gamma}{}_{\alpha} \tag{1.B.9}$$

where

$$\frac{\partial x'^{\gamma}}{\partial x^{\beta}} = \frac{\partial f^{\gamma}(x^{\nu})}{\partial x^{\beta}} = \xi^{\gamma}{}_{\mu}(x'^{\lambda})\, \omega^{\mu}{}_{\beta}(x^{\nu}). \tag{1.B.10}$$

These are the equations of Lie's first fundamental theorem for Lie groups.

Differentiation of the $\boldsymbol{\omega}^{\mu}$ gives the Maurer-Cartan equations

$$d\boldsymbol{\omega}^{\mu} + \frac{1}{2} C^{\mu}{}_{\lambda\nu}\, \boldsymbol{\omega}^{\lambda} \wedge \boldsymbol{\omega}^{\nu} = 0 \tag{1.B.11}$$

and the invariance of the $\boldsymbol{\omega}^{\mu}$ leads to constancy of the "*structure constants*" $C^{\mu}{}_{\lambda\nu}$. Further differentiation makes the first term of the left-hand side of (1.B.11) zero and creates thus the conditions for Lie's third fundamental theorem of Lie groups by the integrability conditions

$$C^{\mu}{}_{\lambda\nu}\, d\boldsymbol{\omega}^{\lambda} \wedge \boldsymbol{\omega}^{\nu} - C^{\mu}{}_{\lambda\nu}\, \boldsymbol{\omega}^{\lambda} \wedge d\boldsymbol{\omega}^{\nu} = 0. \tag{1.B.12}$$

With the Maurer-Cartan equations, we have

$$C^{\mu}{}_{\lambda\nu}\, C^{\nu}{}_{\rho\sigma}\, \boldsymbol{\omega}^{\nu} \wedge \boldsymbol{\omega}^{\rho} \wedge \boldsymbol{\omega}^{\sigma} = 0. \tag{1.B.13}$$

This gives the defining conditions for a Lie algebra

$$C^{\mu}{}_{\lambda\nu}\, C^{\nu}{}_{\rho\sigma} \,+\, C^{\mu}{}_{\rho\nu}\, C^{\nu}{}_{\sigma\lambda} \,+\, C^{\mu}{}_{\sigma\nu}\, C^{\nu}{}_{\lambda\rho} \,=\, 0\,. \tag{1.B.14}$$

By introducing Lie groups through invariant differential forms, Élie Cartan was able to simplify Lie's derivations of his fundamental theorems.

A Riemannian manifold is called *homogeneous* if its metric is invariant under a transitive group of motions. Mathematicians call a transformation group *transitive* if its action maps any point of the manifold into any other point. The $\mathbf{S}^2$, for example, with its standard metric inherited from being rigidly embedded into the 3-dimensional Euclidean space, is homogeneous under the action of the rotation group $\mathbf{O}(3)$.

The homogeneous manifolds we need for gravitational fields are of a special nature. For them, the group of motions has to be *simply transitive*, meaning that there is only one group element that moves a point of the manifold into another given point of the manifold. In this case, a Riemannian manifold is homogeneous if the metric

$$ds^2 \,=\, \eta_{\mu\nu}\, \boldsymbol{\omega}^{\mu}\, \boldsymbol{\omega}^{\nu}\,, \quad \det\left[\,\eta_{\mu\nu}\,\right] \neq 0\,, \quad \eta_{\mu\nu} \,=\, \eta_{\nu\mu} \,=\, constant\,, \tag{1.B.15}$$

is given in terms of invariant differential one-forms $\boldsymbol{\omega}^{\mu}(x^{\lambda})$ for a simply transitive group. These one-forms define in each point of the manifold an orthonormal n-leg of vectors, also known as a "frame" by

$$\boldsymbol{\omega}^{\mu}(\mathbf{e}_{\lambda}) \,=\, \delta^{\mu}{}_{\lambda}\,. \tag{1.B.16}$$

In coordinate components, these vectors are then given by (1.B.9) as

$$\mathbf{e}_{\lambda} \longrightarrow \xi^{\alpha}{}_{\lambda}\,. \tag{1.B.17}$$

The Levi-Civita connection is defined by differential one-forms $\boldsymbol{\omega}^{\mu}{}_{\nu}$, the *connection forms*, that represent the gravitational forces given by

$$d\boldsymbol{\omega}^{\mu} \,=\, \boldsymbol{\omega}^{\mu}{}_{\nu} \wedge \boldsymbol{\omega}^{\nu}\,, \qquad \boldsymbol{\omega}_{\mu\nu} \,=\, -\,\boldsymbol{\omega}_{\nu\mu}\,. \tag{1.B.18}$$

These are Élie Cartan's first structural equations for a Riemannian space with vanishing torsion. Some authors define the connection forms $\boldsymbol{\omega}^{\mu}{}_{\nu}$ with the opposite sign from our convention.

The components of the gravitational field, $g^{\mu}{}_{\nu\lambda}$, are given by

$$\boldsymbol{\omega}^{\mu}{}_{\nu} \,=\, g^{\mu}{}_{\nu\lambda}\, \boldsymbol{\omega}^{\lambda}\,, \qquad g_{\mu\nu\lambda} \,+\, g_{\nu\mu\lambda} \,=\, 0\,. \tag{1.B.19}$$

We have thus

$$d\boldsymbol{\omega}^{\mu} = g^{\mu}{}_{\nu\lambda}\, \boldsymbol{\omega}^{\lambda} \wedge \boldsymbol{\omega}^{\nu} . \tag{1.B.20}$$

Comparing this equation with the Maurer-Cartan equation with the indices lowered by $\eta_{\mu\rho}$ gives

$$d\boldsymbol{\omega}_{\alpha} = \frac{1}{2}\left(g_{\alpha\lambda\nu} - g_{\alpha\nu\lambda}\right) \boldsymbol{\omega}^{\nu} \wedge \boldsymbol{\omega}^{\lambda} = -\frac{1}{2} C_{\alpha\nu\lambda}\, \boldsymbol{\omega}^{\nu} \wedge \boldsymbol{\omega}^{\lambda} . \tag{1.B.21}$$

We obtain, therefore, with even permutations,

$$C_{\alpha\lambda\nu} = g_{\alpha\lambda\nu} - g_{\alpha\nu\lambda} , \tag{1.B.22a}$$

$$C_{\lambda\nu\alpha} = g_{\lambda\nu\alpha} - g_{\lambda\alpha\nu} , \tag{1.B.22b}$$

$$C_{\nu\alpha\lambda} = g_{\nu\alpha\lambda} - g_{\nu\lambda\alpha} . \tag{1.B.22c}$$

Because of the skew-symmetry of $g_{\mu\nu\lambda}$ in its first two indices, according to (1.B.19), adding (1.B.22a) and (1.B.22b) and subtracting (1.B.22c) gives

$$2\, g_{\alpha\lambda\nu} = C_{\alpha\lambda\nu} + C_{\lambda\nu\alpha} - C_{\nu\alpha\lambda} . \tag{1.B.23}$$

This shows that the physical components of a gravitational field's strength are constant in a homogeneous gravitational field.

For the further development it is useful to discuss an example of such a homogeneous gravitational field.

C. A Homogeneous Gravitational Field in the Minkowski Plane

It is easy to find invariant differential forms if one writes the metric for the Minkowski plane as

$$ds^2 = \eta_{\mu\nu}\, dx^{\mu}\, dx^{\nu} = 2\, du\, dv , \qquad \eta_{\mu\nu} = \begin{pmatrix} 0 & 1 \\ 1 & 0 \end{pmatrix} \tag{1.C.1}$$

where we have used the null coordinates

$$u = \frac{1}{\sqrt{2}}(t - x) , \qquad v = \frac{1}{\sqrt{2}}(t + x) , \tag{1.C.2}$$

and the associated vectors

$$\frac{\partial}{\partial u} = \frac{1}{\sqrt{2}}\left(\frac{\partial}{\partial t} - \frac{\partial}{\partial x}\right) , \qquad \frac{\partial}{\partial v} = \frac{1}{\sqrt{2}}\left(\frac{\partial}{\partial t} + \frac{\partial}{\partial x}\right) . \tag{1.C.3}$$

The one-forms

$$\omega^0 = \frac{du}{\gamma u}, \qquad \omega^1 = \gamma u \, dv, \tag{1.C.4a}$$

$$ds^2 = 2\,\omega^0\,\omega^1, \qquad \gamma = constant \neq 0, \tag{1.C.4b}$$

are independent and give us constant structure coefficients. We have

$$d\omega^0 = -\,C^0{}_{01}\,\omega^0 \wedge \omega^1 = 0, \tag{1.C.5}$$

$$d\omega^1 = \gamma\; du \wedge dv = \gamma\;\omega^0 \wedge \omega^1 = -\,C^1{}_{01}\;\omega^0 \wedge \omega^1 \tag{1.C.6}$$

which gives

$$C^0{}_{01} = 0, \qquad C^1{}_{01} = -\,\gamma. \tag{1.C.7}$$

We then have, from (1.B.23), for the gravitational field strength(s)

$$g_{100} = -\,\gamma, \qquad g_{101} = 0. \tag{1.C.8}$$

The frame vectors are given by (1.B.16); in the current situation, they are

$$\mathbf{e}_0 = \gamma\, u\,\frac{\partial}{\partial u}, \qquad \mathbf{e}_1 = \frac{1}{\gamma\, u}\,\frac{\partial}{\partial v}. \tag{1.C.9}$$

A general velocity vector in the Minkowski plane is subject to

$$2\,V^0\,V^1 = 1 \tag{1.C.10}$$

and is thus time-like. It is directed into the future if $V^0 > 0$.

The acceleration vector $\mathbf{g}$ has the covariant components g_j equal to

$$g_j = g_{jkl}\,V^k\,V^l, \tag{1.C.11}$$

which is the geodesic acceleration adapted to our frame. In our case, we obtain

$$g_0 = g_{010}\,V^1\,V^0 = \gamma\,V^1\,V^0, \tag{1.C.12a}$$

$$g_1 = g_{100}\,V^0\,V^0 = -\,\gamma\,\left(V^0\right)^2. \tag{1.C.12b}$$

Evidently, this expression for the acceleration fulfills the requirement that it be orthogonal to the 4-velocity; that is,

$$g_j\,V^j = 0. \tag{1.C.13}$$

If we choose the velocity as a constant, that is, $V^0 = constant > 0$, it follows from (1.C.10) that V^1 is also constant and the velocity vector field

$$\mathbf{V} = V^\mu \, \mathbf{e}_\mu = V^0 \, \gamma \, u \, \frac{\partial}{\partial u} + \frac{1}{2 \, V^0 \, \gamma \, u} \, \frac{\partial}{\partial v} \tag{1.C.14}$$

is invariant in the Minkowski plane. To better visualize the field, we calculate the worldlines that have the vectors of the field as tangents. We have

$$\frac{du}{ds} = V^0 \, \gamma \, u \, , \qquad \frac{dv}{ds} = \frac{1}{2 \, V^0 \, \gamma \, u} \tag{1.C.15}$$

or, with integration constants u_0, v_0, we get

$$u = u_0 \, e^{V^0 \, \gamma \, s} \, , \qquad v - v_0 = - \frac{e^{- V^0 \, \gamma \, s}}{2 \, u_0 \, (V^0 \, \gamma)^2} \, . \tag{1.C.16}$$

By eliminating u_0 and s, we obtain the one-parameter set of curves

$$v - v_0 = - \frac{1}{2 \, u \, (V^0 \, \gamma)^2} \, . \tag{1.C.17}$$

This is a set of identical hyperbolae that are obtained from the hyperbola

$$u \, v + \frac{1}{2 \, (V^0 \, \gamma)^2} = 0 \tag{1.C.18}$$

by translation in the v-direction. With (1.C.2), this hyperbola can also be written as

$$t^2 - x^2 + \frac{1}{(V^0 \, \gamma)^2} = 0 \, . \tag{1.C.19}$$

[See Figure 1.C.1; left and right hyperbolae.]

Next, we calculate the acceleration vector field. We have from (1.C.12)

$$g^0 = - \gamma \, (V^0)^2 \, , \qquad g^1 = \gamma \, V^1 \, V^0 \tag{1.C.20}$$

and thus

$$\mathbf{g} = g^\mu \, \mathbf{e}_\mu = - \gamma^2 \, (V^0)^2 \, u \, \frac{\partial}{\partial u} + \frac{V^1 \, V^0}{u} \, \frac{\partial}{\partial v} \, . \tag{1.C.21}$$

This vector field is tangent to the curves

$$\frac{du}{d\tau} = - (\gamma \, V^0)^2 \, u \, , \qquad \frac{dv}{d\tau} = \frac{V^1 \, V^0}{u} \tag{1.C.22}$$

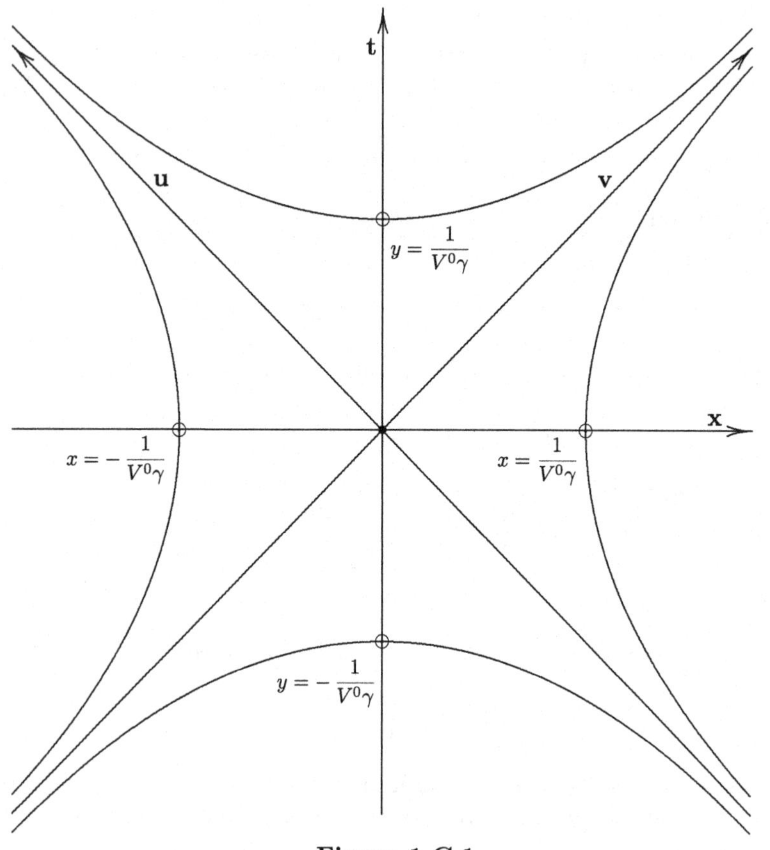

Figure 1.C.1
A Homogeneous Gravitational Field in the Minkowski Plane

described by the parameter τ. We obtain

$$u = u_1 \, e^{-(\gamma V^0)^2 \tau} , \qquad v - v_1 = \frac{V^1 \, V^0}{(\gamma \, V^0)^2 \, u_1} \, e^{(\gamma V^0)^2 \tau} \tag{1.C.23}$$

with integration constants u_1, v_1. By eliminating u_1 and v_1 together with (1.C.10), we obtain

$$v - v_1 = \frac{1}{2 \, u \, (\gamma \, V^0)^2} \, . \tag{1.C.24}$$

These curves are obtained from the hyperbola

$$u \, v = \frac{1}{2 \, (\gamma \, V^0)^2} \tag{1.C.25}$$

by translation in the v-direction.

With (1.C.2), this hyperbola can also be written

$$t^2 - x^2 = \frac{1}{(\gamma \, V^0)^2} \, . \tag{1.C.26}$$

[See Figure 1.C.1; upper and lower hyperbolae.]

By shifting the lower branch of the hyperbola by $\sqrt{2}/V^0\gamma$ in the positive v-direction, we intersect then the time-like hyperbola (1.C.18) at $t = 0$, $x = 1/V^0\gamma$. Tangent to the hyperbolae at their intersection are the two frame vectors of the velocity field and its acceleration. **[See Figure 1.C.2.]** Now, we wish to show that in the neighborhood of the intersection point we are approximating a Newtonian homogeneous gravitational field.

D. The Newtonian Gravitational Field

We study the Newtonian field with acceleration $-\gamma$ in the x-direction of zero velocity and at $t = 0$. In the Minkowski (t-x)-plane, we take parallel worldlines $x = \textit{constant}$ that are geodesics. For $t = 0$ and $x = 1/V^0\gamma$, the parallel to the t-axis through this point is touched by the hyperbola (1.C.19). This hyperbola, parameterized by with proper time s,

$$t = \rho \, \sinh \frac{s}{\rho} , \qquad x = \rho \, \cosh \frac{s}{\rho} , \qquad \rho = \frac{1}{V^0 \, \gamma} , \tag{1.D.1}$$

is the worldline of an observer with constant intrinsic acceleration. Now, let this observer measure the distance to the straight line $x = \rho$ orthogonal

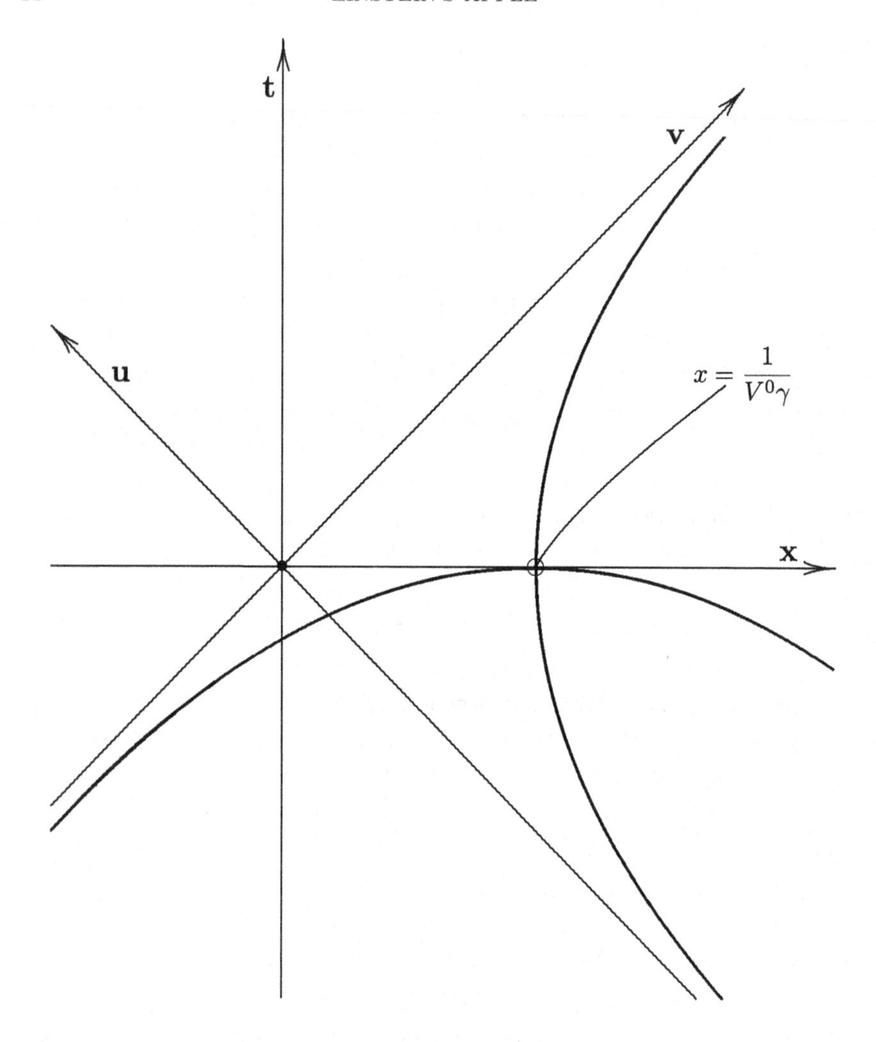

Figure 1.C.2
A Homogeneous Gravitational Field in the Minkowski Plane

The two hyperbolic branches intersecting at $x = 1/(V^0\gamma)$ are the right branch of the hyperbola in Figure D.1 and lower branch of the hyperbola in Figure D.1 displaced in the v-direction. The two curves intersect orthogonally and have as their unit tangent vectors $\boldsymbol{e}_0$ and $\boldsymbol{e}_1$ of a homogeneous gravitational field.

to his worldline. This distance is a function $\eta(s)$. From **Figure 1.D.1**, we read off immediately that

$$\eta = \rho - \frac{\rho}{\cosh(s/\rho)} = \rho\,\frac{\cosh(s/\rho) - 1}{\cosh(s/\rho)} = \frac{1}{2\rho}\,s^2 + \ldots \tag{1.D.2}$$

where higher terms are of fourth-order in time s. The acceleration is given by

$$a = \frac{d^2\eta}{ds^2} = \frac{1}{\rho} = V^0\,\gamma\,. \tag{1.D.3}$$

Since the velocity $d\eta/ds$ increases, the acceleration points into the negative x-direction.

It is for this special case: against absolute space, in the same event, and at the same velocity, an inertial observer and the accelerated one can interpret gravity as acceleration and vice versa. Here the Newtonian equivalence for a homogeneous gravitational field can be made relativistic in a point. This raises the question: Is it possible to save Einstein's equivalence for a homogeneous gravitational field.

E. Relativistic Equivalence

We have applied the calculus of Ricci Curbastro to Minkowski spacetime for the description of a homogeneous gravitational field in Einstein's theory. It turns out that such fields actually exist, without any spacetime curvature, in a flat world. The existence of such a non-vanishing field came as a surprise to us because its analogs on the $\mathbf{S}^2$ or the Euclidean plane do not exist. The reason for its existence on a flat spacetime is that the Poincaré group in (1,1)-dimensions has a simply transitive 2-dimensional subgroup. This is not the case for $\mathbf{SO(3)}$. The Euclidean motions in the plane, $\mathbf{E(2)}$, only have the trivial transformations.

In terms of the light-like coordinates u and v, we can write the Poincaré group of the Minkowski plane as

$$u' = \alpha\,u + a\,, \qquad v' = \frac{1}{\alpha}\,v + b\,, \qquad \alpha \neq 0 \tag{1.E.1}$$

with constants a, b, α. In matrix form, this is

$$\begin{pmatrix} u' \\ v' \\ 1 \end{pmatrix} = \begin{pmatrix} \alpha & 0 & a \\ 0 & \alpha^{-1} & b \\ 0 & 0 & 1 \end{pmatrix} \begin{pmatrix} u \\ v \\ 1 \end{pmatrix}\,. \tag{1.E.2}$$

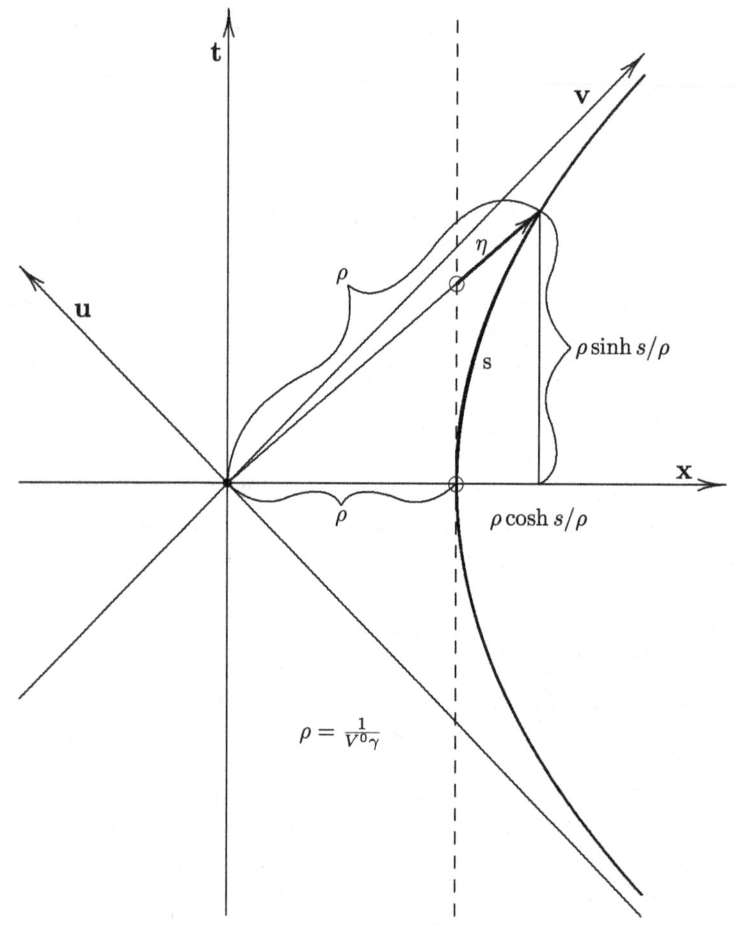

Figure 1.D.1 The Newtonian Gravitational Field

The worldline described by the parametric equations

$$x = \rho \cosh \frac{s}{\rho}, \qquad t = \rho \sinh \frac{s}{\rho}, \qquad \rho = \frac{1}{V^0 \gamma}$$

has acceleration $1/\rho$. At $t = 0$, it osculates the geodesic $\rho =$ *constant* (the vertical dashed line). The vector η determines the distance from that geodesic to the hyperbola in the rest system of the accelerated observer, "orthogonal" to his worldline.

Clearly, the matrices with vanishing a

$$\begin{pmatrix} \alpha & 0 & 0 \\ 0 & \alpha^{-1} & b \\ 0 & 0 & 1 \end{pmatrix} \tag{1.E.3}$$

form a subgroup of the Poincaré group depending on the two parameters α and b. For events with $u' \neq 0$, a unique event with coordinates (u, v) exists and is obtained from the inverse matrix of (1.E.3). This demonstrates that the subgroup acts simply transitively in the two demi-mondes $u \neq 0$. The existence of this subgroup is the source of the homogeneous field. But it is also the source of what is known as "teleparallelism".

If we take an orthonormal frame in one original point with say $u < 0$, the motions of the subgroup move it to all other points with $u < 0$ in a unique fashion. When we assign to a vector $\mathbf{V}$ in the original point the components V^0 and V^1, we can now assign these same components to vectors with respect to the frames that were transported and now define them as parallel to each other. This is clearly what mathematicians call an equivalence relation known as teleparallelism.

Such a more general notion of parallelity, based on simply transitive groups, was introduced by the Princeton geometer and later Dean Luther Pfahler Eisenhart in a brief paper in 1925 [Eisenhart, 1925]. The concept was further developed by Élie Cartan together with the Dutch electrical engineer Jan Arnoldus Schouten who had discovered parallel displacement in Riemannian geometry independently of Levi-Civita [Cartan, 1926].

The homogeneity group is characterized by the invariant differential forms $\boldsymbol{\omega}^\mu$ of (1.B.1) that determine the frames. We get the frames everywhere if we pick a frame in one point and integrate the total differential equations (1.B.11) of Ludwig Maurer and Élie Cartan. We call the vectorial 2-form

$$\boldsymbol{\Theta}^\mu = d\boldsymbol{\omega}^\mu = -\frac{1}{2} C^\mu{}_{\lambda\nu}\, \boldsymbol{\omega}^\lambda \wedge \boldsymbol{\omega}^\nu \tag{1.E.4}$$

the torsion form. In any case, it is simply determined by the structure constants, $C^\mu{}_{\lambda\nu}$, of the homogeneity group.

CHAPTER 2

TORSION AND TELEMOTION

A. Cartan's Torsion

Élie Cartan introduced the notion of torsion in a brief note in 1922 [Cartan, 1922]. He amplified his sketch in a memoir in the following year. It was reprinted in his collected papers in 1955. This paper *"Sur les varietés à connexion affine et la théorie de la relativité généralisée"* is now accessible through its translation in book form and an introduction that facilitates its study [Cartan, 1923][Cartan, 1924].

It is easiest to introduce torsion by using the notion of the exterior covariant derivative of a vector-valued differential one-form. For the general formalism see, for example, [EDM2, 1993].

In the tangent vector space attached to each point of a manifold exists a unit operator I. The torsion is described by the vector-valued 2-form $\boldsymbol{\Theta}$

$$\boldsymbol{\Theta} = \nabla I \,. \tag{2.A.1}$$

Here, ∇ indicates the operator of exterior covariant derivation. One writes

$$I = \mathbf{e}_\mu \otimes \boldsymbol{\omega}^\mu \,, \qquad \boldsymbol{\omega}^\nu (\mathbf{e}_\mu) = \delta^\nu{}_\mu \,. \tag{2.A.2}$$

The connection is then defined by

$$\nabla \mathbf{e}_\mu = -\, \mathbf{e}_\nu \, \boldsymbol{\omega}^\nu{}_\mu \tag{2.A.3}$$

where the $\boldsymbol{\omega}^\nu{}_\mu$ are the connection 1-forms. The torsion $\boldsymbol{\Theta}$ becomes then

$$\begin{aligned} \boldsymbol{\Theta} &= \nabla (\mathbf{e}_\mu \, \boldsymbol{\omega}^\mu) \\ &= -\, \mathbf{e}_\nu \, \boldsymbol{\omega}^\nu{}_\mu \wedge \boldsymbol{\omega}^\mu + \mathbf{e}_\nu \, d\boldsymbol{\omega}^\nu \\ &= \mathbf{e}_\nu \, (d\boldsymbol{\omega}^\nu - \boldsymbol{\omega}^\nu{}_\mu \wedge \boldsymbol{\omega}^\mu) \,. \end{aligned} \tag{2.A.4}$$

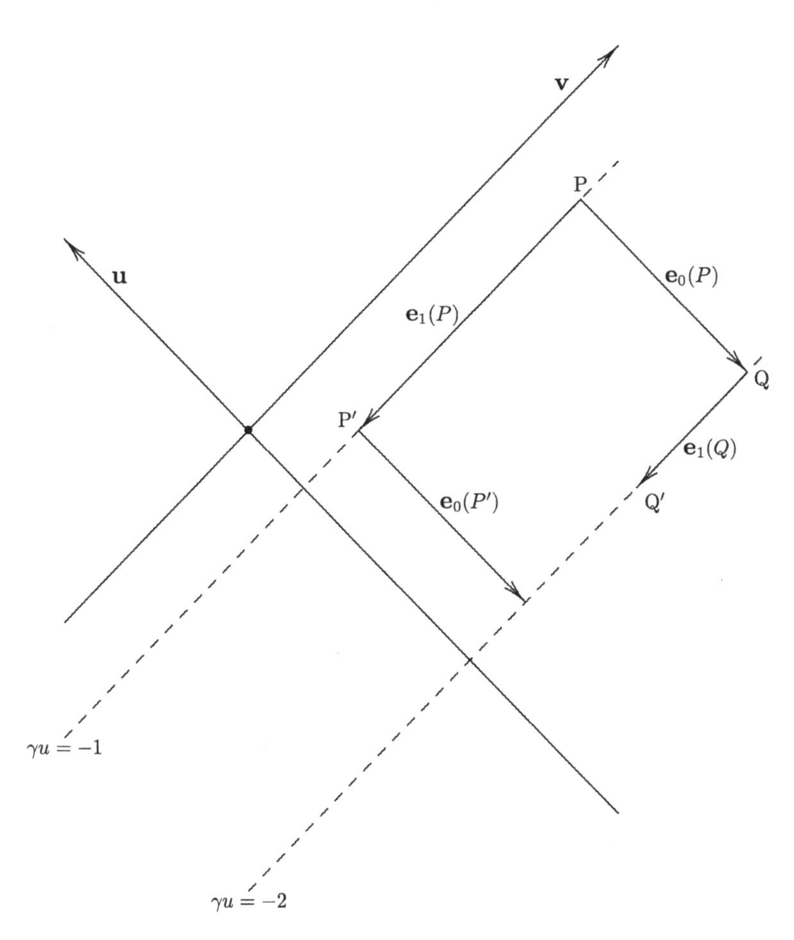

Figure 2.A.1: Cartan's Torsion

Here, we are interested only in affinely connected manifolds for which (2.A.3) holds and where the connection leaves the metric invariant, that is, the scalar product of vectors in the tangent space. We have

$$\begin{aligned}\nabla(\mathbf{e}_\lambda \cdot \mathbf{e}_\mu) &= \nabla\mathbf{e}_\lambda \cdot \mathbf{e}_\mu + \mathbf{e}_\lambda \cdot \nabla\mathbf{e}_\mu \\ &= -\,\mathbf{e}_\nu \cdot \mathbf{e}_\mu\, \boldsymbol{\omega}^\nu{}_\lambda - \mathbf{e}_\lambda \cdot \mathbf{e}_\nu\, \boldsymbol{\omega}^\nu{}_\mu \\ &= -\,(\boldsymbol{\omega}_{\mu\lambda} + \boldsymbol{\omega}_{\lambda\mu}) \\ &= \nabla\eta_{\lambda\mu} \\ &= 0\,. \end{aligned} \tag{2.A.5}$$

This means that the connection form with covariant indices must be skew in these indices.

We now want to study the constant torsion in the Minkowski plane. Writing the vector components of the torsion 2-form

$$\boldsymbol{\Theta}^\mu = \frac{1}{2} T^\mu{}_{\lambda\nu}\, \boldsymbol{\omega}^\lambda \wedge \boldsymbol{\omega}^\nu\,, \tag{2.A.6}$$

we have, from (1.E.4) and (1.C.7)

$$T^0{}_{01} = 0\,, \qquad T^1{}_{01} = \gamma\,. \tag{2.A.7}$$

This shows that the torsion is constant and thus homogeneous.

We now want to look at the basis vectors $\mathbf{e}_0$ and $\mathbf{e}_1$ of our distant parallelism. We have from (1.C.9)

$$\mathbf{e}_0 = \gamma\, u\, \partial_\mu\,, \qquad \mathbf{e}_1 = \frac{1}{\gamma\, u}\, \partial_v\,. \tag{2.A.8}$$

We take a point P on the line $\gamma\, u = -1$ and draw the frame vectors $\mathbf{e}_0(P)$ and $\mathbf{e}_1(P)$. The vector $\mathbf{e}_0(P)$ has its tip in the point Q that lies on the line $\gamma\, u = -2$. According to (2.A.8), the vector $\mathbf{e}_1(Q)$ is only half as long as the vector $\mathbf{e}_1(P)$. [**See Figure 2.A.1.**]

We now assume that the connection coefficients vanish:

The vector $\mathbf{e}_0(P')$, issuing from the tip of the vector $\mathbf{e}_1(P)$, is obtained through parallel transfer from $\mathbf{e}_0(P)$ and has remained unchanged since $\gamma\, u$ was constant. The effect of the torsion is that our square $PQP'Q'$ does not close. There should be a physical explanation for this phenomenon.

B. The Gravitational Frequency Shift

For a first orientation, we draw the two hyperbolic worldlines of two observers who experience a homogeneous gravitational field. [**See Figure 2.B.1.**] Their 4-velocities $\mathbf{V}$ are given by (1.C.14) for negative u

$$\mathbf{V} = V^{\mu}\,\mathbf{e}_{\mu} = -\,V^{0}\,\gamma\,u\,\partial_{u} - \frac{1}{2\,V^{0}\,\gamma\,u}\,\partial_{v}\,. \tag{2.B.1}$$

Since V^0 is constant, the vectors are parallel in the usual affine sense along lines $u = constant$. These are the null lines of rising photons emitted when the lower observer has precisely the same velocity as the higher observer at the reception of the light. There is no frequency shift for a rising photon. It is obvious from the figure that a falling photon is perceived by the lower observer to originate from an approaching source and thus be blue-shifted. To calculate the shift, we remember that the ratio of the frequency ν of the emitter to the frequency ν' of the receiver is given by

$$\frac{\nu'}{\nu} = \frac{\mathbf{k}\cdot\mathbf{V}'}{\mathbf{k}\cdot\mathbf{V}}\,. \tag{2.B.2}$$

Here, $\mathbf{V}$ and $\mathbf{V}'$ are the 4-velocities of emitter and receiver, respectively, while $\mathbf{k}$ is the 4-vector of the photon.

For the falling photon the $\mathbf{k}$ null vector has a u-component only, say,

$$\mathbf{k} \sim (\,1,\,0\,)\,. \tag{2.B.3}$$

This is given, with (2.B.1),

$$\frac{\nu'}{\nu} = \frac{-1}{2\,V^{0}\,\gamma\,u'} : \frac{-1}{2\,V^{0}\,\gamma\,u} = \left|\,\frac{u}{u'}\,\right|\,. \tag{2.B.4}$$

Since $|\,u'\,|$ is smaller than $|\,u\,|$, we obtain a blue-shift. We now also see that this blue-shift for a falling photon is the physical equivalent of the non-closure of our square $PQP'Q'$. The ratio in length of PP'/QQ' is the blue-shift ν'/ν.

As we had seen in (1.C.17), two different observers differ only in their value of the parameter v_0

$$u = -\,\frac{1}{2\,(\,v - v_0\,)\,(\,V^{0}\,\gamma\,)^{2}}\,. \tag{2.B.5}$$

This shows that the ratios of their u-values, $|\,u/u'\,|$, goes towards 1 for $v \to \infty$. The blue-shift vanishes asymptotically.

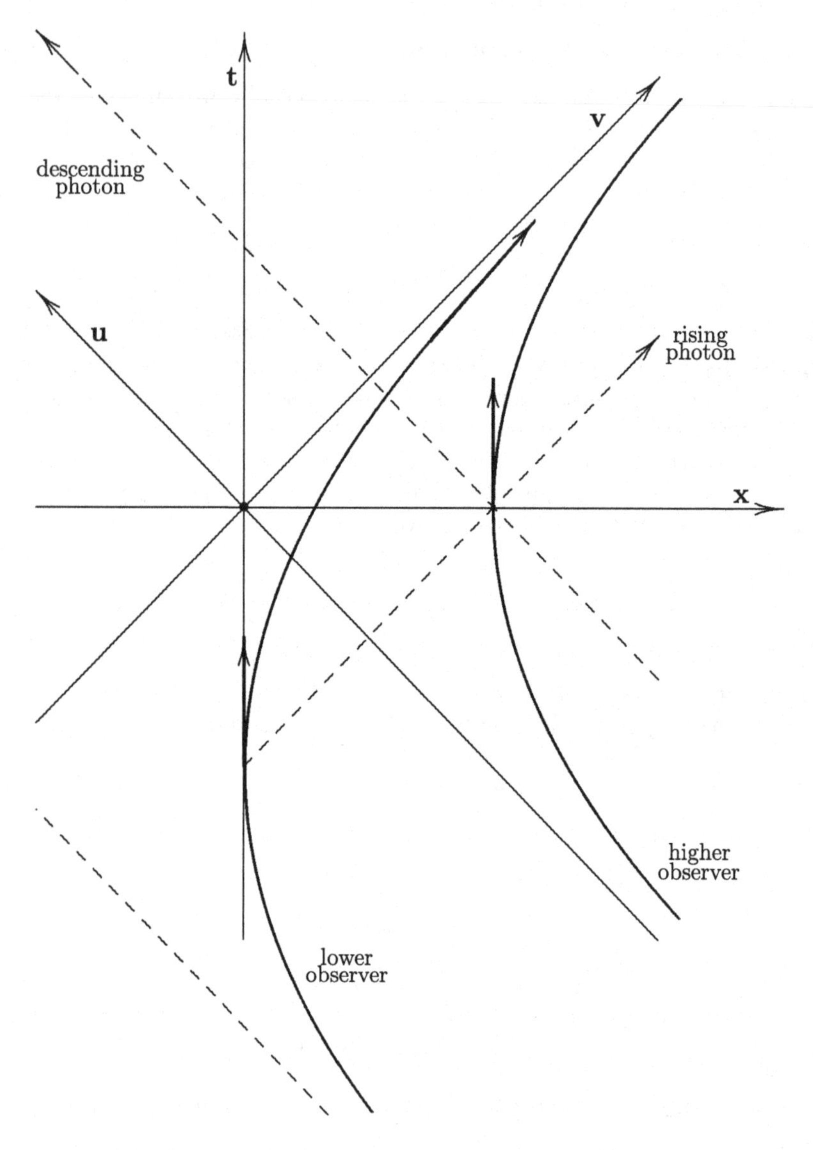

Figure 2.B.1: The Gravitational Frequency Shift

C. Fermi's Torsion

Almost simultaneously with Élie Cartan's 1922 introduction of torsion, this concept and its importance for physics was also discovered by a student at Italy's famous Scuola Normale Superiore in Pisa. Enrico Fermi, then aged 21, had sent a note to the Rendiconti of the Accademia dei Lincei on "*Sopra i fenomeni che avvengono in vicinanza di una linea oraria*" (On the phenomena that occur in the neighborhood of a time-like worldline) [Fermi, 1922]. The short paper, that appeared in three pieces, introduced what became known as Fermi transport of vectors [Walker, 1932].

Let $\mathbf{U}$ be a time-like unit vector field tangent to a bunch of worldlines. We have

$$\mathbf{U} \cdot \mathbf{U} = 1 . \tag{2.C.1}$$

The acceleration along the worldlines is given by

$$\mathbf{A} = \dot{\mathbf{U}} \equiv \frac{D\mathbf{U}}{Ds} , \tag{2.C.2}$$

where $\mathbf{A}$ is the acceleration vector and s measures proper time along the worldlines. Clearly, because of (2.C.1),

$$\dot{\mathbf{U}} \cdot \mathbf{U} = \mathbf{A} \cdot \mathbf{U} = 0 , \tag{2.C.3}$$

the acceleration vector is orthogonal to the four-velocity $\mathbf{U}$. The two vectors $\mathbf{U}$ and $\mathbf{A}$ form a bivector $\mathbf{U} \wedge \mathbf{A}$ that describes a Lorentz boost for the frame that has the four-velocity vector $\mathbf{U}$.

For a vector $\mathbf{V}$ in the frame with four-velocity

$$\mathbf{U} = \mathbf{e}_0 , \tag{2.C.4}$$

the time-part $\mathbf{V}_{||}$ is given by the projection of $\mathbf{V}$ on $\mathbf{U}$

$$\mathbf{V}_{||} \equiv (\mathbf{V} \cdot \mathbf{U})\, \mathbf{U} = V^0 \, \mathbf{e}_0 , \tag{2.C.5}$$

while the space-part $\mathbf{V}_{\perp}$ is defined as

$$\mathbf{V}_{\perp} \equiv V^j \, \mathbf{e}_j = V^\mu \, \mathbf{e}_\mu - V^0 \, \mathbf{e}_0 = \mathbf{V} - (\mathbf{V} \cdot \mathbf{U})\, \mathbf{U} . \tag{2.C.6}$$

Fermi transport for a vector $\mathbf{V}$ along a worldline with unit tangent $\mathbf{U}$ is characterized by a boost that compensates the acceleration $\mathbf{A}$. The time-part of $\mathbf{V}$ remains unchanged, while the space-part becomes subject to a rotation preserving the length of the space-part of the vector $\mathbf{V}$ and, thus, its four-dimensional length. The Fermi connection is thus a metrical connection.

As an example of Fermi transport, we return to Born motion in the Minkowski plane, as described in **Section 1.A**, with the use of "polar coordinates". (They are called "Rindler coordinates" by some authors.) [**See Figure 2.C.1.**]

The unit tangent vector, **U**, along the hyperbolae will remain a unit tangent vector along the hyperbolae under Fermi transport. Along the radial geodesics, however, Fermi transport becomes simply Levi-Civita's parallel transport. It is clear then that we have torsion since parallelograms do not close. It is easy to see that the torsion becomes the cause of frequency shifts of light. We have the metric (1.A.3)

$$ds^2 = \rho^2\, d\tau^2 - d\rho^2 \,. \tag{2.C.7}$$

Null lines are given by $ds = 0$ and thus

$$\pm\, d\tau = \frac{d\rho}{\rho} \,. \tag{2.C.8}$$

A light ray emitted from ρ_1 at time τ_1 will be received at ρ_2 at time τ_2:

$$\pm \int_{\tau_1+d\tau_1}^{\tau_2+d\tau_2} d\tau = \pm \int_{\tau_1}^{\tau_2} d\tau = \int_{\rho_1}^{\rho_2} \frac{d\rho}{\rho} = \ln \frac{\rho_2}{\rho_1} \,. \tag{2.C.9}$$

A second light ray emitted at $\tau_1 + d\tau_1$, will be received at $\tau_2 + d\tau_2$. It follows that

$$d\tau_1 = d\tau_2 \,. \tag{2.C.10}$$

Since frequencies go inversely to proper times, ds, we have the gravitational shift

$$\frac{\nu_1}{\nu_2} = \frac{ds_2}{ds_1} = \frac{\rho_2\, d\tau_2}{\rho_1\, d\tau_1} = \frac{\rho_2}{\rho_1} \,. \tag{2.C.11}$$

Going back to (1.A.3), we can now calculate the torsion. We have

$$\omega^0 = \rho\, d\tau \,, \qquad \omega^1 = d\rho \,, \tag{2.C.12}$$

and

$$ds^2 = \left[\omega^0\right]^2 - \left[\omega^1\right]^2 . \tag{2.C.13}$$

The connection form is zero since vectors do not change their components under Fermi transport in the Minkowski plane. We thus have for the torsion form's vector components, $\boldsymbol{\Theta}^\mu$,

$$\boldsymbol{\Theta}^0 = d\omega^0 = d\rho \wedge d\tau = \frac{1}{\rho}\, \omega^1 \wedge \omega^0 \,, \qquad \boldsymbol{\Theta}^1 = d\omega^1 = 0 \,. \tag{2.C.14}$$

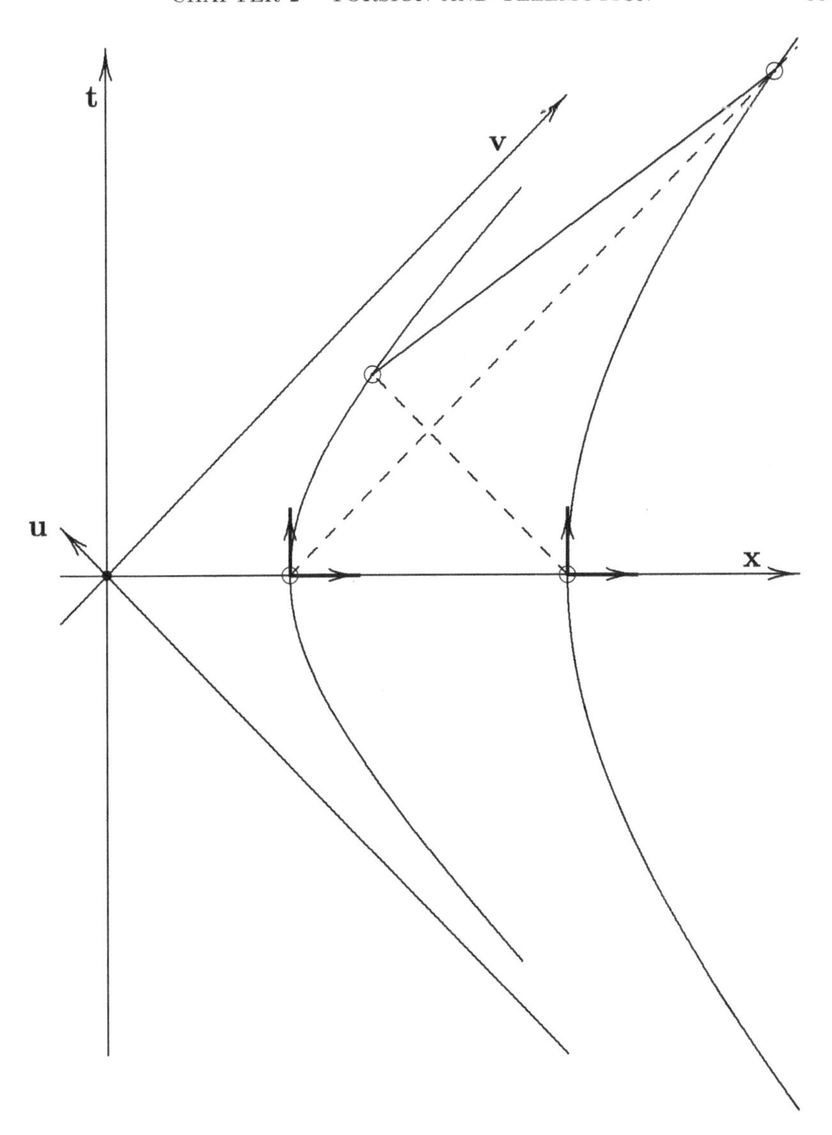

Figure 2.C.1: Fermi's Torsion

With

$$\boldsymbol{\Theta}^{\mu} = \frac{1}{2} T^{\mu}{}_{\lambda\nu}\, \boldsymbol{\omega}^{\lambda} \wedge \boldsymbol{\omega}^{\nu}\,, \tag{2.C.15}$$

we obtain

$$T^{0}{}_{01} = -\frac{1}{\rho}\,, \qquad T^{1}{}_{01} = 0\,. \tag{2.C.16}$$

We can express the gravitational shift in terms of torsion coefficients.

D. Reference Frames

Enrico Fermi had introduced his vector transport in approaching a physical description of a gravitational field. It would seem natural for a physicist to interpret the hyperbolae of Born motion as the worldlines of rigid flat slabs made of incompressible matter. But all attempts to introduce rigid bodies into relativity came to naught. This became particularly clear through a one-page paper by Pawel Sigmundovich Ehrenfest in Leiden that was the death knell for rigidity and led to the introduction of non-Euclidean geometry into relativity theory. The paper posing Ehrenfest's paradox pointed out that a circle of radius R centered on the axis of rotation in a rigid body and orthogonal to it would not show Lorentz's contraction of its radius while its circumference would. The deviation from 2π in the ratio of circumference to radius of the circle would thus indicate the presence of non-Euclidean geometry.

We shall define our reference system abstractly as an orthonormal frame in every point of the spacetime manifold. Nowadays one calls this a section of the frame bundle. How this can be physically realized is another question. Mathematically this section is given by

$$\mathbf{e}_{\nu}(x^{\lambda})\,. \tag{2.D.1}$$

When Einstein wrote about accelerated reference systems, it was assumed that meter sticks and clocks in such systems would be unchanged. Experiments done with muons in synchrocyclotrons have confirmed this assumption to high accuracy. Since clocks and meter sticks can be represented as vectors in four dimensions, it follows that the frames themselves, their metric, and, thus, their orthonormality is not affected by acceleration.

The acceleration of observers or, as Einstein thought equivalent to it, a gravitational field are not due to the curvature of the spacetime manifold. The falling of apples from Earth-bound trees and the weight we are experiencing while standing on the surface of this planet is the result of an acceleration of our reference frame that can be described mathematically as the torsion field of Minkowski spacetime. The presence of curvature near the surface of the Earth, that is, a non-vanishing Riemann tensor, corresponds to second order derivatives of the Newtonian potential and is a small higher order effect. The mantra "gravitation is curvature" is misleading if one associates gravitation with the falling of apples.

CHAPTER 3

INERTIAL AND GRAVITATIONAL FIELDS IN MINKOWSKI SPACETIME

A. Inertial Frames of Reference in Minkowski Spacetime

We begin with a discussion of the Poincaré group. We do this to remind the reader of well known facts and use the occasion to introduce terminology and notation. More on that can be found in **Appendix A** and in the books by Bernard Schutz [Schutz, 1980][Schutz, 2009].

The orthonormal and flat hypersurfaces $x^\mu = constant$ in Minkowski's spacetime do have a physical meaning. Their intersections define orthonormal inertial reference frames with respect to which kinematical and dynamical notions like velocity, angular velocity, acceleration, momentum, or energy of a body can be measured. We describe these distinguished Minkowski coordinates x^μ as a column vector with four entries which we abbreviate by the letter x. Any other set x' of Minkowski coordinates is then related to x by the matrix equation

$$\begin{bmatrix} x \\ 1 \end{bmatrix} = \begin{bmatrix} L & a \\ 0 & 1 \end{bmatrix} \begin{bmatrix} x' \\ 1 \end{bmatrix} . \tag{3.A.1}$$

Here a is another column vector with four entries while L is the 4×4 matrix of a Lorentz transformation

$$L^T \eta L = \eta \tag{3.A.2}$$

where η is the 4×4 matrix

$$\eta = \begin{bmatrix} 1 & 0 & 0 & 0 \\ 0 & -1 & 0 & 0 \\ 0 & 0 & -1 & 0 \\ 0 & 0 & 0 & -1 \end{bmatrix} \tag{3.A.3}$$

and the superscript T indicates transposition. The set of these transformations form the ten-dimensional Poincaré group. The principle of special relativity expresses the covariance of the equations of physics under the Poincaré group. By covariance we mean that physical quantities described by tensors transform as representations of the Lorentz group. The worldline of a force-free particle remains a straight line in Minkowski spacetime under all transformations of the Poincaré group. [If one deals with spinors—as we shall not—one has to go to a covering group of the Poincaré group.] We introduce the metric through the four differential forms

$$\boldsymbol{\omega}^{\mu} = dx^{\mu} \tag{3.A.4}$$

by

$$ds^2 = \eta_{\mu\nu}\, \boldsymbol{\omega}^{\mu}\, \boldsymbol{\omega}^{\nu} \tag{3.A.5}$$

and the frame vectors, also known as four-legs or *vierbeine*, by

$$\boldsymbol{\omega}^{\mu}(\mathbf{e}_{\nu}) = \delta^{\mu}{}_{\nu}\,, \qquad \mathbf{e}_{\nu} = \frac{\partial}{\partial x^{\nu}}\,. \tag{3.A.6}$$

An energy-momentum-stress tensor $\mathbf{T}(x)$, to give an example, would be written

$$\mathbf{T}(x) = T^{\mu\nu}(x)\, \mathbf{e}_{\mu} \otimes \mathbf{e}_{\nu} \tag{3.A.7}$$

with components $T^{\mu\nu}(x)$. These components are known as physical components or leg components. Let us now carry out a Poincaré transformation as a coordinate transformation. We have then

$$\boldsymbol{\omega}^{\mu} = dx^{\mu} = L^{\mu}{}_{\rho}\, dx'^{\rho} = L^{\mu}{}_{\rho}\, \boldsymbol{\omega}'^{\rho} \tag{3.A.8}$$

and correspondingly,

$$\frac{\partial}{\partial x'^{\lambda}} = \mathbf{e}'_{\lambda} = L^{\nu}{}_{\lambda} \left(\frac{\partial}{\partial x^{\nu}} \right) = L^{\nu}{}_{\lambda}\, \mathbf{e}_{\nu}\,. \tag{3.A.9}$$

With this transformation of the leg vectors, we have

$$\boldsymbol{\omega}^{\mu}(\mathbf{e}'_{\lambda}) = L^{\mu}{}_{\rho}\, \boldsymbol{\omega}'^{\rho}(\mathbf{e}'_{\lambda}) = \boldsymbol{\omega}^{\mu}(\mathbf{e}_{\nu} L^{\nu}{}_{\lambda}) = L^{\mu}{}_{\lambda}\,. \tag{3.A.10}$$

This gives again

$$\boldsymbol{\omega}'^{\rho}(\mathbf{e}'_{\lambda}) = \delta^{\rho}{}_{\lambda}\,. \tag{3.A.11}$$

For the tensor $\mathbf{T}$ we get

$$\mathbf{T}(Lx' + a) = T'^{\alpha\lambda}(Lx' + a)\; \mathbf{e}'_{\alpha} \otimes \mathbf{e}'_{\lambda}\,. \tag{3.A.12}$$

This gives the transformation law for the leg components

$$T^{\mu\nu}\,(Lx' + a\,) \;=\; L^{\mu}{}_{\alpha}\; L^{\nu}{}_{\lambda}\; T'^{\alpha\lambda}\,(Lx' + a\,)\;. \tag{3.A.13}$$

We stress that under this coordinate transformation the coefficients $L^{\mu}{}_{\nu}$ of the Lorentz transformation do not depend on position in spacetime, nor do the legs $\mathbf{e}_{\lambda}$ or $\mathbf{e}'_{\mu}$. We call the inertial frames $\mathbf{e}_{\nu}$, introduced above by the Minkowski coordinates x^{μ}, a natural inertial frame. The Poincaré group transforms natural inertial frames into natural inertial frames.

B. General Leg Transformations

It is often convenient in physics to use accelerated or rotating frames for the description of phenomena. Such non-inertial frames can be obtained from inertial frames by a general leg transformation given by

$$\boldsymbol{\omega}^{\mu} \;=\; L^{\mu}{}_{\rho}(x)\;\boldsymbol{\omega}'^{\rho}\,, \qquad \mathbf{e}'_{\lambda} \;=\; \mathbf{e}_{\nu}\,L^{\nu}{}_{\lambda}(x)\,. \tag{3.B.1}$$

The $L^{\mu}{}_{\lambda}(x)$ are differentiable functions of x which fulfill (3.A.2). They can be parameterized in terms of six functions of the four variables x^{μ}. A tensor $\mathbf{T}(x)$ under such a general leg transformation gives

$$\mathbf{T}\,(x) \;=\; T^{\mu\nu}(x)\;\mathbf{e}_{\mu}\otimes\mathbf{e}_{\nu} \;=\; T'^{\alpha\lambda}(x)\;\mathbf{e}'_{\alpha}\otimes\mathbf{e}'_{\lambda} \tag{3.B.2}$$

with

$$T^{\mu\nu}(x) \;=\; L^{\mu}{}_{\alpha}(x)\,L^{\nu}{}_{\lambda}(x)\,T'^{\alpha\lambda}(x)\;. \tag{3.B.3}$$

If the $L^{\mu}{}_{\alpha}$ are constants, the general leg transformation leads to a new set of inertial frames in the Minkowski spacetime. But it is no longer true that

$$\mathbf{e}_{\rho} \;=\; \frac{\partial}{\partial x^{\rho}}\,. \tag{3.B.4}$$

The frames then no longer coincide with the natural inertial frames given by the intersection of the flat coordinate hypersurfaces. For the general leg transformation it is often convenient to work with the differential leg forms $\boldsymbol{\omega}^{\mu}$ which have simpler transformation properties than the leg vectors $\mathbf{e}_{\rho}$. We have introduced a general leg through a transformation from a natural frame. This means that the general leg is described in terms of the Minkowski coordinates x^{μ}. But that is not essential for the notion of a general leg. We are free to use other coordinates. The reader will not be offended, we hope, if we give a simple example of a general leg. Let

$$\boldsymbol{\omega}^{0} \;=\; dt\,, \quad \boldsymbol{\omega}^{1} \;=\; dx\,, \quad \boldsymbol{\omega}^{2} \;=\; dy\,, \quad \boldsymbol{\omega}^{3} \;=\; dz \tag{3.B.5}$$

be a natural frame in Minkowski spacetime. We combine $\boldsymbol{\omega}^1$ and $\boldsymbol{\omega}^2$ into $\boldsymbol{\omega}^\pm$

$$\boldsymbol{\omega}^\pm = dx \pm i\, dy = d\left(r\, e^{\pm i\phi}\right) = \left(dr \pm i\, r\, d\phi\right) e^{\pm i\phi} . \tag{3.B.6}$$

This is still a natural frame, but now in polar coordinates. If we carry out the position-dependent leg transformation

$$\begin{aligned} \boldsymbol{\omega}^{+\prime} &= \boldsymbol{\omega}^+ e^{-i\phi} = dr + i\, r\, d\phi , \\ \boldsymbol{\omega}^{-\prime} &= \boldsymbol{\omega}^- e^{+i\phi} = dr - i\, r\, d\phi \end{aligned} \tag{3.B.7}$$

we have

$$\boldsymbol{\omega}^{\pm\prime} = \boldsymbol{\omega}^{1\prime} \pm i\, \boldsymbol{\omega}^{2\prime} , \quad \boldsymbol{\omega}^{1\prime} = dr , \quad \boldsymbol{\omega}^{2\prime} = r\, d\phi \tag{3.B.8}$$

and obtain the general leg

$$\mathbf{e}'_0 = \frac{\partial}{\partial t} , \quad \mathbf{e}'_1 = \frac{\partial}{\partial r} , \quad \mathbf{e}'_2 = \frac{1}{r}\frac{\partial}{\partial \phi} , \quad \mathbf{e}'_3 = \frac{\partial}{\partial z} . \tag{3.B.9}$$

A simple way to get a general leg is to start from the metric in the form

$$ds^2 = \left(\boldsymbol{\omega}^0\right)^2 - \left(\boldsymbol{\omega}^1\right)^2 - \left(\boldsymbol{\omega}^2\right)^2 - \left(\boldsymbol{\omega}^3\right)^2 . \tag{3.B.10}$$

In our case this would have been

$$\boldsymbol{\omega}'^0 = dt , \quad \boldsymbol{\omega}'^1 = dx , \quad \boldsymbol{\omega}'^2 = dy , \quad \boldsymbol{\omega}'^3 = dz . \tag{3.B.11}$$

For a curved spacetime where natural inertial frames no longer exist, this method is the only one available.

C. Inertial Fields

For the following discussion it is useful to introduce the concept of a directional derivative. Suppose, we have in our natural inertial frame $\{x^\mu, \mathbf{e}_\nu\}$ vector fields $\mathbf{A}$ and $\mathbf{B}$

$$\mathbf{A} = A^\mu\left(x^\lambda\right) \mathbf{e}_\mu , \quad \mathbf{B} = B^\nu\left(x^\lambda\right) \mathbf{e}_\nu . \tag{3.C.1}$$

We define then the directional derivative of $\mathbf{A}$ with respect to $\mathbf{B}$

$$\boldsymbol{\nabla}_{\mathbf{B}}\, \mathbf{A} = A^\mu{}_{,\nu}\, B^\nu\, \mathbf{e}_\mu . \tag{3.C.2}$$

This operation creates, from the vector fields $\mathbf{A}$ and $\mathbf{B}$, the new vector field $\boldsymbol{\nabla}_{\mathbf{B}}\mathbf{A}$. Koszul used it to introduce an affine connection or covariant differentiation for a differential manifold [Koszul, 1950]. For differentiable vector fields $\mathbf{A}, \mathbf{B}, \mathbf{C}$ and a function f he postulated the following properties

$$\boldsymbol{\nabla}_{(\mathbf{B}+f\mathbf{C})}\,\mathbf{A} = \boldsymbol{\nabla}_{\mathbf{B}}\,\mathbf{A} + f\,\boldsymbol{\nabla}_{\mathbf{C}}\,\mathbf{A} \tag{3.C.3}$$

and

$$\boldsymbol{\nabla}_{\mathbf{A}}\,(\mathbf{B} + f\,\mathbf{C}) = \boldsymbol{\nabla}_{\mathbf{A}}\,\mathbf{B} + f\,\boldsymbol{\nabla}_{\mathbf{A}}\,\mathbf{C} + df(\mathbf{A})\,\mathbf{C}\,. \tag{3.C.4}$$

If one has a general affine leg $\mathbf{e}_k$, the operation of covariant differentiation for vector fields is completely determined if one knows the directional derivatives of the leg vectors $\mathbf{e}_k$ with respect to all leg vectors $\mathbf{e}_j$

$$\boldsymbol{\nabla}_{\mathbf{e}_j}\mathbf{e}_k = -\,g^m{}_{kj}\,\mathbf{e}_m\,. \tag{3.C.5}$$

The $g^m{}_{kj}$ are known as the connection coefficients. If the manifold has a metric, we take the leg vectors to be orthonormal

$$\mathbf{e}_j\boldsymbol{\cdot}\mathbf{e}_l = \eta_{kl} = \mathit{constant} \tag{3.C.6}$$

and get then

$$\begin{aligned} 0 &= \boldsymbol{\nabla}_{\mathbf{e}_j}\,(\mathbf{e}_k\boldsymbol{\cdot}\mathbf{e}_l) \\ &= \left(\boldsymbol{\nabla}_{\mathbf{e}_j}\mathbf{e}_k\right)\boldsymbol{\cdot}\mathbf{e}_l + \mathbf{e}_k\boldsymbol{\cdot}\left(\boldsymbol{\nabla}_{\mathbf{e}_j}\mathbf{e}_l\right) \\ &= -\,g^m{}_{kj}\,\mathbf{e}_m\boldsymbol{\cdot}\mathbf{e}_l - \mathbf{e}_k\boldsymbol{\cdot}g^m{}_{lj}\,\mathbf{e}_m \\ &= -\,g_{lkj} - g_{klj}\,. \end{aligned} \tag{3.C.7}$$

This means that the g_{lkj} with all indices downstairs are skew-symmetric in the first two indices. For a Riemannian space they are known as the Ricci coefficients. For a natural inertial frame $\{x^\mu, \mathbf{e}_\nu\}$, all Ricci coefficients vanish since the vectors $\mathbf{e}_\nu$ are constants

$$\boldsymbol{\nabla}_{\mathbf{e}_\tau}\mathbf{e}_\nu = -\,g^\alpha{}_{\nu\tau}\,\mathbf{e}_\alpha = 0\,. \tag{3.C.8}$$

If we go now from a natural inertial frame in Minkowski spacetime to an arbitrary orthonormal leg using the leg transformation

$$\mathbf{e}'_j = \mathbf{e}_\nu\,L^\nu{}_j\,(x) \tag{3.C.9}$$

we have

$$\begin{aligned} \nabla_{\mathbf{e}'_j}\mathbf{e}'_k \; - \; \nabla_{\mathbf{e}'_j}\mathbf{e}_\nu \, L^\nu{}_k(x) \\ &= dL^\nu{}_k(\mathbf{e}'_j)\,\mathbf{e}_\nu \\ &= \left(L^{-1}\right)^m{}_\nu \, dL^\nu{}_k(\mathbf{e}'_j)\,\mathbf{e}'_m \,. \end{aligned} \tag{3.C.10}$$

For the general leg, we now use Latin indices to distinguish it from the natural inertial leg. Comparison with (3.C.5) now gives

$$-\,g^m{}_{kj}(x) \;=\; \left(L^{-1}\right)^m{}_\nu \, dL^\nu{}_k(\mathbf{e}'_j) \;. \tag{3.C.11}$$

We now call the Ricci coefficients the components of an inertial field. To see that non-vanishing Ricci coefficients give rise to inertial forces, we study geodesic motion in a general frame $\mathbf{e}'$. Let the timelike vector field $\mathbf{U}$ belong to a geodesic congruence

$$\mathbf{U} = U^k\,\mathbf{e}'_k\,, \qquad \mathbf{U}\cdot\mathbf{U} = 1\,, \qquad \nabla_{\mathbf{U}}\mathbf{U} = 0\,. \tag{3.C.12}$$

We have then

$$\begin{aligned} 0 &= \nabla_{\mathbf{U}}\mathbf{U} \\ &= \nabla_{\mathbf{U}}\left(U^k\,\mathbf{e}'_k\right) \\ &= dU^l(\mathbf{U})\,\mathbf{e}'_l + U^k\,U^m\,\nabla_{\mathbf{e}'_m}\mathbf{e}'_k \\ &= \left[dU^l(\mathbf{U}) - U^k\,U^m\,g^l{}_{km}\right]\mathbf{e}'_l\,. \end{aligned} \tag{3.C.13}$$

This gives

$$dU^l(\mathbf{U}) \;=\; g^l{}_{km}\,U^k\,U^m\,. \tag{3.C.14}$$

The U^l are the components of the four-velocity. The expression on the left-hand side gives the components of the four-acceleration in the leg. The right-hand side is the specific inertial force, that is, the inertial force per unit mass.

D. The Rigidly Rotating Frame in Minkowski Spacetime

It may be useful to look at an example that should appear in every undergraduate text on Mechanics, but doesn't. We choose a set of rigidly rotating frames for special relativity. We obtain such a leg $\mathbf{e}''_j$ by subjecting the leg of (3.B.9) to the following Lorentz transformation

$$\begin{aligned} \mathbf{e}''_0 &= \gamma\,(\mathbf{e}'_0 + \nu\,\mathbf{e}'_2)\,, \qquad \mathbf{e}''_1 = \mathbf{e}'_1\,, \\ \mathbf{e}''_2 &= \gamma\,(\mathbf{e}'_2 + \nu\,\mathbf{e}'_0)\,, \qquad \mathbf{e}''_3 = \mathbf{e}'_3\,, \end{aligned} \tag{3.D.1}$$

$$\gamma = \left(1 - \nu^2\right)^{-1/2}, \qquad \nu = \alpha\,r\,, \qquad \alpha = constant$$

with a constant α that has the dimension of an inverse length

$$\alpha = R^{-1} . \tag{3.D.2}$$

It is clear that this transformation is only possible within the "light cylinder"

$$r < R \tag{3.D.3}$$

since at a distance R from the rotation axis, the frame would move with the speed of light. To calculate the Ricci coefficients, it is easier to use differential forms. For the necessary formulae, see **Appendix A.**

We start with the metric

$$ds^2 = dt^2 - dr^2 - r^2 d\phi^2 - dz^2 \tag{3.D.4}$$

and carry out the Lorentz transformation for its differential forms that corresponds to (3.D.1).

$$\begin{aligned} \omega^0 &= \gamma\left(dt - \alpha\, r^2 d\phi\right) , \quad & \omega^1 &= dr , \\ \omega^2 &= \gamma\,(r\, d\phi - \alpha\, r\, dt) , \quad & \omega^3 &= dz , \\ \gamma &= \left(1 - \alpha^2 r^2\right)^{-1/2} . \end{aligned} \tag{3.D.5}$$

We then have

$$ds^2 = \left(\omega^0\right)^2 - \left(\omega^1\right)^2 - \left(\omega^2\right)^2 - \left(\omega^3\right)^2 . \tag{3.D.6}$$

To calculate the connection forms we obtain

$$\begin{aligned} d\omega^0 &= \alpha^2\gamma^2 r \;\; \omega^0 \wedge \omega^1 \; - \; 2\,\alpha\,\gamma^2 \;\; \omega^1 \wedge \omega^2 \\ d\omega^1 &= 0 , \\ d\omega^2 &= \frac{\gamma^2}{r} \; \omega^1 \wedge \omega^2 , \\ d\omega^3 &= 0 . \end{aligned} \tag{3.D.7}$$

The formula

$$d\omega_{jk} = \omega_{jk} \wedge \omega^k \tag{3.D.8}$$

gives then the unique solution for skew-symmetric $\boldsymbol{\omega}_{jk}$

$$\begin{aligned} \boldsymbol{\omega}_{01} &= \alpha^2\gamma^2\, r\, \boldsymbol{\omega}^0 \,+\, \alpha\,\gamma^2\, \boldsymbol{\omega}^2\,, \\ \boldsymbol{\omega}_{02} &= -\,\alpha\,\gamma^2\, \boldsymbol{\omega}^1\,, \\ \boldsymbol{\omega}_{12} &= -\,\alpha\,\gamma^2\, \boldsymbol{\omega}^0 \,-\, \frac{\gamma^2}{r}\, \boldsymbol{\omega}^2\,. \end{aligned} \tag{3.D.9}$$

All others containing the index "3" vanish. From

$$\boldsymbol{\omega}_{jk} \,=\, g_{jkl}\, \boldsymbol{\omega}^l \tag{3.D.10}$$

we now obtain the Ricci coefficients

$$\begin{aligned} g_{010} &= \alpha^2\gamma^2\, r\,, \\ g_{012} &= \alpha\,\gamma^2\,, \quad g_{021} \,=\, -\,\alpha\,\gamma^2\,, \quad g_{120} \,=\, -\,\alpha\,\gamma^2\,, \\ g_{122} &= -\,\frac{\gamma^2}{r}\,. \end{aligned} \tag{3.D.11}$$

This gives, for instance, for the acceleration of a test particle at rest

$$\dot{U}^0 = 0\,, \quad \dot{U}^1 \,=\, \alpha^2\gamma^2\, r\,, \quad \dot{U}^2 \,=\, \dot{U}^3 \,=\, 0\,. \tag{3.D.12}$$

And for the centrifugal force, bringing back the velocity, of light, c,

$$F^0 = 0\,, \quad F^1 \,=\, \frac{m\,\omega^2 r}{1-(\omega^2 r^2/c^2)}\,, \quad F^2 \,=\, F^3 \,=\, 0 \tag{3.D.13}$$

where $\omega \equiv \alpha\, c$ is the angular velocity and m is the mass of the particle at rest in the rotating frame. For arbitrary four-velocity, we have

$$\dot{U}^0 \,=\, \alpha^2\gamma^2\, r\, U^0\, U^1\,, \tag{3.D.14a}$$

$$\dot{U}^1 \,=\, \alpha^2\gamma^2\, r\,(U^0)^2 \,+\, 2\,\alpha\,\gamma^2\, U^0\, U^2 \,+\, \frac{\gamma^2}{r}\,(U^2)^2\,, \tag{3.D.14b}$$

$$\dot{U}^2 \,=\, -\,2\,\alpha\,\gamma^2\, U^0\, U^1 \,-\, \frac{\gamma^2}{r}\, U^1\, U^2\,, \tag{3.D.14c}$$

$$\dot{U}^3 \,=\, 0\,. \tag{3.D.14d}$$

E. Gravitational Fields

It is clear that the mysterious strict proportionality of inertial mass with passive gravitational mass allows us to interpret the inertial fields of **Section 3.C** as gravitational fields. These gravitational fields are described by a tensor $g^l{}_{km}$. In first approximation for non-relativistic motion, we have, from (3.C.14), for the acceleration a^j of a particle

$$a^j = g^j{}_{00}\,, \qquad j \in \{1, 2, 3\}\,. \tag{3.E.1}$$

By equating the two kinds of masses m one has, then, for the gravitational force

$$F^j = m\, g^j{}_{00}\,. \tag{3.E.2}$$

Through the equivalence principle, a special relativistic theory of gravitation can now be formulated by interpreting gravitational forces as inertial forces. However, Einstein's original 1908 formulation used non-inertial coordinate systems instead of orthonormal frames in arbitrary coordinate systems. Since coordinate systems have no physical meaning beyond identifying events, this introduced confusion into the interpretation of the theory. So defined, the gravitational fields are source-free fields in Minkowski spacetime. This means that both the curvature form and, equivalently, the Riemann tensor, calculated with the Ricci coefficients $g^j{}_{kl}$, vanish. General fields can then be discussed by dropping this requirement.

CHAPTER 4

THE NOTION OF TORSION

A. The Meaning of Einstein's First Principle of Equivalence

> "The worldline of a spinless test particle, moving under the influence of gravitational fields only, depends on its initial position and velocity, but not its mass and composition."

This is a modern statement of what we now call the "weak" principle of equivalence. Apart from its fanciful spacetime formulation, Isaac Newton could have put that into his *"Principia Philosophiae Naturalis"*. What happened to the equivalence of acceleration and gravitation?

For a Riemannian spacetime, using Élie Cartan's first fundamental equation, we can formulate a kind of equivalence of inertial and gravitational acceleration. The equation

$$\boldsymbol{\Theta}^{\mu} = d\boldsymbol{\omega}^{\mu} - \boldsymbol{\omega}^{\mu}{}_{\nu} \wedge \boldsymbol{\omega}^{\nu} \tag{4.A.1}$$

can be written down twice: for the same frame $\boldsymbol{\omega}^{\nu}$ and as

$$\boldsymbol{\Theta}'^{\mu} = d\boldsymbol{\omega}^{\mu} - \boldsymbol{\omega}'^{\mu}{}_{\nu} \wedge \boldsymbol{\omega}^{\nu}. \tag{4.A.2}$$

If we put the torsion, $\boldsymbol{\Theta}^{\mu}$, equal to zero in the first equation, we have

$$d\boldsymbol{\omega}^{\mu} = \boldsymbol{\omega}^{\mu}{}_{\nu} \wedge \boldsymbol{\omega}^{\nu} \tag{4.A.3}$$

where the connection forms, $\boldsymbol{\omega}^{\mu}{}_{\nu}$, describe a gravitational field in a spacetime without torsion. If we put the connection forms, $\boldsymbol{\omega}'^{\mu}{}_{\nu}$, in the second equation equal to zero, we have

$$d\boldsymbol{\omega}^{\mu} = \boldsymbol{\Theta}'^{\mu} \tag{4.A.4}$$

and thus

$$\boldsymbol{\Theta}'^{\mu} = \boldsymbol{\omega}^{\mu}{}_{\nu} \wedge \boldsymbol{\omega}^{\nu}\,. \tag{4.A.5}$$

The vanishing connection forms $\boldsymbol{\omega}'^{\mu}{}_{\nu}$ imply teleparallelism in spacetime. The tensor character of the torsion follows from the fact that

$$\boldsymbol{\Theta}'^{\mu} - \boldsymbol{\Theta}^{\mu} = \boldsymbol{\Theta}'^{\mu} = -\left(\boldsymbol{\omega}'^{\mu}{}_{\nu} - \boldsymbol{\omega}^{\mu}{}_{\nu}\right) \wedge \boldsymbol{\omega}^{\nu} = \boldsymbol{\omega}^{\mu}{}_{\nu} \wedge \boldsymbol{\omega}^{\nu}, \qquad (4.A.6)$$

is obtained through the difference of two connection forms. The justification for the choice of a vanishing connection in the accelerated frame is suggested by the way in which measurements are carried out.

Encouraged by Ernst Mach's critique of Isaac Newton's mechanics, Einstein tried to argue that the equivalence of inertia and gravitation made acceleration against absolute space obsolete. Only relative acceleration of bodies was supposed to be observable. All inertial forces could then also be interpreted as gravitational ones. However, these gravitational fields had no sources and were generated by coordinate transformations.

For the modern physicist, acceleration against Newton's absolute space is no longer so implausible as it appeared a century ago. The vacuum of spacetime is filled with the fluctuations of all fields and houses a non-zero Higgs field and a metric too. Absolute space is realized by a local inertial system.

The modern view of Einstein's theory of gravitation was stated by Sir Hermann Bondi at the occasion of the Centenary of Einstein's birthday in 1979. He wrote:

> "From this point of view, Einstein's elevators have nothing to do with gravitation; they simply analyze inertia in a perfectly Newtonian way. Thus, the notion of general relativity does not in fact introduce any post-Newtonian physics; it simply deals with coordinate transformations. Such a formalism may have some convenience, but physically it is wholly irrelevant. It is rather late to change the name of Einstein's theory of gravitation, but general relativity is a physically meaningless phrase that can only be viewed as a historical memento of a curious philosophical observation." [Bondi, 1979]

Physically, it is a clearcut case whether the accelerometer of an observer shows zero or not; this holds too for the mathematics. Whether two events in Minkowski spacetime are connected by a straight timelike line or by a hyperbolic path can easily be distinguished because the hyperbolic path has a smaller proper time. So it is clear that the accelerated frame has to be described as the frame that has torsion and there is nothing left of the so-called equivalence or relativity.

What has been gained, however, is an extension of the class of reference systems from the inertial frames of Minkowski spacetime to all smooth sections of the frame bundle.

This does not lead to the most general kind of torsion since

$$\boldsymbol{\Theta}'^{\mu} = d\boldsymbol{\omega}^{\mu} \tag{4.A.7}$$

gives that

$$d\boldsymbol{\Theta}'^{\mu} = 0\,. \tag{4.A.8}$$

Because the curvature, $\boldsymbol{\Omega}'^{\mu}{}_{\nu}$, of the connection, $\boldsymbol{\omega}'^{\mu}{}_{\nu}$, vanishes, the integrability conditions of the torsion

$$\nabla\boldsymbol{\Theta}'^{\mu} = d\boldsymbol{\Theta}'^{\mu} - \boldsymbol{\omega}'^{\mu}{}_{\nu} \wedge \boldsymbol{\Theta}^{\nu} = -\boldsymbol{\Omega}'^{\mu}{}_{\nu} \wedge \boldsymbol{\omega}^{\nu} \tag{4.A.9}$$

give also that

$$\nabla\boldsymbol{\Theta}'^{\mu} = 0\,. \tag{4.A.10}$$

It is a remarkable fact that the frames expressing linear and rotational acceleration can be interpreted—via torsion—as an invariant property of spacetime.

B. The Use of Frames

Gregorio Ricci-Curbastro used frames in his calculus and so did the French geometer Gaston Darboux. His technique of moving frames was developed to perfection by his student Élie Cartan. Students of general relativity followed Einstein in the unphysical use of coordinate bases. In 1929, Eugene Wigner, Hermann Weyl, and Vladimir Fock introduced orthonormal frames into general relativity to formulate the Dirac equation for curved spacetimes.

While frames are important for the introduction of a spin structure and the four-dimensional Gauss-Bonnet theorem, there are critical physical reasons for their use. Since physics generally deals with spatially extended objects (that is, fields) by working with their frame components, we can also consider global Lorentz transformations for a section of the frame bundle.

The most important aspect of the frame is that we can consider it as a simplified geometric representation of the quantized measuring instruments (Bohr radius of the hydrogen atom, Cesium hyperfine structure line defining the second, etc.) that determine the metric. Mathematicians often seem to ignore that all measurements are relative.

C. Einstein's Torsion

Torsion, invented by Élie Cartan and announced in a three page note to the French Academie, was introduced by Emile Borel in the session of February 27, 1922. Its publication in the Comptes Rendus bore the title *"Sur une généralisation de la notion de courbure de Riemann et les espaces à torsion"*. Einstein learned about this new revolutionary concept in geometry only four weeks later. His friend Paul Langevin had invited him to give a lecture at the Collège de France in Paris on March 31, 1922.

In the aftermath of World War I, the first lecture by a professor from the archenemy country was a highly charged political affair. To cut down on demonstrations, it was by invitation only and the French Prime Minister, Paul Painlevé, stood at the door checking the invitations. During this lecture week, Jacques Hadamard, professor at the Collège de France, gave a party for Einstein. Among his guests was Élie Cartan who was to meet Einstein there. Cartan thought torsion might have important physical applications and used the occasion to tell Einstein about his recent discovery. He tried to explain the novel concept to him using the simplest example well known to mapmakers and navigators. [**See Figure 4.C.1.**]

If one takes a sphere, removes the North and South poles, and puts unit vectors pointing North along meridians and along latitude circles pointing East, one has created an orthonormal frame everywhere. If one now moves vectors tangent to the sphere by keeping their frame components constant, one has a connection with torsion. The lines that have such teleparallel vectors as tangents have been long known as loxodromes or rhumb lines, discovered in 1537 as *"curvas dos rumos"* by a *marraño*, the Portuguese geometer Pedro Nuñez. Cartan's example of torsion was the exact analog of Fermi transport for a two dimensional spacetime that had been published on January 22, 1922, a month before Cartan. It was uncanny that seven years later Einstein wrote to Cartan:

> "I didn't at all understand the explanations you gave me in Paris; still less was it clear to me how they might be made useful for physical theory." [Einstein, 1929]

The occasion of this remark was a letter by Cartan pointing out that Einstein's teleparallelism was a special case of Cartan's torsion. But since publication of his first paper on teleparallelism in the year before, Einstein had learned of Eisenhart's work and that of Roland Weitzenböck. In fact, Weitzenböck had given a supposedly complete bibliography of papers on torsion with fourteen references without mentioning Élie Cartan. Since Weitzenböck, of Amsterdam University in the Netherlands, had written the

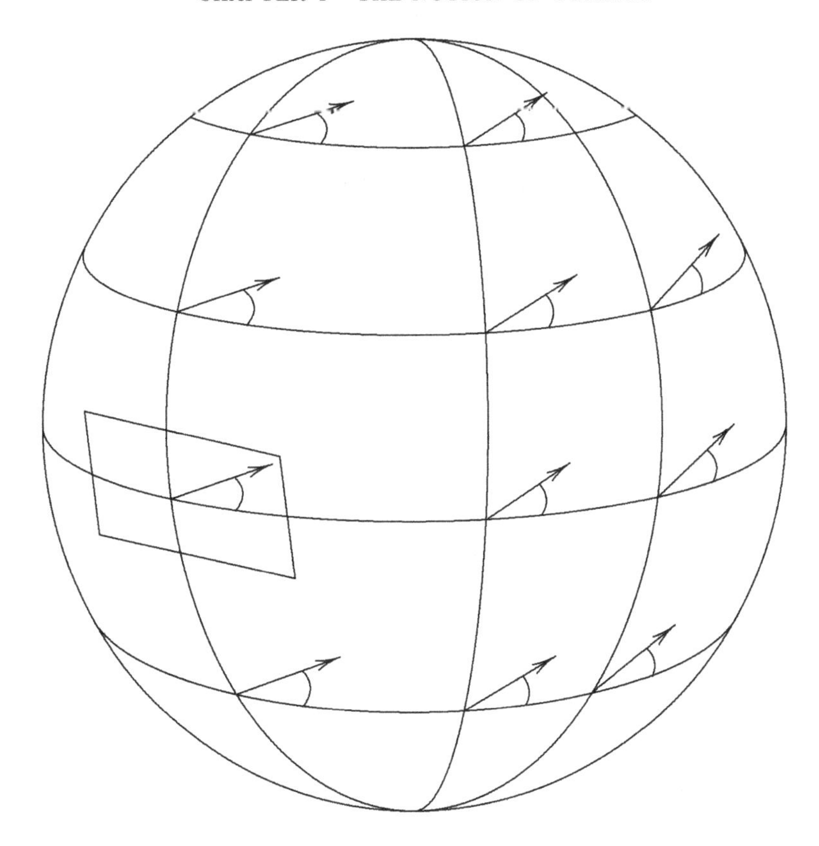

Figure 4.C.1 Einstein's Torsion

Cartan's example of a connection with torsion that confused Einstein. The arrows are the same length and make constant angles with their respective latitude lines. They lie in tangent planes located at their origins.

review article on Differential Invariants for the Encyclopedia of Mathematical Science, he certainly would have been aware of Cartan's activity in this field. A bizarre circumstance suggested that this omission may have been deliberate: In 1923 Weitzenböck published a modern monograph on the Theory of Invariants that included tensor calculus. In the innocent looking Preface, we find that the first letter of the first word in each of its twenty-one sentences spell out:

"NIEDER MIT DEN FRANZOSEN" (Down with the French!)

Weitzenböck, a Prussian-born army officer of World War I, had in fact mentioned his own kind of torsion as a possible connection in the review article of the Encyclopedia in 1923. But his silence on the French work on torsion in his 1928 review appeared to be a subliminal intellectual continuation of warfare. To set the record straight, Einstein asked Cartan to describe the history of torsion in an article in the leading German mathematics journal *Mathematische Annalen* as *"Notice historique sur la notion de parallélisme absolu"* [Cartan, 1930].

Einstein had introduced frames for a generalized theory using teleparallelism in 1928. In fact, Wigner's paper was based on Einstein's tetrad formalism. Einstein was using torsion for the electromagnetic potentials in his unified field theory. The resulting opus provoked Wolfgang Pauli's scathing critique in a letter to Einstein of December 19, 1929. Concluding his letter, Pauli wrote:

> "I am also not so naif as to believe you would change your mind because of criticism by others. But I would bet you any odds that in a year at the latest you will have dropped the whole teleparallelism as you earlier dropped your affine theory. And I do not wish to provoke your spirit of contradiction by continuing this letter to not delay the approaching natural demise of the theory of teleparallelism. In this definite hope of winning the bet, I wish you a merry Christmas affectionately, truly yours, W.Pauli" [Pauli, 1929]

Einstein did not accept Pauli's criticism or his bet. It took him two year's to drop teleparallelism and to come up with a new variant of Kaluza's five-dimensional theory.

D. The Notion of Torsion

Looking back, we may wonder why the concept of torsion and its importance for physics were apparently so difficult to grasp. There are, in fact, more examples in the literature. The gravitational *"Handbuch"* by

Charles Misner, Kip Thorne, and John Archibald Wheeler deals with the subject in their **Box 10.20** where the torsion is set to zero by a circular argument [Misner, 1973]. This was discovered by Borut Gogala [Gogala, 1980] and James Nester [Nester, 1984]. But then even expert differential geometers could have trouble with the concept: in his "Notes on Differential Geometry," Noel J. Hicks declares

> "As far as we know, there is no motivation for the word 'torsion' to describe the above tensor." [Hicks, 1965]

It may thus be useful to discuss another example of torsion.

The simplest non-trivial case of torsion in two-dimensional manifolds with a positive definite metric is the space of constant negative curvature viewed as Poincaré's upper half-plane. There, the torsion is constant everywhere and homogeneous.

The metric of the space is given by

$$ds^2 = \frac{1}{y^2}\left(dx^2 + dy^2\right), \qquad y > 0. \tag{4.D.1}$$

The geodesics appear as semi-circles that intersect the x-axis orthogonally. The upper y-axis and all its analogs, $x = constant$, $y > 0$, are special cases of the semi-circles and the geodesics. In the hyperbolic geometry, they are known as a pencil of parallels that all intersect in a point at $y = \infty$. All this can be derived from the metric.

The lines $y = constant$ are, however, not of this type. They are a special case of "horocycles" that appear in Poincaré's map as full circles that touch the x-axis or the point $y = \infty$. In the latter case, these horocycles are the lines $y = constant$, orthogonal to the pencil of parallel rays $x = constant$.

We now introduce the frame $\{\mathbf{e}_1, \mathbf{e}_2\}$ by

$$\mathbf{e}_1 = y\,\frac{\partial}{\partial x}, \qquad \mathbf{e}_2 = y\,\frac{\partial}{\partial y}, \tag{4.D.2}$$

or the corresponding differential forms

$$\omega_1 = \frac{1}{y}\,dx, \qquad \omega_2 = \frac{1}{y}\,dy, \qquad ds^2 = (\omega_1)^2 + (\omega_2)^2. \tag{4.D.3}$$

(For convenience, we have given the ω's lower indices.) **[See Figure 4.D.1.]** We then have

$$d\omega_1 = -\frac{1}{y^2}\,dy \wedge dx = \omega_1 \wedge \omega_2 = \omega_{12} \wedge \omega_2, \tag{4.D.4}$$

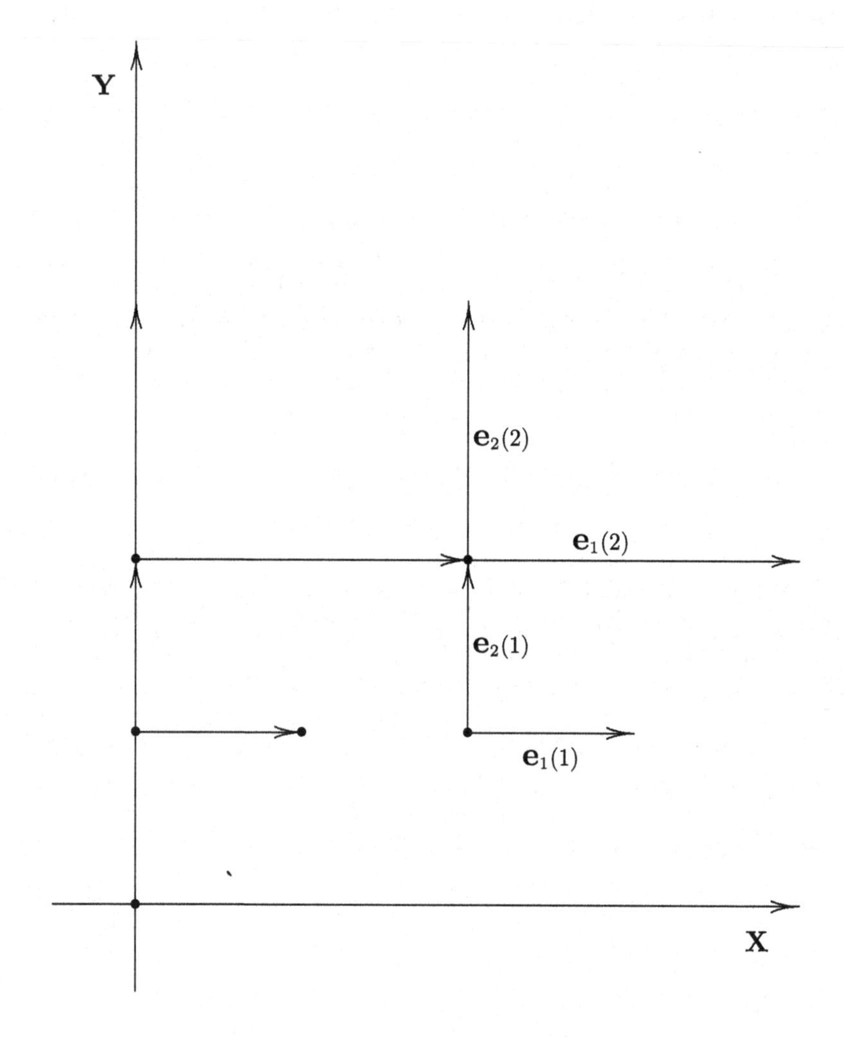

Figure 4.D.1 The Notion of Torsion (1)

$$d\boldsymbol{\omega}_2 = 0 = \boldsymbol{\omega}_{21} \wedge \boldsymbol{\omega}_1 \,. \tag{4.D.5}$$

It follows that the Levi-Civita connection form is given by

$$\boldsymbol{\omega}_{12} = \boldsymbol{\omega}_1 \,, \qquad \boldsymbol{\omega}_{12} = g_{12\mu}\,\boldsymbol{\omega}_\mu \,, \tag{4.D.6}$$

which gives

$$g_{121} = 1 \,, \qquad g_{122} = 0 \,. \tag{4.D.7}$$

The connection field is constant.

We obtain, as the only non-zero component of the curvature form,

$$\boldsymbol{\Omega}_{12} = -\,d\boldsymbol{\omega}_{12} = -\,d\boldsymbol{\omega}_1 = -\,\boldsymbol{\omega}_1 \wedge \boldsymbol{\omega}_2 \,, \tag{4.D.8}$$

which tells us that the curvature is minus one (that is, -1).

The torsion of our frame appears if we try to create a rectangle through parallel displacements. [**See Figure 4.D.2.**]

Let us start at (0,1), moving up to (0,2), moving a distance one unit right to (2,2), moving down to (2,1), and then moving a distance one unit left to (1,1). We are left with the gap from (0,1) to (1,1).

In this example, we study torsion on a manifold with curvature under the Levi-Civita connection without torsion. Our frame that gives teleparallelism with a flat (zero) connection and torsion gives rise to a gap in the parallelogram. The analog of this gap for the Levi-Civita connection is *not* the phenomenon of curvature but rather the shape of the geodesics and the parallel propagation of vectors one-way (not around closed loops).

Clarifying that through geometry, we look at two points (x_1, y_1) and (x_2, y_1) equidistant from the x-axis. [**See Figure 4.D.3.**] The circle through these two points that meets the x-axis orthogonally is the Levi-Civita geodesic. The horocycle $y_1 = constant$ is the auto-parallel connection for the teleparallelism. The horocycle is not a geodesic; it is a circle with curvature $1/y_1$ and infinite radius. Parallel transfer of a vector, $\mathbf{V}$, initially at (x_1, y_1) and pointing in the opposite of the y-direction, changes its orientation by the angle $-(x_2 - x_1)/y_1$. That results from the Levi-Civita connection. In teleparallelism, there is no change: $\mathbf{V} = \overline{\mathbf{V}}$.

We get the Gauss-Riemann curvature from the Levi-Civita connection if we consider the rectangle formed by two parallels to the y-axis and their orthogonal horocycles. [**See Figure 4.D.4.**] Parallel to the y-axis the vector $\mathbf{V}$ does not change its angle; but along the horocycle $y_2 = constant$, it changes its angle by $(x_2 - x_1)/y_2$ which is less for $y_2 > y_1$ than the change in the opposite direction along $y_1 = constant$. The overall change in angle is

$$\Delta\phi = (x_2 - x_1)\left(\frac{1}{y_1} - \frac{1}{y_2}\right) = -\int\!\!\int \frac{dx \wedge dy}{y^2} \,. \tag{4.D.9}$$

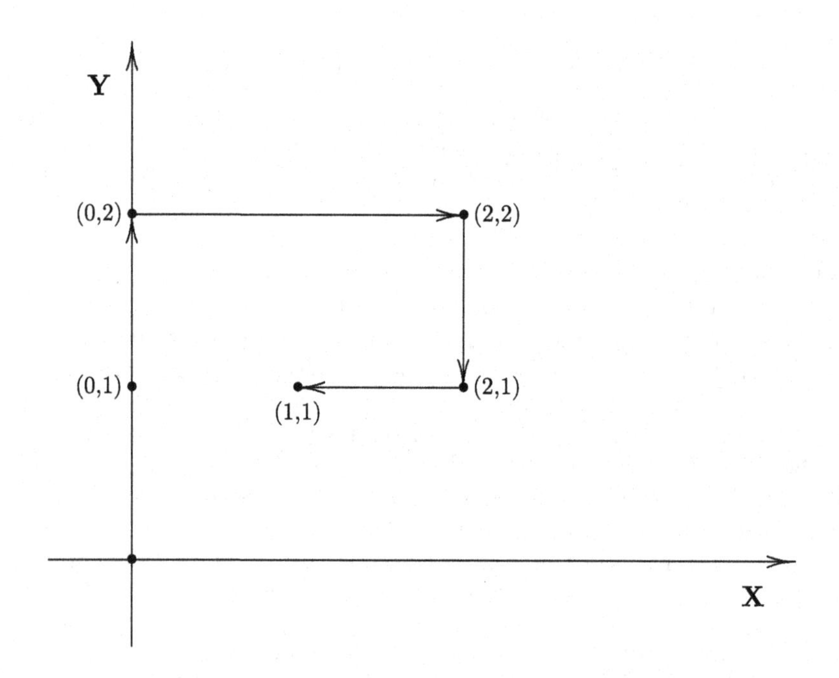

Figure 4.D.2 The Notion of Torsion (2)

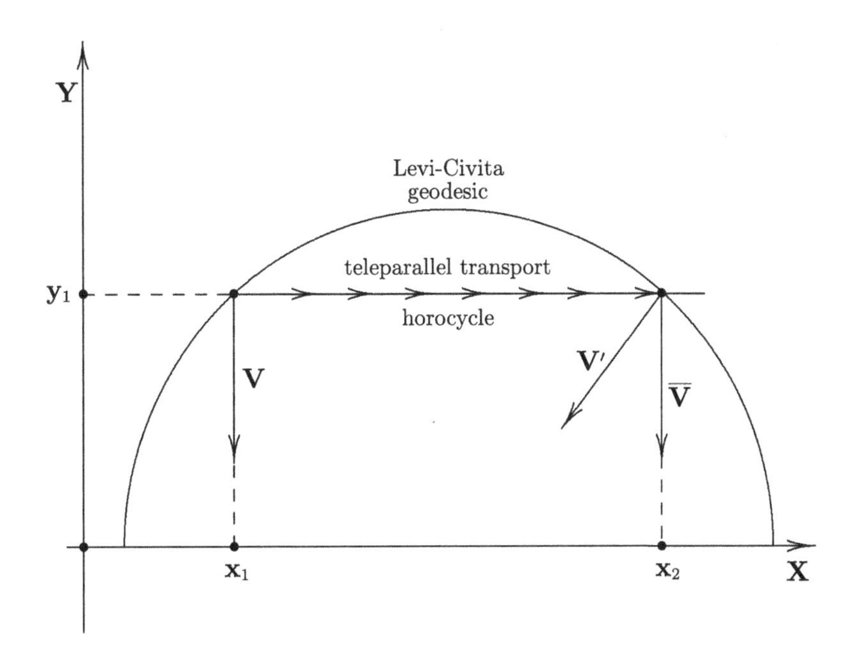

Figure 4.D.3 The Notion of Torsion (3)

When the vector $\mathbf{V}$ is parallel transported to (x_2, y_1) along the Levi-Civita geodesic it becomes the vector $\mathbf{V}'$. When the vector $\mathbf{V}$ is teleparallel transported to (x_2, y_1) along the horocycle $y_1 = constant$, it becomes $\overline{\mathbf{V}}$.

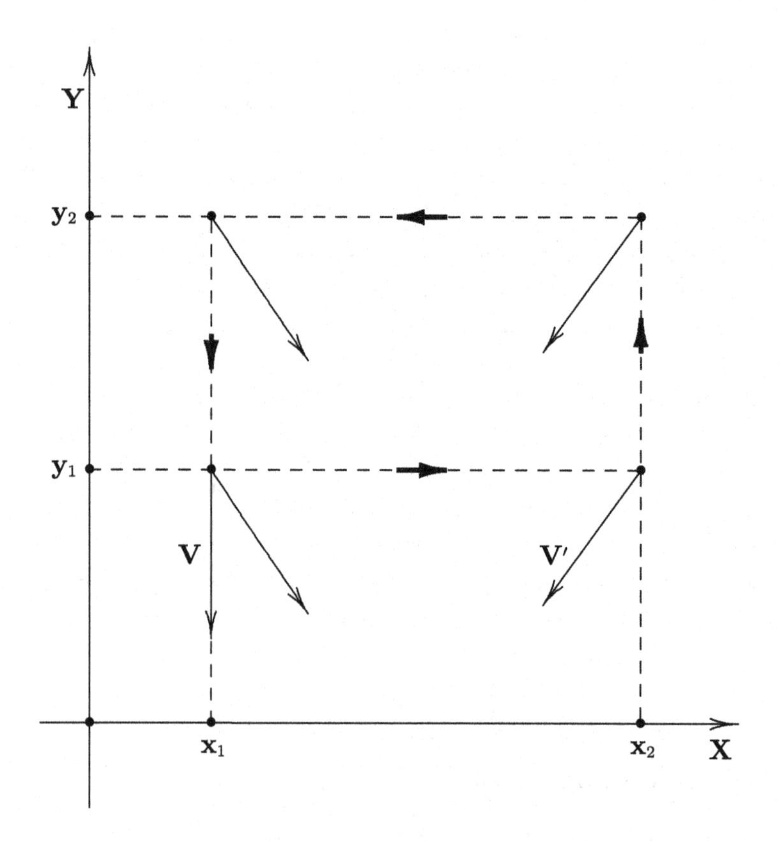

Figure 4.D.4 The Notion of Torsion (4)

The change of angle by parallel transferring a vector **V** in the positive sense divided by the area gives the curvature as a constant equal to minus one.

We see that the curvature is a higher order effect that does not appear at the level of the torsion. It appears as the difference of two parallel transfers. Although we are dealing with a curved manifold, all is in perfect analogy to the flat Minkowski plane.

E. Conclusion

In 1907 Einstein did not show the equivalence of acceleration and gravitation described by spacetime curvature. He did not show either the equivalence of geodesics and non-geodesics or the equivalence of rotating and non-rotating systems.

What he did, we now can see more clearly, was the introduction of accelerated reference systems exhibiting torsion through distant parallelism. There were physical consequences that needed to be checked for these systems, like the constancy of the speed of light independent of acceleration, no influence of acceleration on the rate of clocks and the length of standards. As far as these assumptions have been tested, they appear to be in order.

In 1911 Einstein formulated an equivalence principle that involved relative acceleration in an attempt to introduce ideas of Ernst Mach into his theory [Einstein, 1911]. This did not prove to be a happy idea since this notion makes mathematical sense only for bodies having the same 4-velocity and from a physical point of view accelerations are absolute. These ideas gave the theory its name.

However, as John Stachel [Stachel, 1980] pointed out, it was the older idea of 1907 that guided him through Ehrenfest's paradox of the rotating disc to Riemannian geometry. We can now see that going from Levi-Civita's connection in Minkowski's spacetime to teleparallelism opens the door to going back from torsion to a Levi-Civita connection with curvature.

CHAPTER 5

HOMOGENEOUS FIELDS ON TWO-DIMENSIONAL RIEMANNIAN MANIFOLDS

A. Outline

For a study of homogeneous fields it is useful to initially explore this concept in the simplest case where we have a non-trivial example. That occurs on two-dimensional surfaces. Although we are mainly interested in homogeneous fields on Lorentz manifolds, especially Minkowski space, we begin with the case of a positive definite metric since its differential geometry is simpler and more familiar. We will then take up the manifolds with indefinite metrics.

B. Definition of the Frame Bundle for a Surface

We shall make the usual assumptions about our surface M. We request that M be a separable Hausdorff space covered by overlapping two-dimensional coordinate charts. In the overlap, the two sets of coordinates should be related by regular transformations which are three times continuously differentiable. We shall further assume that M is orientable. This is equivalent to saying that there is an atlas for M with positive Jacobians for the coordinate transformations.

A Riemannian structure can be given on M by specifying, on each coordinate chart, $\{x, y\}$, an ordered pair of linearly independent differential one-forms

$$\omega_1(x, y)\,, \quad \omega_2(x, y)\,. \tag{5.B.1}$$

They are given by

$$\omega_1(x, y) = A_{11}(x, y)\,dx + A_{12}(x, y)\,dy\,, \tag{5.B.2a}$$

$$\omega_2(x, y) = A_{21}(x, y)\,dx + A_{22}(x, y)\,dy\,. \tag{5.B.2b}$$

Linear independence means that

$$\boldsymbol{\omega}_1 \wedge \boldsymbol{\omega}_2 = \det \begin{bmatrix} A_{11} & A_{12} \\ A_{21} & A_{22} \end{bmatrix} dx \wedge dy \neq 0 . \tag{5.B.3}$$

This is equivalent to requesting that the determinant

$$\det \begin{bmatrix} A_{11} & A_{12} \\ A_{21} & A_{22} \end{bmatrix} \neq 0 . \tag{5.B.4}$$

The differential forms (5.B.1) are supposed to be two-times continuously differentiable. This means that we assume for the functions

$$A_{jk}(x, y) , \qquad j, k \in \{1, 2\} , \tag{5.B.5}$$

two-times continuous differentiability. We may combine the one-forms into a single complex valued differential one-form defined by

$$\boldsymbol{\kappa} \equiv \boldsymbol{\omega}_1 + i\, \boldsymbol{\omega}_2 . \tag{5.B.6}$$

On the same coordinate chart we could choose a different complex valued differential one-form defined by

$$\boldsymbol{\kappa}_\phi \equiv \boldsymbol{\kappa}\, e^{i\phi(x,y)} \tag{5.B.7}$$

with a real function ϕ that is two times differentiable. If ϕ is constant on the chart, we call the transition from $\boldsymbol{\kappa}$ to $\boldsymbol{\kappa}_\phi$ a "local gauge transformation of the first kind". If ϕ is not constant, the transformation is said to be "of the second kind".

The form $\boldsymbol{\kappa}$ is also known as a local orthonormal differential one-form when we impose the additional restriction of orthonormality. For brevity, it is then designated as a "local frame". The set of all the frames $\boldsymbol{\kappa}_\phi$ on the coordinate chart S is called the "local frame bundle", $B(S)$. With a choice of $\boldsymbol{\kappa}$ on the chart we can now introduce a local Riemannian metric by

$$ds^2 = \boldsymbol{\kappa}\, \bar{\boldsymbol{\kappa}} = [\boldsymbol{\omega}_1]^2 + [\boldsymbol{\omega}_2]^2 = \boldsymbol{\kappa}_\phi\, \bar{\boldsymbol{\kappa}}_\phi \tag{5.B.8}$$

which is not affected by the choice of the function ϕ. The bar indicates the complex conjugate.

Before we discuss the introduction of a global Riemannian metric, we use a more intuitive and geometric description of the local frame bundle. Clearly, it depends on three parameters, namely x, y, and ϕ. We can, therefore, think of it as a three-dimensional space. The coordinate chart is some open set S in the x-y–plane, while the coordinate ϕ is circular. The local frame bundle, $B(S)$, is thus a three-dimensional space of topology $S\times\mathbb{S}^1$, where $\mathbb{S}^1$ is the unit circle parameterized by ϕ. The product structure shows that the local frame bundle is a fiber space where fibers are circles and whose base manifold is S.

Definition: The frame $\boldsymbol{\kappa}$ is a *reference section* in $S \times \mathbb{S}^1$ corresponding to $\phi = 0$ on S.

A gauge transformation of the second kind is the choice of a new reference section. This, naturally, does not affect the metric.

We are now well prepared to define the global Riemannian metric on M. Suppose we have two coordinate charts, which cover sets on M, with a non-empty intersection. We then have local frame bundles with local frames

$$\boldsymbol{\kappa}'_{\phi'}(x', y'), \qquad \boldsymbol{\kappa}_{\phi}(x, y) \tag{5.B.9}$$

on the respective charts. We postulate, in the overlapping region of the two charts, the existence of a two-times continuously differentiable function

$$\boldsymbol{\Phi}(\mathbf{p}) \tag{5.B.10}$$

where $\mathbf{p}$ designates a point of the manifold in the overlap without regard to any specific chart or atlas. $\boldsymbol{\Phi}$ designates the function, which has its support on the intersection of the charts. The overlap is thus promoted to a chart with $\boldsymbol{\Phi}$ as its coordinate function. When $\boldsymbol{\Phi}$ is "interpreted" with respect to each of the charts in the pair, $\boldsymbol{\Phi}$ becomes a chart specific function, Φ and Φ', on each chart such that the functions are invariant

$$\Phi'\,(x', y') \equiv \Phi\,(x, y) \tag{5.B.11}$$

and also satisfy the requirement that the corresponding local frames also be invariant

$$\boldsymbol{\kappa}'_{\phi'}(x', y') = \boldsymbol{\kappa}_{\phi}(x, y)\,. \tag{5.B.12}$$

Coordinate maps on the respective charts are related by

$$x' = x'(x, y)\,, \qquad y' = y'(x, y)\,, \tag{5.B.13}$$

and the gauge functions for the differential one-forms have the transformations

$$\phi'(x', y') = \phi(x, y) \quad + \Phi(x, y)\,, \tag{5.B.14a}$$

$$\phi(x, y) = \phi'(x', y') - \Phi'(x', y')\,. \tag{5.B.14b}$$

A non-trivial example of this transformation for the two-sphere is given in **Appendix B**.

The function $\mathbf{\Phi}$ relates the two reference sections of the two local frame bundles in the overlap. By means of the transformations (5.B.11) one can "glue" together the local frame bundles into a global frame bundle. This global bundle for the manifold M will in general no longer have the product topology $M \times \mathbb{S}^1$. For the two-sphere it has the topology of the projective three-sphere. One has to stress also that the one-forms may not exist globally, which amounts to saying that global sections cannot be found.

C. Local Homogeneous Fields

The connection form is defined as the unique solution of the Maurer-Cartan equations

$$d\omega_1 - \boldsymbol{\omega} \wedge \omega_2 = 0\,, \qquad d\omega_2 + \boldsymbol{\omega} \wedge \omega_1 = 0\,. \tag{5.C.1}$$

On a domain of M, an open and connected subset M', we have:

Definition: A frame bundle $B(M')$ is called a *local homogeneous field* if the independent one-forms (5.B.1) are defined on M' and the connection form $\boldsymbol{\omega}$ is also developed in the basis forms

$$\boldsymbol{\omega} = g_1\,\omega_1 + g_2\,\omega_2 \tag{5.C.2}$$

with the additional requirement on the scalar coefficients that they be constants, that is, we have

$$g_1 = constant\,, \qquad g_2 = constant\,. \tag{5.C.3}$$

In general, these scalars would be functions of position, rather than constants. Thus, their constancy is the decisive condition for a local homogeneous field.

We define the strength of the field by

$$g \equiv \left[[g_1]^2 + [g_2]^2\right]^{\frac{1}{2}}\,. \tag{5.C.4}$$

The notion of a homogeneous field is clearly independent of coordinates. We do not assume that the domain M can be covered by one coordinate chart; it may need several, but it is decisive that at least one of the differential one-forms (5.B.1) be defined throughout M. (The other and the complex combination are then trivially defined.) The components, g_1 and

g_2, transform like components of a vector under a uniform rotation of all frames by a constant angle, ϕ. This will leave the field strength, g, the magnitude of the vector, constant. The field strength will, however, not be invariant under general gauge transformations of the second kind.

Now, we want to find conditions for the existence of local homogeneous fields. For such a field we have from (5.C.1), (5.C.2), and (5.C.3)

$$d\boldsymbol{\omega} = g_1\, d\omega_1 + g_2\, d\omega_2 \tag{5.C.5a}$$

$$= \left[[g_1]^2 + [g_2]^2 \right] \omega_1 \wedge \omega_2 \,. \tag{5.C.5b}$$

Since the intrinsic curvature of a 2-surface can be calculated from the differential of the connection form $\boldsymbol{\omega}$ [Flanders, 1963] [Misner, 1973], we have

$$d\boldsymbol{\omega} = -K\, \omega_1 \wedge \omega_2 \,, \tag{5.C.6}$$

where K is the Gaussian curvature of M. We immediately see that K is always negative since

$$K = -g^2 \,. \tag{5.C.7}$$

This shows clearly that homogeneous fields do not exist where the Gaussian curvature is positive. We have learned that homogeneous fields exist only on surfaces of zero curvature and constant negative curvature. We discuss the case of vanishing curvature first.

D. Surfaces of Zero Curvature

The field strength, g, vanishes for surfaces with vanishing curvature. This observation leads us to adopt a special name for these fields:

Definition: We call a homogeneous field of zero strength a "*vacuum*".

There are only three geodesically complete orientable surfaces of vanishing curvature (up to affine transformations). These have the topologies $\mathbb{R} \times \mathbb{R}$, $\mathbb{R} \times \mathbb{S}^1$, and $\mathbb{S}^1 \times \mathbb{S}^1$. They are defined by means of quotient maps of the Euclidean plane with respect to properly discrete subgroups of the group of translations of the plane.

We take Cartesian coordinates in the plane and select for our linearly independent one-forms

$$\omega_1 = dx, \qquad \omega_2 = dy \tag{5.D.1}$$

since the connection form $\boldsymbol{\omega}$ vanishes. The complete, orientable manifolds of zero curvature are now the plane, the cylinder, and the torus. The invariant of the cylinder is the circumference. The invariants of the torus are the two circumferences. By a transformation with constant ϕ we can arrange that the coordinates lie along the circumferences if the complete manifold is not the plane. These manifolds comprise global vacua. Any domains of them are local vacua. We have thus found a construction for the general local vacuum for zero curvature. One might stress that such a local vacuum is perfectly well defined although the topology of the manifold M might be extremely complicated with many different holes in it. This could not affect the vacuum at all.

E. Surfaces of Constant Negative Curvature

Having disposed of the two-dimensional surfaces of positive curvature as carriers of homogeneous fields and explored the situation on the corresponding surfaces of zero curvature, we now pass on to two-dimensional surfaces of negative curvature.

We choose a coordinate chart for which the upper half of the complex plane is given by

$$y > 0 \,. \tag{5.E.1}$$

The points $y > 0$ are then those of the Poincaré model for the space of constant negative curvature. We obtain for the curvature

$$K = -\frac{1}{R^2}, \qquad R > 0 \,. \tag{5.E.2}$$

The metric is given by means of the one-forms

$$\omega_1 = \frac{R}{y}\,dx, \qquad \omega_2 = \frac{R}{y}\,dy \tag{5.E.3}$$

when combined as in (5.B.8). The connection form $\boldsymbol{\omega}$, defined by the equations (5.C.1), becomes

$$\boldsymbol{\omega} = \frac{1}{y}\,dx = \frac{1}{R}\,\omega_1 \tag{5.E.4}$$

with the Ricci coefficients and field strength given by

$$g_1 = \frac{1}{R}, \qquad g_2 = 0, \qquad g = \frac{1}{R}. \tag{5.E.5}$$

We say, therefore, the space of constant negative curvature gives rise to a constant field. The field strength is determined by the curvature and is the inverse of its radius of curvature. The metric is conformally Euclidean

$$ds^2 = \frac{R^2}{y^2}\left[\,dx^2 + dy^2\,\right] \tag{5.E.6}$$

with the conformal factor R^2/y^2.

We now want to describe the homogeneous field in invariant terms. Its frame is based on a pencil of geodesics with unit tangent vectors **T** such that

$$\mathbf{T} = \frac{y}{R}\frac{\partial}{\partial y}, \qquad \omega_1(\mathbf{T}) = 0, \qquad \omega_2(\mathbf{T}) = 1 \tag{5.E.7}$$

and on their orthogonal trajectories, a pencil of horocycles. [See **Appendix C** for a discussion of the Poincaré half-plane.]

The field given by (5.E.4) is a pseudovector (as g_1 does not change sign under the reflection $\omega_1 \longrightarrow -\omega_1,\ \omega_2 \longrightarrow \omega_2$) along the horocycles, oriented from left to right, that is, "down" to "up" in the upper half-plane of Poincaré's plane. To visualize the nature of the field one can describe it as follows: In the Euclidean plane parallel geodesics keep a constant distance. In the hyperbolic plane they do not. The distance between parallels increases steadily as one runs along these parallels from their meeting point at infinity while they diverge further and further. One can describe this by saying that "new" space is being created or is opening up as they spread out. This space is created in the opposite direction of the sideways orientation of the pseudovector. The "field strength", which has dimension cm^{-1}, is a measure of the relative rate for the creation "down" to "up" of new space. A brief calculation shows this as follows: the distance L between geodesics $x = 0$ and $x = x_0$ measured along the horocycle $y = constant > 0$ is according to (5.E.5)

$$L = \frac{R}{y}\, x_0\,. \tag{5.E.8}$$

For the relative change of length L for fixed x_0, we have

$$\frac{1}{L}\frac{dL}{ds} = \left[\frac{R}{y}\,x_0\right]^{-1}\left[-\frac{R}{y^2}\,x_0\,dy\right]\left[-\frac{R}{y}\,dy\right]^{-1} = \frac{1}{R}\,. \tag{5.E.9}$$

This relative rate is independent of position (namely, coordinates x and y) and, as we learn from **Appendix D**, also independent of direction. For further remarks about "space creation," see **Appendix E**.

F. Remarks on Curvature

Before leaving the discussion of the hyperbolic plane a brief remark about topological questions. It appears that none of the finite hyperbolic space-forms can carry a global homogeneous field. But an infinite space-form with a homogeneous field can be constructed by taking a strip of the Poincaré half-plane bounded by the geodesics $x = 0$ and $x = x_0$ and identifying points on the boundary with the same y. It appears that this sort of hyperbolic cone is the only other global homogeneous field. A piece of this hyperbolic cone can be imbedded isometrically as a surface of revolution. [**See Figure 5.F.1.**]

The advantage of this formulation of curvature is its introduction at a lower level of differentiation. This follows from the fact that the quantity defined is $1/R$ and not $1/R^2$. The Gaussian curvature has here been integrated once. This definition might be useful from a physicist's point of view since the dimension cm^{-1} is also the dimension of acceleration. The Riemann tensor has dimension cm^{-2} and corresponds to the gradient of accelerations in Einstein's theory of gravitation. [**See Appendix K.**]

There are also disadvantages to the introduction of curvature as sketched above. First of all it applies only to homogeneous spaces. This was the reason why one could integrate the Gaussian curvature since it was constant. However, it is only through the study of homogeneous spaces that one can hope to gain a better understanding of the geometry in more general spaces. The homogeneous spaces are models for the local behavior of spaces in which the curvature varies from point to point.

A definite disadvantage of our formulation is the fact that the procedure does not work for 2-surfaces of positive curvature and there not even locally. The reason is that homogeneous fields do not exist on any domain of a two-sphere. This is a peculiar phenomenon of the two-dimensional case (the three-dimensional sphere $\mathbb{S}^3$ allows homogeneous fields). Such exceptional cases will also occur in higher dimensions, for example, for manifolds which are products containing two-spheres. A possible cure of the disease might be the admission of complex frames.

Another consideration might be mentioned here, although it does not apply to the two-dimensional case with a definite metric: For a definite metric in two dimensions, the field strength of a homogeneous field determines the curvature. This is no longer so for indefinite metrics or higher dimensions.

Non-vanishing homogeneous fields exist in spaces with vanishing Riemann tensors. This creates the task of distinguishing between "genuine" fields (occurring in spaces with Riemann tensors different from zero) and "fictitious" fields occurring in flat spaces. It is precisely this phenomenon which is of interest to the physicist. We shall see that fictitious homogeneous fields exist in Minkowski space. The fact that these fields can masquerade as gravitational fields has led to the new invariant formulation of the principle of equivalence. [**See Chapter 4, Section A.**]

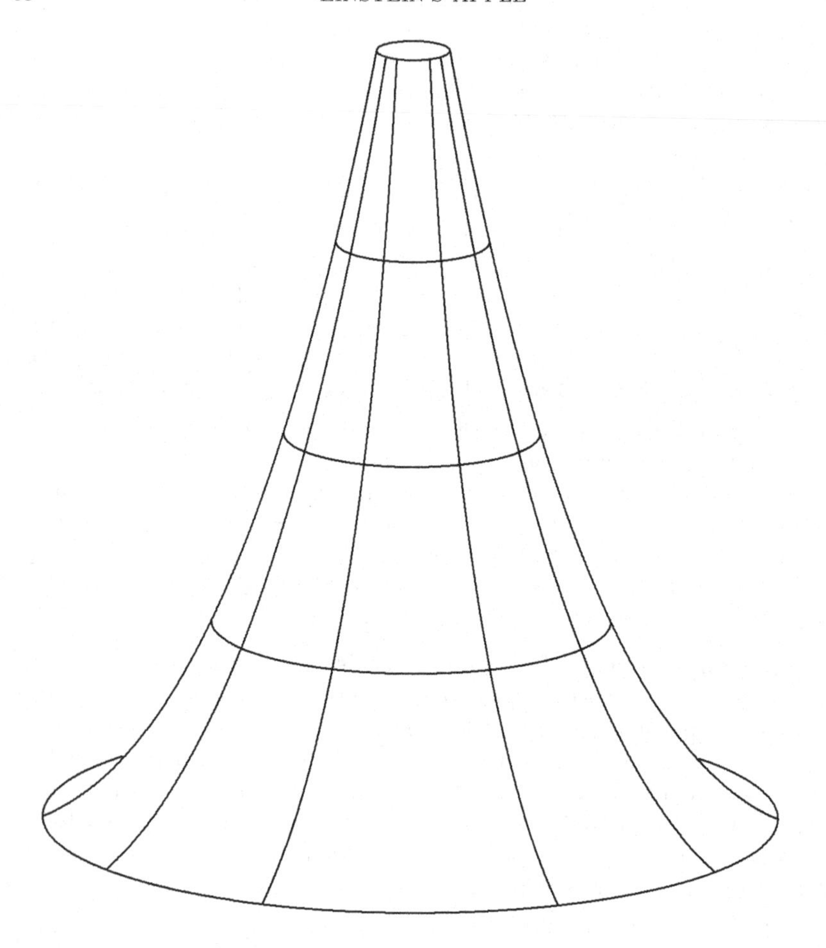

Figure 5.F.1 Two-Dimensional Space of Constant Curvature Embedding A piece of the "pseudosphere", the two-dimensional space of constant negative curvature can be isometrically embedded as a surface of revolution. Every meridian of the surface is a plane curve known as a tractrix. The meridians are truncated left parallels of each other. They meet at infinity and correspond to lines of constant x in the Poincaré half-plane. The horocycles of $y = constant$ appear, in this model, as circles orthogonal to the meridians. The largest circle, where the surface has a singular edge and the meridians are truncated, has the circumference $2\pi R$. If we place it at $y = 1$, the lines $x = 0$ and $x = 2\pi$ have been identified to create a surface of revolution. The complete pseudosphere cannot be imbedded isometrically as a surface in three-dimensional space without singularities.

G. Indefinite Two-Dimensional Metrics

For definite two-dimensional metrics we found that there were no such fields on spheres, only the trivial vacuum field on the plane, and fields of fixed strength (determined by the curvature) on the hyperbolic plane.

We find a different situation for the indefinite metrics. Again, we distinguish between the cases of negative, positive, and vanishing curvature. The corresponding spacetimes are known as de Sitter, anti-de Sitter, and Minkowski spaces. Here we shall find non-trivial homogeneous fields in all three cases. These fields will be characterized by a pseudo-two-vector (the connection form) which will be spacelike, timelike, or null according to the curvature of the spacetime.

The relation between curvature and field strength can again be easily derived. We put

$$ds^2 = [\omega_0]^2 - [\omega_1]^2 \tag{5.G.1a}$$

$$= M^{AB}\,\omega_A\,\omega_B \tag{5.G.1b}$$

where

$$M^{AB} = \begin{bmatrix} 1 & 0 \\ 0 & -1 \end{bmatrix}, \qquad A, B \in 0, 1\,. \tag{5.G.2}$$

The connection form is defined by the equations

$$d\omega_0 = \omega_{01} \wedge \omega^1 = -\,\omega_{01} \wedge \omega_1\,, \tag{5.G.3a}$$

$$d\omega_1 = \omega_{10} \wedge \omega^0 = -\,\omega_{01} \wedge \omega_0\,, \tag{5.G.3b}$$

where we have used

$$\omega^0 = \omega_0\,, \qquad \omega^1 = -\,\omega_1\,, \qquad \omega_{01} = -\,\omega_{10}\,. \tag{5.G.4}$$

For a homogeneous field, we have the development of the connection form

$$\omega_{01} = g^0\,\omega_0 + g^1\,\omega_1\,, \tag{5.G.5}$$

with the side condition on the connection coefficients that they must be constant:

$$g^0 = constant\,, \qquad g^1 = constant. \tag{5.G.6}$$

We define the Gaussian curvature, K, through the expression

$$d\omega_{01} = -\,K\,\omega_0 \wedge \omega_1 \tag{5.G.7}$$

and obtain from equations (5.G.3) and (5.G.5)

$$d\omega_{01} = g^0\, d\omega_0 + g^1\, d\omega_1 \tag{5.G.8a}$$

$$= \left[-\left[g^0\right]^2 + \left[g^1\right]^2 \right] \omega_0 \wedge \omega_1 . \tag{5.G.8b}$$

$$= \widetilde{g}^2\, \omega_0 \wedge \omega_1 . \tag{5.G.8c}$$

Comparing (5.G.7) and (5.G.8b), we obtain

$$K = -\,\widetilde{g}^2 = \left[g^0\right]^2 - \left[g^1\right]^2 = M_{AB}\, g^A\, g^B . \tag{5.G.9}$$

This equation establishes the relation between the sign of the curvature of a two-dimensional homogeneous spacetime and the character of the pseudo-two-vector field strength g^A. One sees that the components of the connection form, by themselves, without the use of their derivatives, determine the curvature invariants of the metric. This is a special feature of homogeneous spacetimes.

We can now go to a new frame

$$\omega_0' , \quad \omega_1' \tag{5.G.10}$$

through a Lorentz transformation with rapidity χ,

$$\omega_0 = ch\,\chi\, \omega_0' + sh\,\chi\, \omega_1' \tag{5.G.11a}$$

$$\omega_1 = sh\,\chi\, \omega_0' + ch\,\chi\, \omega_1' \tag{5.G.11b}$$

where χ is constant. We also allow space and time reflections of the frame. The three cases above can then be normalized to

$$\omega_{01} = \frac{1}{R}\, \omega_1 \qquad \text{de Sitter}, \qquad K < 0, \tag{5.G.12a}$$

$$\omega_{01} = \frac{1}{R}\, \omega_0 \qquad \text{anti-de Sitter}, \qquad K > 0, \tag{5.G.12b}$$

$$\omega_{01} = \widehat{g}\,[\,\omega_0 + \omega_1\,] \qquad \text{Minkowski}, \qquad K = 0. \tag{5.G.12c}$$

We do not normalize the constant $\widehat{g}$ in the Minkowski case since we can still vary it for a given frame by letting it go to zero without changing the curvature. This makes the Minkowski case different from the other two.

Another feature that occurs for an indefinite metric is the horizon phenomenon. We found a homogeneous field for the whole hyperbolic plane which is geodesically complete. When we study the individual indefinite metrics, we shall see that homogeneous fields do not cover geodesically complete regions of the covering manifolds except for the vacuum field in Minkowski space. As we shall see later, this can also be expressed by saying that although the frame bundle is trivial, it does not allow homogeneous sections.

H. The Field Concept

Before we go on to discussions of the specific indefinite metrics, it might be useful to justify the terminology invoking the "field" concept. We do this in the familiar four-dimensional spacetime of relativity. Let $\mathbf{T}$ be the tangent vector of a timelike worldline normalized to length one for a particle of mass m and charge e. We write

$$\mathbf{T} = u^a \, \mathbf{e}_a \,, \qquad \mathbf{e}_a \cdot \mathbf{e}_b = \eta_{ab} \,, \qquad u^a \, u_a = 1 \,. \tag{5.H.1}$$

We have for the scalar product of the orthonormal vectors $\mathbf{e}_a$ and $\mathbf{e}_b$

$$\eta_{ab} = \begin{bmatrix} 1 & 0 & 0 & 0 \\ 0 & -1 & 0 & 0 \\ 0 & 0 & -1 & 0 \\ 0 & 0 & 0 & -1 \end{bmatrix} . \tag{5.H.2}$$

The $\mathbf{e}_a$ define an orthonormal frame in a Lorentz manifold with

$$\boldsymbol{\omega}^b \left(\mathbf{e}_a \right) = \delta^b{}_a \,. \tag{5.H.3}$$

As Einstein's theory of gravitation is based in a geometrical description, the specific solutions of his field equations provide expressions for line elements easily written in terms of orthonormal frames. Therefore, the frames of (5.H.3) can carry information about the local gravitational field when chosen appropriately. Let

$$\mathbf{F} = \frac{1}{2} F_{ab} \, \boldsymbol{\omega}^a \wedge \boldsymbol{\omega}^b \,, \qquad F_{ab} = - F_{ba} \tag{5.H.4}$$

be the electromagnetic field tensor and

$$\boldsymbol{\nabla}_{\mathbf{T}} \tag{5.H.5}$$

the operator of covariant differentiation in the direction $\mathbf{T}$. We have then as the Lorentz equation of motion for the particle (assuming its moments are vanishing)

$$\boldsymbol{\nabla}_{\mathbf{T}} \left[u^a \mathbf{e}_a \right] = \frac{e}{m} F^a{}_b \, u^b \, \mathbf{e}_a \,. \tag{5.H.6}$$

Working on the left-hand side of this equation, we have

$$\boldsymbol{\nabla}_{\mathbf{T}} \left[u^a \mathbf{e}_a \right] = \boldsymbol{\nabla}_{\mathbf{T}} \left[u^a \right] \mathbf{e}_a + u^a \, \boldsymbol{\nabla}_{\mathbf{T}} \left[\mathbf{e}_a \right] \,. \tag{5.H.7}$$

The four-acceleration is

$$\dot{u}_a \equiv \frac{d}{ds}[u_a] \quad (= \boldsymbol{\nabla}_{\mathbf{T}}[u_a]\,)\,, \tag{5.H.8}$$

and, furthermore, we have as definition of the connection form

$$\boldsymbol{\nabla}_{\mathbf{T}}[\mathbf{e}_a] = \boldsymbol{\omega}_a{}^b(\mathbf{T})\,\mathbf{e}_b\,. \tag{5.H.9}$$

Leaving out the common vector $\mathbf{e}_a$, we get from (5.H.7) the components

$$\dot{u}_a + \boldsymbol{\omega}_{ba}(\mathbf{T})\,u^b = \frac{e}{m}F_{ab}\,u^b. \tag{5.H.10}$$

For a homogeneous field, we first write the connection forms, as applied to the tangent vector, in terms of the Ricci coefficients

$$\boldsymbol{\omega}_{ba}(\mathbf{T}) = u^c\,g_{bac}\,, \qquad g_{bac} = -\,g_{abc} \tag{5.H.11}$$

and then require constant Ricci coefficients, that is, $g_{bac} = constant$. We finally obtain

$$\dot{u}_a = \frac{e}{m}F_{ab}\,u^b + g_{abc}\,u^b\,u^c\,. \tag{5.H.12}$$

The first term on the right hand side of the equation is the acceleration due to the Lorentz force. The second term, which is also orthogonal to the four velocity, u^a, is due to gravitational and inertial forces. It describes the acceleration of a neutral, freely falling particle, that is, one moving on a geodesic in spacetime.

Specializing to the case of two-dimensional Minkowski spacetime, where there is only one space dimension (and no electromagnetic field), we have a 'two-velocity' with

$$[u^0]^2 - [u^1]^2 = 1\,, \qquad u^0 = \gamma\,, \qquad u^1 = \gamma\,\beta\,, \tag{5.H.13a}$$

$$\gamma = \frac{1}{\sqrt{1-\beta^2}}\,, \qquad \beta = \frac{v}{c}\,, \tag{5.H.13b}$$

where v is the spatial velocity. Again, the connection form(s) for a homogeneous field must be

$$g_0 \equiv g_{010} = constant\,, \qquad g_1 \equiv g_{011} = constant\,. \tag{5.H.14}$$

The equation of motion to be solved is the geodesic equation

$$\boldsymbol{\nabla}_{\mathbf{T}}[u^a\mathbf{e}_a] = 0\,, \tag{5.H.15}$$

which is the left hand side of (5.H.10) when no other physical field, such as the electromagnetic field, is present. For a neutral free particle in such an inertial or gravitational field (which are equivalent by Einstein's first principle [**See Chapter 0, Section A.**]), we have

$$\dot{u}_0 = (\gamma)^{\cdot} = g_0\left[u^0 u^1\right] + g_1\left[u^1\right]^2 \tag{5.H.16a}$$

$$= \beta\,\gamma^2\left[g_0 + \beta\, g_1\right], \tag{5.H.16b}$$

and also

$$\dot{u}_1 = -(\beta\,\gamma)^{\cdot} = -g_0\left[u^0\right]^2 - g_1\left[u^0 u^1\right] \tag{5.H.17a}$$

$$= -\gamma^2\left[\,g_0 + \beta\, g_1\right]. \tag{5.H.17b}$$

The accelerations do not depend on the spacetime coordinates. This independence of position gives us the right to talk of homogeneous fields. The accelerations, do however, depend on velocity just as in the electromagnetic case.

We obtain from equations (5.H.13), in normalized frames,

$$(\beta\,\gamma)^{\cdot} = \frac{1}{R}\beta\,\gamma^2 \quad \text{de Sitter}, \qquad K < 0, \tag{5.H.18a}$$

$$(\beta\,\gamma)^{\cdot} = \frac{1}{R}\gamma^2 \quad \text{anti-de Sitter}, \quad K > 0, \tag{5.H.18b}$$

$$(\beta\,\gamma)^{\cdot} = \frac{g}{1-\beta} \quad \text{Minkowski}, \qquad K = 0. \tag{5.H.18c}$$

Only in the de Sitter model is $\beta = 0$ a solution of the equation of motion.

I. The de Sitter Metric ($K < 0$)

The de Sitter metric can be normalized as we know from (5.H.13a) to

$$\omega_{01} = \frac{1}{R}\omega_1, \qquad R = \textit{constant} > 0. \tag{5.I.1}$$

We then have, from equations (5.F.3)

$$d\omega_0 = 0, \qquad d\omega_1 = \frac{1}{R}\omega_0 \wedge \omega_1. \tag{5.I.2a,b}$$

The first equation of (5.I.2) shows that ω_0 is a closed form and therefore at least locally exact which enables us to introduce a variable t by putting

$$\omega_0 = R\,dt\,. \tag{5.I.3}$$

The second equation of (5.I.2) gives then

$$d\omega_1 - dt \wedge \omega_1 = e^t\, d\big[e^{-t}\omega_1\big] = 0\,. \tag{5.I.4}$$

This again enables us to introduce a second variable σ by means of the closed nature of the form $e^{-t}\omega_1$. We thus are able to write it as

$$e^{-t}\,\omega_1 = R\,d\sigma\,, \tag{5.I.5}$$

which immediately gives

$$\omega_1 = R\,e^t\,d\sigma\,. \tag{5.I.6}$$

According to equations (5.G.1), we then have the metric,

$$ds^2 = [\omega_0]^2 - [\omega_1]^2 = R^2\left[dt^2 - e^{2t}\,d\sigma^2\right]. \tag{5.I.7}$$

With (5.I.1), the connection form becomes

$$\omega_{01} = e^t\,d\sigma\,. \tag{5.I.8}$$

The two-dimensional de Sitter spacetime is often represented as a one-sheeted hyperboloid in a three-dimensional Minkowski space with coordinates T, X, Y and metric

$$ds^2 = dT^2 - dX^2 - dY^2\,. \tag{5.I.9}$$

The surface is given by the parameter representation

$$T = R\,sh\,\tau\,, \qquad X + iY = R\,ch\,\tau\,e^{i\phi} \tag{5.I.10}$$

with equation

$$T^2 - X^2 - Y^2 = -R^2\,. \tag{5.I.11}$$

The metric is then given by this imbedding as

$$ds^2 = R^2\left[d\tau^2 - ch^2\tau\;d\phi^2\right]. \tag{5.I.12}$$

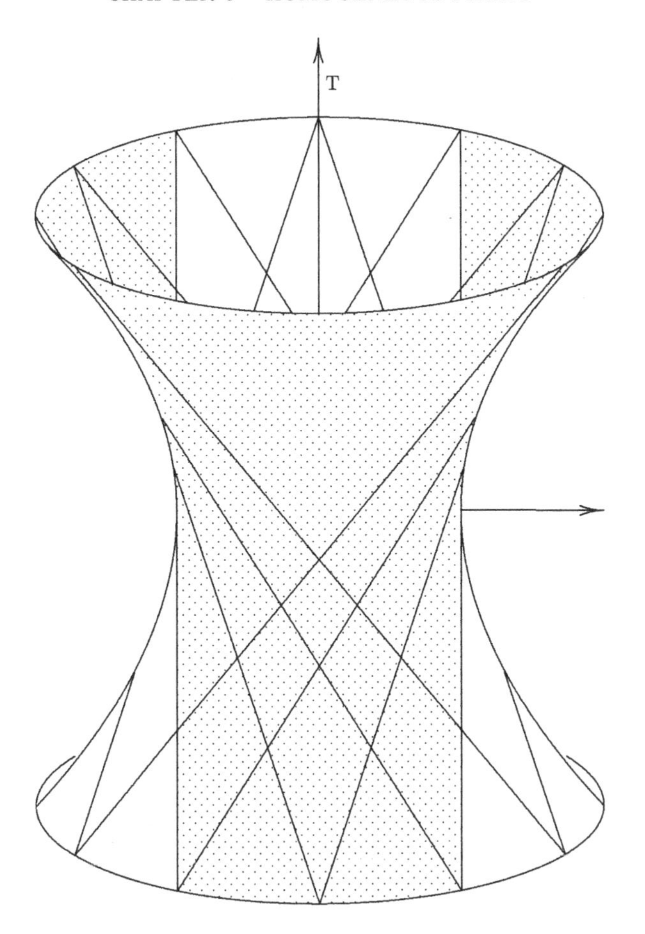

Figure 5.I.1 Hyperboloid of Steady State Model The hyperboloid $T^2 - X^2 - Y^2 = -R^2$ is shown with the Y–axis pointing towards the viewer. The homogeneous field on the de Sitter hyperboloid is the shaded area bounded by two generators. The two-dimensional surface model is not simply connected, as can be seen by a circle around the waist of the hyperboloid; it cannot be shrunk to a point.

This spacetime is complete. One can now also go to the covering surface by dropping the assumption of periodicity in the coordinate ϕ modulo 2π.

We want to see now which part of the manifold is covered by the coordinates t and σ of the homogeneous field. We put

$$X = R\sigma e^t\,, \tag{5.I.13a}$$

$$T + Y = R e^t\,, \tag{5.I.13b}$$

$$T - Y = R\sigma^2 e^t - R e^{-t}\,. \tag{5.I.13c}$$

This clearly satisfies equation (5.I.11) and gives rise to the metric (5.I.7).

We see from equations (5.I.13) that the homogeneous field with coordinates t and σ is restricted to the region of the hyperboloid where $T+Y > 0$. This covers only half of the hyperboloid, also known as the Steady State Model of cosmology. [**See Figure 5.I.1.**]

The connection form of the homogeneous field becomes

$$\omega_{01} = e^t\, d\sigma = -\frac{\cos\phi\; d\tau + [\mathit{ch}\,\tau + \mathit{sh}\,\tau\,\sin\phi]\; \mathit{ch}\,\tau\, d\phi}{\mathit{sh}\,\tau + \mathit{ch}\,\tau\,\sin\phi} \tag{5.I.14}$$

in the τ-ϕ–coordinates (which cover the whole manifold). The denominator vanishes where $T + Y$ vanishes. This shows that the homogeneous field cannot be extended over the whole manifold. It is bounded by two horizons. The tangents of these geodesics are the timelike vectors of our homogeneous frame. They form a pencil whose common origin has been removed to $t = -\infty$. This is analogous to the homogeneous field on the hyperbolic plane. Again, the curvature of spacetime can be described as the creation of space between the geodesics of the pencil. What is new is the appearance of horizons.

J. Geodesics of the de Sitter Spacetime ($K < 0$)

It is easy to check that the lines

$$\sigma = \frac{X}{T+Y} = \mathit{constant} \tag{5.J.1}$$

are geodesics being the analogs of great circles on a sphere obtained through intersection of an equilateral hyperboloid with a plane through its center.

The geodesics can be obtained from the Euler-Lagrange equations of a variational principle with Lagrange function

$$L = \frac{1}{2} R^2 \left(\dot{t}^2 - e^{2t}\,\dot{\sigma}^2 \right)\,. \tag{5.J.2}$$

A dot indicates derivation with respect to proper time. We have the two integrals

$$\frac{\partial L}{\partial \dot{\sigma}} = R^2 \, e^{2t} \, \dot{\sigma} = -\alpha = constant \tag{5.J.3}$$

and

$$R^2 \left(\dot{t}^2 - e^{2t} \, \dot{\sigma}^2 \right) = 1 \tag{5.J.4}$$

for timelike geodesics. The frame vectors $\mathbf{e}_0$ and $\mathbf{e}_1$ of the homogeneous field are defined by

$$\boldsymbol{\omega}^j \left(\mathbf{e}_k \right) = \delta^j{}_k \,, \tag{5.J.5}$$

which gives

$$\mathbf{e}_0 = \frac{1}{R} \, \partial_t \,, \qquad \mathbf{e}_1 = -\frac{1}{R} \, e^{-t} \, \partial_\sigma \,. \tag{5.J.6}$$

We obtain for the unit tangent vector $\mathbf{T}$ of the timelike geodesic

$$\mathbf{T} = T^0 \, \mathbf{e}_0 + T^1 \, \mathbf{e}_1 \,, \tag{5.J.7}$$

with

$$T^0 = \frac{1}{R} \left[R^2 + \alpha^2 \, e^{-2t} \right]^{\frac{1}{2}} \,, \qquad T^1 = \frac{\alpha}{R \, e^t} \,. \tag{5.J.8}$$

We have

$$\left(T^0 \right)^2 - \left(T^1 \right)^2 = 1 \,. \tag{5.J.9}$$

The equation of the geodesic is

$$\boldsymbol{\nabla}_{\mathbf{T}} \mathbf{T} = 0 \,. \tag{5.J.10}$$

With (5.J.7) and (A.Q.2), from **Appendix A**,

$$\mathbf{e}_j \cdot \mathbf{e}_k = \eta_{jk} \,, \tag{5.J.11}$$

we obtain

$$\mathbf{A} = A^m \, \mathbf{e}_m = dT^m(\mathbf{T}) \, \mathbf{e}_m = \mathbf{e}_m \, g^m{}_{kj} \, T^k \, T^j \,, \tag{5.J.12}$$

where $\mathbf{A}$ is the leg-acceleration vector. From

$$\boldsymbol{\omega}_{01} = \frac{1}{R} \, \boldsymbol{\omega}_1 \,, \tag{5.J.13}$$

we obtain

$$g^0{}_{10} = g^1{}_{00} = 0 \,, \qquad g^0{}_{11} = g^1{}_{01} = \frac{1}{R} \,. \tag{5.J.14}$$

This gives

$$A^0 = \frac{1}{R}\left(T^1\right)^2, \qquad A^1 = \frac{1}{R}T^0 T^1. \tag{5.J.15}$$

We obtain thus for the value of the leg-acceleration

$$\left[\left(A^1\right)^2 - \left(A^0\right)^2\right]^{\frac{1}{2}} = \frac{T^1}{R}. \tag{5.J.16}$$

The geodesics with vanishing T^1

$$\mathbf{T} = \mathbf{e}_0 \tag{5.J.17}$$

define a congruence of observers adapted to the homogeneous field. They have vanishing leg-acceleration.

K. Anti-de Sitter Spacetime ($K > 0$)

A discussion of this case proceeds along the same lines as that of de Sitter spacetime except that the roles of space and time are interchanged. We obtain

$$\omega_0 = R\,e^{\sigma}\,dt\,, \qquad \omega_1 = R\,d\sigma\,, \qquad \omega_{01} = e^{\sigma}\,dt\,, \tag{5.K.1}$$

and have as the metric

$$ds^2 = R^2\left[e^{2\sigma}\,dt^2 - d\sigma^2\right]. \tag{5.K.2}$$

The hyperboloid is the same, namely,

$$T^2 - X^2 - Y^2 = -R^2, \tag{5.K.3}$$

but the three-dimensional Minkowski space has now two time and one space dimensions with the metric

$$ds^2 = -\,dT^2 + dX^2 + dY^2. \tag{5.K.4}$$

We have then, with (5.H.10),

$$ds^2 = R^2\left[ch^2\tau\,d\phi^2 - d\tau^2\right] \tag{5.K.5}$$

where ϕ is a time coordinate and τ a space coordinate. Everything goes now in analogy. The curvature is here the phenomenon that the pencil of geodesics issuing from spacelike infinity expands time and not space.

Homogeneous fields in the Minkowski plane will be discussed in the more general setting of *homogeneous vector fields.*

CHAPTER 6

HOMOGENEOUS VECTOR FIELDS IN N-DIMENSIONS

A. Existence

We will now extend the study of homogeneous fields to manifolds of an arbitrary number of dimensions. After demonstrating the existence of homogeneous fields on n-dimensional manifolds, we will again examine their specialized forms for positive definite and indefinite metrics.

We assume an n-dimensional metric, ds^2, with the n independent differential one-forms $\boldsymbol{\omega}^j$, which is constructed as

$$ds^2 = \eta_{jk}\,\boldsymbol{\omega}^j\,\boldsymbol{\omega}^k\,. \tag{6.A.1}$$

The $n \times n$ square matrix η_{jk} is non-singular, symmetric, and constant:

$$\det[\,\eta_{jk}\,] \neq 0\,, \tag{6.A.2a}$$

$$\eta_{jk} = \eta_{kj} = constant\,. \tag{6.A.2b}$$

In general, we shall assume η_{jk} to be in diagonal form with elements ± 1. The independence of the differential one-forms is expressed by demanding that their n-fold wedge product be different from zero:

$$\boldsymbol{\omega}^1 \wedge \ldots \wedge \boldsymbol{\omega}^n \neq 0\,. \tag{6.A.3}$$

Latin indices like j, k, l range from 1 to n. We use the matrix coefficients, η_{jk} to lower upper indices, for example:

$$\boldsymbol{\omega}_j = \eta_{jk}\,\boldsymbol{\omega}^k\,. \tag{6.A.4}$$

We now define the connection one-forms $\boldsymbol{\omega}_{jk}$ by the Cartan structure equations for the case of zero torsion

$$d\boldsymbol{\omega}_j - \boldsymbol{\omega}_{jk} \wedge \boldsymbol{\omega}^k = 0\,, \tag{6.A.5a}$$

where we have

$$\boldsymbol{\omega}_{jk} + \boldsymbol{\omega}_{kj} = 0 . \tag{6.A.5b}$$

The $\boldsymbol{\omega}_{jk}$ thus form a skew-symmetric matrix of differential one-forms that is uniquely determined through the differentials of the one-forms $\boldsymbol{\omega}_j$. The connection forms have the components $g_{jk}{}^l$ with respect to the forms $\boldsymbol{\omega}_l$ such that

$$\boldsymbol{\omega}_{jk} = g_{jk}{}^l\,\boldsymbol{\omega}_l = g_{jkl}\,\boldsymbol{\omega}^l , \tag{6.A.6a}$$

$$g_{jk}{}^l = -\,g_{kj}{}^l, \qquad g_{jkl} = -\,g_{kjl} . \tag{6.A.6b}$$

We now introduce our fundamental definitions:

Definition: If the connection coefficients are constant, that is,

$$g_{jk}{}^l = \mathit{constant}, \tag{6.A.7}$$

we have a *homogeneous field.*

Definition: A homogeneous field is called a *homogeneous vector field* if the field components can be represented by a vector g_j such that

$$g_{jk}{}^l = g_j\,\delta_k{}^l - g_k\,\delta_j{}^l . \tag{6.A.8}$$

Here, Kronecker's $\delta_j{}^l$ is the unit matrix, a matrix of constants. We have then, by contraction,

$$g_j = \frac{1}{n-1}\,g_{jk}{}^k . \tag{6.A.9}$$

From (6.A.6a), One obtains for the connection form

$$\boxed{\boldsymbol{\omega}_{jk} = g_j\,\boldsymbol{\omega}_k - g_k\,\boldsymbol{\omega}_j} . \tag{6.A.10}$$

For constant g_j this is the defining equation for a homogeneous vector field.

A cautionary remark is, perhaps, helpful: In the two-dimensional case we saw that the field is described by a pseudovector. The generalization of this to n dimensions is contained in equation (6.A.10). The vector g_j should not be blithely identified with the field itself, but should be seen as an essential ingredient in the construction of a third rank tensor.

It should further be pointed out that our situation is quite specialized. It is only for $n = 2$ that this assumption deals with the general case. For $n > 2$ there are many other homogeneous fields. In the following chapter an example for $n = 3$ will be given for Euclidean and for Minkowski geometry.

Curvature Form

We now compute the matrix of the curvature form's two-form components as

$$\boldsymbol{\Omega}_{jk} = d\boldsymbol{\omega}_{jk} - \boldsymbol{\omega}_{jl} \wedge \boldsymbol{\omega}^{l}{}_{k}\,, \tag{6.A.11}$$

where we have raised the index l in the last connection form with the inverse of the matrix η_{jk}. We have with (6.A.5), (6.A.6), and (6.A.10)

$$\boldsymbol{\Omega}_{jk} = g_j\, d\boldsymbol{\omega}_k - g_k\, d\boldsymbol{\omega}_j - \left[\, g_j\, \boldsymbol{\omega}_l - g_l\, \boldsymbol{\omega}_j \,\right] \wedge \left[\, g^l\, \boldsymbol{\omega}_k - g_k\, \boldsymbol{\omega}^l \,\right]\,. \tag{6.A.12}$$

Upon substituting for the differentials of the one-forms, this gives

$$\begin{aligned}\boldsymbol{\Omega}_{jk} = {} & g_j\, \left[\, g_k\, \boldsymbol{\omega}_l - g_l\, \boldsymbol{\omega}_k \,\right] \wedge \boldsymbol{\omega}^l - g_k\, \left[\, g_j\, \boldsymbol{\omega}_l - g_l\, \boldsymbol{\omega}_j \,\right] \wedge \boldsymbol{\omega}^l \\ & - g_j\, g^l\, \boldsymbol{\omega}_l \wedge \boldsymbol{\omega}_k - g^l\, g_k\, \boldsymbol{\omega}_j \wedge \boldsymbol{\omega}_l \\ & + g_l\, g^l\, \boldsymbol{\omega}_j \wedge \boldsymbol{\omega}_k\,,\end{aligned} \tag{6.A.13}$$

or, after removing terms that annihilate each other,

$$\boldsymbol{\Omega}_{jk} = g_l\, g^l\, \boldsymbol{\omega}_j \wedge \boldsymbol{\omega}_k\,. \tag{6.A.14}$$

The preceding equation can also be written as

$$\boldsymbol{\Omega}_{jk} = \frac{1}{2}\, R_{jk}{}^{lm}\, \boldsymbol{\omega}_l \wedge \boldsymbol{\omega}_m\,, \tag{6.A.15}$$

where $R_{jk}{}^{lm}$ is the Riemann tensor. For homogeneous vector fields, we have it decomposed as

$$\boxed{R_{jk}{}^{lm} = g_p\, g^p \left[\, \delta_j{}^l\, \delta_k{}^m - \delta_k{}^l\, \delta_j{}^m \,\right]}\,. \tag{6.A.16}$$

This shows that homogeneous vector fields can only occur in spaces of constant curvature K. Further, we have from (6.A.14) that the Gaussian curvature, K, is given as

$$K = -\, g_p\, g^p\,. \tag{6.A.17}$$

This generalizes exactly the situation that we discussed in the previous chapter where we studied the general case for $n = 2$.

We find again the following results:

For definite metrics:

(1) No homogeneous vector fields exist for spaces of positive curvature.
(2) The only homogeneous vector field for zero curvature is the vacuum.
(3) Homogeneous vector fields exist for spaces of constant negative curvature. One has for them

$$g_p\, g^p \;=\; -\,K \;=\; \frac{1}{R^2}\,,$$

where R is the radius of curvature.

For Lorentz manifolds, we have the results that:

(1) n-dimensional de Sitter space has a homogeneous field for a timelike vector g_j (in two dimensions the timelike vector corresponds to a spacelike pseudovector);
(2) n-dimensional Minkowski space has a homogeneous vector field for a null vector g_j;
(3) n-dimensional anti-de Sitter space has a homogeneous vector field for a spacelike vector g_j.

These results settle the existence question for homogeneous vector fields in n-dimensional Lorentz manifolds.

B. Positive Definite Metrics

We now want to construct the homogeneous vector fields mentioned in the previous section. For positive definite metrics, the only non-trivial case occurs for negative curvature K. By a suitable rotation of the frame, we arrange that

$$g_j \;=\; -\,\frac{1}{R}\,\delta_j{}^n\,, \qquad R > 0, \tag{6.B.1}$$

where n is a constant, fixed index. Introducing the convention that Greek indices run from 1 to $n-1$, we can now write the connection form (6.A.10) as

$$\omega_{jk} \;=\; -\,\frac{1}{R}\,[\,\delta_j{}^n\,\omega_k \;-\; \delta_k{}^n\,\omega_j\,]\;. \tag{6.B.2}$$

We then have from (6.A.4) and (6.A.5)

$$d\omega_\alpha \;=\; \omega_{\alpha n} \wedge \omega_n \;=\; \frac{1}{R}\,\omega_\alpha \wedge \omega_n\,, \tag{6.B.3a}$$

$$d\omega_n \;=\; \omega_{n\alpha} \wedge \omega_\alpha \;=\; 0\,. \tag{6.B.3b}$$

The last equation asserts ω_n to be a closed form and gives, with the introduction of a variable y, we have

$$\omega_n = \frac{R}{y}\,dy\,. \tag{6.B.4}$$

The first $n-1$ equations lead to

$$\frac{1}{y}\,d(y\,\omega_\alpha) = d\omega_\alpha + \frac{1}{y}\,dy \wedge \omega_\alpha = 0\,. \tag{6.B.5}$$

We have, therefore, with variables x_α,

$$\omega_\alpha = \frac{R}{y}\,dx_\alpha\,, \qquad \omega_n = \frac{R}{y}\,dy\,. \tag{6.B.6a,b}$$

This gives the metric

$$ds^2 = \frac{R^2}{y^2}\left[[dx_1]^2 + \ldots + [dx_{n-1}]^2 + [dy]^2\right]. \tag{6.B.7}$$

Here is the n-dimensional generalization of the Poincaré half-plane. The metric is built around an $(n-1)$-dimensional bundle of geodesics with equation

$$x_\alpha = \mathit{constant} \tag{6.B.8}$$

which originate at $y = +\infty$ where they have a common point of intersection. The horo-hyperspheres given by $y = \mathit{constant}$ are orthogonal to these geodesics and carry a Euclidean metric

$$dx_\alpha\,dx_\alpha\,. \tag{6.B.9}$$

This is the geometric background for the homogeneous vector field.

The space with positive y-coordinate is geodesically complete. The simply transitive group of motions operating in it is given by

$$x'_\alpha = \sigma\,x_\alpha + \mu_\alpha\,, \qquad y' = \sigma\,y\,, \tag{6.B.10a,b}$$

where $\sigma > 0$ and $\mu_\alpha \in \mathbb{R}$. Just as in the two-dimensional case, where a transformation exists from the upper half-plane to the interior of the unit disc, we have a similar map in the n-dimensional case which also gives a conformal metric, namely,

$$ds^2 = 4\,[1 - y_k\,y_k]^{-2}\,dy_j\,dy_j \tag{6.B.11}$$

with $j, k \in 1, \ldots, n$. This metric puts the isotropic nature of the space into evidence.

C. Lorentz Manifolds

Everything here is similar to the situation in two dimensions. We can distinguish between the three cases where the vector g_j is timelike, spacelike, or null. This can occur, respectively for the de Sitter, the anti-de Sitter, and Minkowski space.

D. De Sitter Space

For de Sitter, we assume

$$g_j\, g^j = \frac{1}{R^2} \tag{6.D.1}$$

and after a suitable rotation and possible reflection

$$g_j = \frac{1}{R}\, \delta_j{}^0\,, \qquad j \in 0, \ldots, n-1\,. \tag{6.D.2}$$

We now let the Greek index α run from 1 to $n-1$; from (6.A.10), we have

$$\omega_{0\alpha} = \frac{1}{R}\, \omega_\alpha\,, \qquad \omega_{\alpha\beta} = 0\,. \tag{6.D.3}$$

This gives

$$d\omega_0 = 0\,, \qquad d\omega_\alpha = \omega_{\alpha 0} \wedge \omega_0 = \frac{1}{R}\, \omega_0 \wedge \omega_\alpha\,. \tag{6.D.4}$$

We write the solution as

$$\omega_0 = R\, dt\,, \qquad \omega_\alpha = R\, e^t dx^\alpha\,. \tag{6.D.5}$$

This gives the metric

$$ds^2 = \omega_0\, \omega_0 - \omega_\alpha\, \omega_\alpha \tag{6.D.6a}$$

$$= R^2 \left[dt^2 - e^{2t}\, dx^\alpha\, dx^\alpha \right]. \tag{6.D.6b}$$

For $n > 2$ the complete manifold is the hyperboloid itself because it is simply connected, which was not the case for $n = 2$. The form of the de Sitter spacetime (6.D.6a,b) is known as the Steady State universe and covers exactly half of the total manifold. A further discussion can be carried out along the lines developed in the description of the two-dimensional case.

E. Anti-de Sitter Space

For anti-de Sitter, we take

$$g_j\, g^j = -\frac{1}{R^2}\,, \qquad g_j = \frac{1}{R}\,\delta_j{}^1\,, \qquad j \in 0, \ldots, n-1\,. \tag{6.E.1}$$

We now let Greek indices run from 2 to $n-1$ and have from (6.A.10)

$$\omega_{01} = -\frac{1}{R}\,\omega_0\,, \qquad \omega_{1\alpha} = \frac{1}{R}\,\omega_\alpha\,, \qquad \omega_{\alpha 0} = \omega_{\alpha\beta} = 0\,. \tag{6.E.2}$$

This gives

$$d\omega_0 = \frac{1}{R}\,\omega_0 \wedge \omega_1\,, \qquad d\omega_1 = 0\,, \qquad d\omega_\alpha = \frac{1}{R}\,\omega_\alpha \wedge \omega_1\,. \tag{6.E.3}$$

We write the solution as

$$\omega_1 = -R\,dx\,, \qquad \omega_0 = +R\,e^x\,dt\,, \qquad \omega_\alpha = -R\,e^x\,dx^\alpha\,. \tag{6.E.4}$$

This gives the metric

$$ds^2 = \omega_0^2 - \omega_1^2 - \omega_\alpha\,\omega_\alpha \tag{6.E.5a}$$

$$= R^2\left[e^{2x}\,dt^2 - dx^2 - e^{2x}\,dx^\alpha\,dx^\alpha\right]. \tag{6.E.5b}$$

For $n > 2$ the complete manifold is the hyperboloid itself. The metric (6.E.5) covers only half of the hyperboloid and we have, as in the de Sitter case, the horizon phenomenon. The space can be described as a system of parallel spacelike geodesics, which expand from their common origin at infinity and create increasingly more time between them as they run on.

F. Minkowski Space

Finally, we deal with the n-dimensional Minkowski space. We take

$$g_j\, g^j = 0\,, \qquad g_j = -g\left[\delta^0{}_j - \delta^1{}_j\right], \tag{6.F.1a,b}$$

$$M_{jk} = \begin{bmatrix} 1 & 0 & 0 \\ 0 & -1 & 0 \\ 0 & 0 & -\delta_{\alpha\beta} \end{bmatrix}. \tag{6.F.1c}$$

This gives, with (6.A.8) and (6.A.10) when $\alpha, \beta \in 2, ..., n-1$,

$$\omega_{01} = -g\,[\,\omega_0 + \omega_1\,]\,, \qquad \omega_{0\alpha} = -g\,\omega_\alpha\,, \tag{6.F.2a,b}$$

$$\omega_{1\alpha} = g\,\omega_\alpha\,, \qquad \omega_{\alpha\beta} = 0\,. \tag{6.F.2c,d}$$

The Maurer-Cartan structure equations become

$$d\omega_0 = \omega_{01} \wedge \omega^1 + \omega_{0\alpha} \wedge \omega^\alpha\,, \tag{6.F.3a}$$

$$d\omega_1 = \omega_{10} \wedge \omega^0 + \omega_{1\alpha} \wedge \omega^\alpha\,, \tag{6.F.3b}$$

$$d\omega_\alpha = \omega_{\alpha 0} \wedge \omega^0 + \omega_{\alpha 1} \wedge \omega^1 + \omega_{\alpha\beta} \wedge \omega^\beta\,. \tag{6.F.3c}$$

We obtain then with the expressions (6.F.2)

$$d\omega_0 = g\,\omega_0 \wedge \omega_1\,, \tag{6.F.4a}$$

$$d\omega_1 = g\,\omega_1 \wedge \omega_0\,, \tag{6.F.4b}$$

$$d\omega_\alpha = g\,\omega_\alpha \wedge [\,\omega_0 + \omega_1\,]\,. \tag{6.F.4c}$$

These explicit expressions for the differentials give

$$d\,(\omega_0 + \omega_1) = 0\,, \tag{6.F.5a}$$

$$d\,(\omega_0 - \omega_1) = -g\,[\,\omega_0 + \omega_1\,] \wedge [\,\omega_0 - \omega_1\,]\,. \tag{6.F.5b}$$

We now put, with a variable u,

$$\omega_0 + \omega_1 = \frac{1}{g}\frac{du}{u}\,, \tag{6.F.6a}$$

$$d(\,u\,[\,\omega_0 - \omega_1\,]\,) = 0\,, \tag{6.F.6b}$$

$$d(\,u\,\omega_\alpha\,) = 0\,. \tag{6.F.6c}$$

Introducing the variables v and σ^α by

$$u\,[\,\omega_0 - \omega_1\,] \equiv 2\,g\,dv\,, \qquad u\,\omega_\alpha \equiv d\sigma^\alpha\,. \tag{6.F.7a,b}$$

We obtain as the metric

$$ds^2 = \frac{1}{u^2}\,[\,2\,du\,dv - d\sigma^\alpha\,d\sigma^\alpha\,]\,. \tag{6.F.8}$$

With the substitution

$$u = \frac{1}{\sqrt{2}}\,[\,\tau - \sigma\,]\,, \qquad v = \frac{1}{\sqrt{2}}\,[\,\tau + \sigma\,]\,, \tag{6.F.9a,b}$$

the metric becomes

$$ds^2 \;=\; \frac{2}{[\,\tau\,-\,\sigma\,]^2}\left[\,d\tau^2 \,-\, d\sigma^2 \,-\, d\sigma^\alpha\,d\sigma^\alpha\,\right]. \tag{6.F.10}$$

It is now of interest to find out what all this means in terms of Minkowski coordinates. We introduce the transformation

$$u \;=\; -\frac{1}{u'}\,, \qquad\qquad v \;=\; v' \,-\, \frac{1}{2u'}\,x^\alpha\,x^\alpha\,, \tag{6.F.11a,b}$$

$$\sigma^\alpha \;=\; +\frac{1}{u'}\,x^\alpha\,, \qquad u' \neq 0\,, \tag{6.F.11c,d}$$

along its with inverse

$$u' \;=\; -\frac{1}{u}\,, \qquad v' \;=\; v \,-\, \frac{1}{2u}\,\sigma^\alpha\,\sigma^\alpha\,, \tag{6.F.12a,b}$$

$$x^\alpha \;=\; -\frac{1}{u}\,\sigma^\alpha\,, \quad u \neq 0\,. \tag{6.F.12c,d}$$

This transformation turns the metric (6.F.8) into pseudo-Euclidean form

$$ds^2 \;=\; 2\,du'\,dv' \,-\, dx^\alpha\,dx^\alpha\,. \tag{6.F.13}$$

The transformation (6.F.11) maps the half-spaces $u' > 0$ and $u' < 0$ onto the half-spaces $u < 0$ and $u > 0$ one-to-one, respectively.

We observe here the horizon phenomenon again: the coordinates become singular for $u' = 0$ which is the same as the null hyperplane $x = t$. The Minkowski space is split into the two regions of positive and negative u' separated by the horizon $u' = 0$.

The group of isometries

$$u' \;=\; c^0\,u\,, \quad v' \;=\; c^0\,v \,+\, c^1\,, \tag{6.F.14a,b}$$

$$\sigma'^\alpha \;=\; c^0\,\sigma^\alpha + c^\alpha\,, \quad c^0 > 0\,, \tag{6.F.14c,d}$$

operates transitively in each half-space. The hyperplane $u' = 0$ goes into itself by all transformations. The transformations (6.F.14) form the same group as those of (6.B.10) for the de Sitter spaces. This result is remarkable because the isometry groups of de Sitter and Minkowski space of equal dimensions are quite different, but the n-dimensional subgroups are identical.

The easiest way to describe the homogeneous vector field in Minkowski space is in terms of “potentials”. We now turn to this procedure.

G. Potentials in Minkowski Space

Potentials in pseudo-Riemannian manifolds are best discussed in terms of $F(M)$, the frame bundle of the manifold M. Only orthonormal frames are admitted. By a potential on M, we mean a cross-section of $F(M)$.

Suppose the metric on M is given by

$$ds^2 = M_{jk}\, \boldsymbol{\omega}^j\, \boldsymbol{\omega}^k\,, \tag{6.G.1a}$$

where

$$\det\left[\, M_{jk} \,\right] \neq 0\,, \qquad M_{jk} = M_{kj} = constant\,. \tag{6.G.1b,c}$$

We call the $\boldsymbol{\omega}(x)$ the reference frame. Another frame $\boldsymbol{\omega}'^b$ is related to $\boldsymbol{\omega}^a$ by means of the pseudo-orthogonal transformation

$$\boldsymbol{\omega}^a = A^a{}_b \left(x^j\right) \boldsymbol{\omega}'^b\,, \qquad A^a{}_b\, M_{ac}\, A^c{}_d = M_{bd}\,. \tag{6.G.2a,b}$$

Thinking of $\boldsymbol{\omega}^a$ as a column vector $\boldsymbol{\omega}$ and of A and M as matrices, we might also formulate (6.G.2) as

$$\boldsymbol{\omega} = A\, \boldsymbol{\omega}'\,, \qquad A^t\, M\, A = M\,, \tag{6.G.3a,b}$$

where the superscript "t" indicates the transpose of the matrix.

The potentials, $A^a{}_b$, as understood here, are the elements of an n-dimensional pseudo-orthogonal matrix. There are only $n(n-1)/2$ independent elements among the n^2 forming the entire matrix. The potentials, $A^a{}_b$, are scalars with respect to coordinate transformations. We have in matrix notation

$$d\boldsymbol{\omega} = \boldsymbol{\theta} \wedge \boldsymbol{\omega}\,, \tag{6.G.4}$$

where $\boldsymbol{\theta}$ is the matrix of the connection forms. We obtain from (6.G.3) and (6.G.4)

$$d\boldsymbol{\omega} = dA \wedge \boldsymbol{\omega}' + A\, d\boldsymbol{\omega}'\,, \tag{6.G.5a}$$

$$\boldsymbol{\theta} \wedge \boldsymbol{\omega} = dA\, A^{-1} \wedge \boldsymbol{\omega} + A\, \theta' \wedge \boldsymbol{\omega}'\,, \tag{6.G.5b}$$

$$\boldsymbol{\theta} \wedge \boldsymbol{\omega} = \left[dA\, A^{-1} + A\, \boldsymbol{\theta}'\, A^{-1}\right] \wedge \boldsymbol{\omega}\,. \tag{6.G.5c}$$

For a gauge transformation, this gives a transformation of the connection form matrix $\boldsymbol{\theta}$:

$$\boldsymbol{\theta} = dA\, A^{-1} + A\, \boldsymbol{\theta}' A^{-1}\,. \tag{6.G.6}$$

H. Vacua Defined

For pseudo-Riemannian spaces with a flat metric, that is, Minkowski space, we have distinguished reference sections of the frame bundle. These are the sections for which the connection vanishes. We have

$$d\omega^j = 0 \longrightarrow \omega^j = dx^j \,, \tag{6.H.1}$$

or, in matrix notation,

$$d\boldsymbol{\omega} = \boldsymbol{\theta} \wedge \boldsymbol{\omega} = 0 \longrightarrow \boldsymbol{\theta} = 0 \,. \tag{6.H.2}$$

We call such sections "vacua". This is the generalization of the vacua defined for two dimensions in **Section 5.D**.

I. Force Fields

One obtains, from (6.G.3) and using (6.H.2),

$$\boldsymbol{\theta} = 0 \longrightarrow A\,\boldsymbol{\theta}'\,A^{-1} + dA\,A^{-1} = 0 \,, \tag{6.I.1}$$

where $\boldsymbol{\theta}'$ is the new matrix of connection forms. We learn from this equation, by multiplication from the left with A^{-1} and from the right with A, that

$$\boldsymbol{\theta}' = -\,A^{-1}\,dA \,. \tag{6.I.2}$$

This equation relates the potential A to the force field $\boldsymbol{\theta}'$.

J. Conservation Laws

Before we explicitly give the potentials for the homogeneous vector fields in Minkowski space, we mention an important application of these potentials. Namely, the existence of readily derived conservation laws.

Let us define

$$\mathbf{U} = u^a\,\mathbf{e}_a \,, \qquad u_a\,u^a = 1 \tag{6.J.1a,b}$$

to be the unit tangent vector of a timelike geodesic. The components u^a will be written as the column vector $\mathbf{u}$. They transform under gauge transformations like the vector $\boldsymbol{\omega}$.

In a vacuum, we have

$$\eta_{ab}\,u^a\,u^b = 1, \qquad \eta_{ab}\,u'^a\,u'^b = 1 \,, \tag{6.J.2a,b}$$

$$u^a = A^a{}_b\,u'^b = constant \,. \tag{6.J.2c}$$

Therefore, we have in a space with a vacuum the n conservation laws

$$A^a{}_b \, u'^b = constant \tag{6.J.3}$$

for any geodesic and any section described by the matrix of potentials A.

K. The Minkowski Potentials

We now proceed with the calculation of the potentials for homogeneous vector fields in Minkowski space. We take M_{jk} in the form (6.F.1c) and have

$$ds^2 = \left[\omega^0\right]^2 - \left[\omega^1\right]^2 - \omega^\alpha \, \omega^\alpha \,, \tag{6.K.1a}$$

$$= dt^2 - dx^2 - dx^\alpha \, dx^\alpha \,. \tag{6.K.1b}$$

With (6.F.13) and the transformation

$$u' = \frac{1}{\sqrt{2}} \left[t - u \right] , \qquad v' = \frac{1}{\sqrt{2}} \left[t + u \right] , \tag{6.K.2a,b}$$

this gives

$$\omega^0 = \frac{1}{\sqrt{2}} \left[du' + dv' \right] , \tag{6.K.3a}$$

$$\omega^1 = \frac{1}{\sqrt{2}} \left[du' - dv' \right] , \tag{6.K.3b}$$

$$\omega^\alpha = dx^\alpha \,, \qquad \alpha \in 2, ..., n-1 \,. \tag{6.K.3c}$$

By differentiation of (6.F.12), and using definitions (6.F.7) we obtain

$$\omega^0 = \frac{1}{\sqrt{2}} \left[\frac{1}{u^2} \left[\frac{1}{2} \sigma^\alpha \sigma^\alpha + 1 \right] du + dv - \frac{1}{u} \sigma^\alpha \, d\sigma^\alpha \right] , \tag{6.K.4a}$$

$$\omega^1 = \frac{1}{\sqrt{2}} \left[\frac{1}{u^2} \left[\frac{1}{2} \sigma^\alpha \sigma^\alpha - 1 \right] du + dv - \frac{1}{u} \sigma^\alpha \, d\sigma^\alpha \right] , \tag{6.K.4b}$$

$$\omega^\alpha = \frac{1}{u^2} \sigma^\alpha \, du - \frac{1}{u} \, d\sigma^\alpha \,. \tag{6.K.4c}$$

Here we have simply expressed the frame of the vacuum in new coordinates u, v, and σ^α. We call ω' the new section of the frame bundle which gives the homogeneous vector field of (6.F.6) and (6.F.7). We have

$$\omega'^0 - \omega'^1 = + \frac{1}{g\,u}\, du\,, \tag{6.K.5a}$$

$$\omega'^0 + \omega'^1 = + 2\,\frac{g}{u}\, dv\,, \tag{6.K.5b}$$

$$\omega'^\alpha = - \frac{1}{u}\, d\sigma^\alpha\,. \tag{6.K.5c}$$

Inserting this into equations (6.K.3), we obtain

$$\omega^0 = \frac{1}{\sqrt{2}}\left[\frac{g}{u}\left[\frac{1}{2}\sigma^\alpha\sigma^\alpha + 1\right]\left[\omega'^0 - \omega'^1\right]\right. \\ \left. + \frac{1}{\sqrt{2}}\left[\frac{u}{2g}\left[\omega'^0 + \omega'^1\right] + \sigma^\alpha\,\omega'^\alpha\right]\right., \tag{6.K.6a}$$

$$\omega^1 = \frac{1}{\sqrt{2}}\left[\frac{g}{u}\left[\frac{1}{2}\sigma^\alpha\sigma^\alpha - 1\right]\left[\omega'^0 - \omega'^1\right]\right. \\ \left. + \frac{1}{\sqrt{2}}\left[\frac{u}{2g}\left[\omega'^0 + \omega'^1\right] + \sigma^\alpha\,\omega'^\alpha\right]\right., \tag{6.K.6b}$$

$$\omega^\alpha = \frac{g}{u}\sigma^\alpha\left[\omega'^0 - \omega'^1\right] + \omega'^\alpha\,. \tag{6.K.6c}$$

These expressions can be rewritten

$$\omega^0 = + \left[\frac{1}{2\sqrt{2}}\frac{u}{g} + \frac{1}{\sqrt{2}}\frac{g}{u}\left[\frac{1}{2}\sigma^\alpha\sigma^\alpha + 1\right]\right]\omega'^0 \\ + \left[\frac{1}{2\sqrt{2}}\frac{u}{g} - \frac{1}{\sqrt{2}}\frac{g}{u}\left[\frac{1}{2}\sigma^\alpha\sigma^\alpha + 1\right]\right]\omega'^1 \\ + \frac{1}{\sqrt{2}}\sigma^\alpha\,\omega'^\alpha\,, \tag{6.K.7a}$$

$$\begin{aligned}
\boldsymbol{\omega}^1 &= + \left[\frac{1}{2\sqrt{2}} \frac{u}{g} + \frac{1}{\sqrt{2}} \frac{g}{u} \left[\frac{1}{2} \sigma^\alpha \sigma^\alpha - 1 \right] \right] \boldsymbol{\omega}'^0 \\
&\quad + \left[\frac{1}{2\sqrt{2}} \frac{u}{g} - \frac{1}{\sqrt{2}} \frac{g}{u} \left[\frac{1}{2} \sigma^\alpha \sigma^\alpha - 1 \right] \right] \boldsymbol{\omega}'^1 \\
&\quad + \frac{1}{\sqrt{2}} \sigma^\alpha \boldsymbol{\omega}'^\alpha ,
\end{aligned} \tag{6.K.7b}$$

$$\boldsymbol{\omega}^\alpha = + \frac{g}{u} \sigma^\alpha \boldsymbol{\omega}'^0 - \frac{g}{u} \sigma^\alpha \boldsymbol{\omega}'^1 + \boldsymbol{\omega}'^\alpha . \tag{6.K.7c}$$

We have from (6.F.11) and the text following (6.K.1) that

$$u = -\frac{1}{u'} = \frac{\sqrt{2}}{x-t}, \qquad \sigma^\alpha = -u\, x^\alpha = \frac{\sqrt{2}}{x-t} x^\alpha . \tag{6.K.8a,b}$$

Equations (6.K.7) give the matrix A for

$$\boldsymbol{\omega} = A \boldsymbol{\omega}' . \tag{6.K.9}$$

By means of (6.K.7) we can read off the matrix elements and express them in Minkowski coordinates (t, x, x^α), with $\alpha \in 2, ..., n-1$. Upon defining

$$\left[\Sigma^+\right]^2 \equiv \frac{x^\alpha x^\alpha}{(t-x)^2} + 1 , \qquad \left[\Sigma^-\right]^2 \equiv \frac{x^\alpha x^\alpha}{(t-x)^2} - 1 , \tag{6.K.10a,b}$$

we get explicit expressions for the Minkowski potential matrix elements

$$A^0{}_0 = + \frac{g\,(x-t)}{2} \left[\Sigma^+\right]^2 + \frac{1}{2g\,(x-t)} , \tag{6.K.11a}$$

$$A^0{}_1 = - \frac{g\,(x-t)}{2} \left[\Sigma^+\right]^2 + \frac{1}{2g\,(x-t)} , \tag{6.K.11b}$$

$$A^0{}_\alpha = + \frac{x^\alpha}{t-x} , \tag{6.K.11c}$$

$$A^1{}_0 = + \frac{g\,(x-t)}{2} \left[\Sigma^-\right]^2 + \frac{1}{2g\,(x-t)} , \tag{6.K.11d}$$

$$A^1{}_1 = - \frac{g\,(x-t)}{2} \left[\Sigma^-\right]^2 + \frac{1}{2g\,(x-t)} , \tag{6.K.11e}$$

$$A^1{}_\alpha = + \frac{x^\alpha}{t-x} , \tag{6.K.11f}$$

$$A^{\alpha}{}_{1} = +\, g\, x^{\alpha}\,, \tag{6.K.11g}$$

$$A^{\alpha}{}_{0} = -\, g\, x^{\alpha}\,, \tag{6.K.11h}$$

$$A^{\alpha}{}_{\beta} = +\, \delta^{\alpha}{}_{\beta}\,. \tag{6.K.11i}$$

CHAPTER 7

HOMOGENEOUS FIELDS ON THREE-DIMENSIONAL SPACETIMES: ELEMENTARY CASES

X. Overview

In this chapter we will examine the existence of some homogeneous fields in three-dimensional spaces. First, we will examine the positive definite Euclidean space where we will exercise the mathematical techniques of transformation groups in a familiar context. We analyze a transformation group and an immediate generalization that includes translations. Next, we will pass into (2+1)-Minkowski spacetime while studying a closely analogous group of transformations.

A. Homogeneous Fields on Positive Definite Spaces

The simplest three-dimensional positive definite space is Euclidean space which has the line element

$$ds^2 = dx^2 + dy^2 + dz^2 . \tag{7.A.1}$$

The line element is a differential invariant regardless of coordinates. We can facilitate the analysis if we define a complex coordinate and its conjugate

$$\Sigma \equiv x + i\,y , \qquad \bar{\Sigma} \equiv x - i\,y , \tag{7.A.2}$$

thus shifting the coordinate realization from $\mathbb{R}^3$ to $\mathbb{C}\times\mathbb{R}$ and rewrite the line element as

$$ds^2 = d\Sigma\, d\bar{\Sigma} + dz^2 , \tag{7.A.3}$$

where the z coordinate remains real. In both expressions, (7.A.1) and (7.A.3), the right-hand-side is a coordinate representation of the invariant. We are better served by a representation in terms of the invariant properties of Euclidean space. We should prefer to use other differential invariants, specifically those of the transformation groups of the congruences of Euclidean space, that is, their left-invariant differential forms.

A Simply Transitive Subgroup

There is a subgroup of the Euclidean congruence transformation group which we can now exhibit by a change of coordinates

$$\mathrm{T}(\phi;\beta): \quad \Sigma' = e^{i\phi}\,\Sigma\,, \qquad z' = z + \beta\,\phi \tag{7.A.4}$$

where ϕ is real, and β is a fixed non-zero real constant. The subgroup's single parameter is ϕ (1 real parameter) and β is a coupling and scaling factor. Notice that when $\beta \neq 0$ the translation in the z-direction results in rotations of the $(x$-$y)$-plane and, conversely, rotation yields a z-translation. This is a helical screwing motion and the orbits of the subgroup are helices with the motions of the subgroup on a surface known as a helicoid. The helices themselves are one-dimensional submanifolds of the three-dimensional Euclidean space; they are topologically equivalent to the real line.

The Euclidean space's transformation group (and its subgroups) may themselves have local rotations applied to them which involve only constants related to the Euler angles of the group **SO(3)**. This is suppressed for the present, but its significance will be provided at the end of this section when the irrelevance of the direction of our coordinate axes gets further comment.

For the same β, we have

$$\Sigma'' = e^{i\phi'}\,\Sigma'\,, \qquad z'' = z' + \beta\,\phi'\,, \tag{7.A.5a,b}$$

as a subsequent transformation. When we compose the successive transformations, we get

$$\Sigma'' = e^{i\phi'}\left[e^{i\phi}\Sigma\right], \qquad z'' = z + \beta\,[\,\phi' + \phi\,]\,, \tag{7.A.6a,b}$$

which is itself a coordinate transformation of type $\mathrm{T}(\phi;\beta)$ that gives

$$\Sigma'' = e^{i\phi''}\,\Sigma\,, \qquad z'' = z + \beta\,\phi''\,, \tag{7.A.7a,b}$$

where we have now identified

$$\phi'' \equiv \phi' + \phi \tag{7.A.8}$$

as the new subgroup parameters for the net transformation that results from the composition of the two individual transformations.

Computing the inverse transformation requires that we arrive back at the original point, so we must satisfy

$$\phi'' = 0 = \phi' + \phi \quad \Longrightarrow \quad \phi' = -\phi \tag{7.A.9}$$

for the parameter of the inverse transformation.

Furthermore, as required by the comments in **Section 1.B**, before equation (1.B.15), our present Riemannian manifold is homogeneous when it is invariant under the action of a transitive group. Our subgroup is also *simply transitive* (or *regular*) with respect to the subgroup's helical orbits since the equations (7.A.4) have a unique solution for the subgroup's parameter

$$\phi = \frac{z' - z}{\beta}, \tag{7.A.10}$$

for given z and z' and $|\Sigma'| = |\Sigma|$ and values of β (the subgroup's coupling constant) not equal to zero. This satisfies the additional requirement for the presence of a homogeneous gravitational field.

Left-invariant Differential Forms

A differential form is said to be left-invariant when the operations of a group leave it unchanged

$$\omega = A\,\omega \tag{7.A.11}$$

where A is an element of a group $\mathfrak{G}$. In a matrix representation, the components of A are analytic functions which the d operation on these 0-forms transform into one-forms such as

$$dA_{ij} = \frac{\partial A_{ij}}{\partial \phi} d\phi\,. \tag{7.A.12}$$

This action on the matrix representation of an element of $\mathfrak{G}$ results in the differential matrix dA. Its elements are functions of the differentials of the group parameters occurring in A; the partial derivatives also contain functions of the group parameters. However, the differential forms dA_{ij} are not necessarily left-invariant in the sense of (7.A.11).

When any element of the transformation group (in matrix representation) is multiplied (in the sense of group multiplication) on the left by a specific member of the group $\mathfrak{G}$, we have

$$A \to R\,A \qquad dA \to R\,dA\,, \tag{7.A.13}$$

since R only contains unvarying constants, no term involving dR appears. We therefore have, given that $(R\,A)^{-1} = A^{-1}\,R^{-1}$,

$$A^{-1}dA \to \left(A^{-1}R^{-1}\right)(R\,dA) = A^{-1}dA. \tag{7.A.14}$$

Thus the expression $A^{-1}dA$ is unchanged by transformation by a constant matrix R; it is invariant and its elements $\left[A^{-1}dA\right]_{ij}$ are now invariant differential forms satisfying (7.A.11).

The Subgroup's Left-invariant Differential Forms

Our next step is the computation of the left-invariant differential forms. This is easiest in a matrix representation of $\mathrm{T}(\phi;\beta)$, that is, of (7.A.4) which is given by

$$\begin{bmatrix}\Sigma'\\ z'\\ 1\end{bmatrix} = \begin{bmatrix}e^{i\phi}\,\Sigma\\ z+\beta\,\phi\\ 1\end{bmatrix} = \begin{bmatrix}e^{i\phi} & 0 & 0\\ 0 & 1 & \beta\,\phi\\ 0 & 0 & 1\end{bmatrix}\begin{bmatrix}\Sigma\\ z\\ 1\end{bmatrix} = A\begin{bmatrix}\Sigma\\ z\\ 1\end{bmatrix}. \tag{7.A.15}$$

The transformation is given by the non-orthogonal matrix

$$A = \begin{bmatrix}e^{i\phi} & 0 & 0\\ 0 & 1 & \beta\,\phi\\ 0 & 0 & 1\end{bmatrix}, \quad \text{where} \quad A^{-1} = \begin{bmatrix}e^{-i\phi} & 0 & 0\\ 0 & 1 & -\beta\,\phi\\ 0 & 0 & 1\end{bmatrix}, \tag{7.A.16}$$

the inverse has been written down immediately with the aid of (7.A.9). The matrix A is non-orthogonal since it does not satisfy $A\,A^t = I$. The differential of the transformation matrix, with constant β, is

$$dA = \begin{bmatrix}e^{i\phi}\,i\,d\phi & 0 & 0\\ 0 & 0 & \beta\,d\phi\\ 0 & 0 & 0\end{bmatrix} \tag{7.A.17}$$

and the left-invariant differential forms are obtained from the elements of

$$A^{-1}\,dA = \begin{bmatrix}i\,d\phi & 0 & 0\\ 0 & 0 & \beta\,d\phi\\ 0 & 0 & 0\end{bmatrix}. \tag{7.A.18}$$

Thus, we have only one left-invariant differential form, that is, multiples of $d\phi$ such as

$$\omega_3 = \beta\,d\phi, \tag{7.A.19}$$

which is unsurprising since the transformations $\mathrm{T}(\phi;\beta)$ have only one parameter, ϕ, for the subgroup. At the neighborhood of the coordinate origin, where $\Sigma' = 0$ and $\phi' = 0$, we have $\phi = -z/\beta$ from (7.A.10) and hence

$$\omega_3 = -dz\,. \tag{7.A.20}$$

In point of fact, the group we have just examined is a special case of another subgroup of the Euclidean congruence group. It will depend on two more parameters, to be written as a single complex number. With the aid of the left-invariant differential form (7.A.20) we can exhibit the more general subgroup.

Exhibit the Larger Subgroup

We will uncover the parameter(s) of the new subgroup by integrating the invariance requirement

$$\omega_3' = -dz' = -dz = \omega_3\,, \tag{7.A.21}$$

which immediately yields

$$z' = z + \gamma\,, \qquad \text{with} \qquad \gamma = constant\,. \tag{7.A.22}$$

We naturally require that the line element (7.A.3) also be invariant under any permissible transformation; that is, we must have

$$ds^2 = d\Sigma\, d\bar{\Sigma} + dz^2 = d\Sigma'\, d\bar{\Sigma}' + dz'^2\,. \tag{7.A.23}$$

The invariance of the line element and the further invariance requirement (7.A.21) demands that we also require

$$d\Sigma'\, d\bar{\Sigma}' = d\Sigma\, d\bar{\Sigma}\,, \tag{7.A.24}$$

or, re-specifying this last condition, we have

$$e^{i\Phi'}\, d\Sigma' = e^{i\Phi}\, d\Sigma\,, \tag{7.A.25}$$

for some real parameter Φ (Φ') that we will now identify with the parameter $\phi\,(\phi')$ of (7.A.4) and (7.A.10) at $z' = 0\;(z'' = 0)$ and obtain

$$e^{-iz'/\beta}\, d\Sigma' = e^{-iz/\beta}\, d\Sigma\,. \tag{7.A.26}$$

Substituting (7.A.22) into (7.A.26) yields up

$$e^{\,i\gamma/\beta}\, d\Sigma' \;=\; d\Sigma\,. \tag{7.A.27}$$

After moving the exponential factor to the right-hand-side and integrating this last expression, we obtain

$$\Sigma' \;=\; e^{i\gamma/\beta}\,\Sigma \,+ \alpha \tag{7.A.28}$$

and where α is a complex constant of integration. To complete our task, we just replace the exponentiated constant γ/β by ϕ, a new real parameter for the generalized subgroup of coordinate transformations. We will now examine what role (7.A.28) plays in the generalized subgroup.

Examine the Larger Subgroup

The larger subgroup of the Euclidean congruence group is shown in the coordinate transformations

$$\mathrm{T}\,(\alpha, \phi; \beta)\;:\quad \Sigma' \;=\; e^{i\phi}\,\Sigma \,+\, \alpha\,, \qquad z' \;=\; z \,+\, \beta\,\phi \tag{7.A.29}$$

where ϕ is real, α is complex, and β is a fixed non-zero real constant. The subgroup parameters are α (2 real parameters) and ϕ (1 real parameter). The new complex parameter provides translations of the (x-y)-plane. Notice that translation in the z-direction results in pointwise rotations of the (x-y)-plane.

For the same β, we have

$$\Sigma'' \;=\; e^{i\phi'}\,\Sigma' \,+\, \alpha'\,, \qquad z'' \;=\; z' \,+\, \beta\,\phi'\,, \tag{7.A.30}$$

as a subsequent transformation. When we compose the successive transformations, we get

$$\Sigma'' \;=\; e^{i\phi'}\left[\,e^{i\phi}\Sigma \,+\, \alpha\,\right] \,+\, \alpha'\,, \tag{7.A.31a}$$

$$z'' \;=\; z \,+\, \beta\,[\,\phi' \,+\, \phi\,]\,, \tag{7.A.31b}$$

which is itself a coordinate transformation of type $\mathrm{T}(\alpha, \phi; \beta)$ that gives

$$\Sigma'' \;=\; e^{i\phi''}\,\Sigma \,+\, \alpha''\,, \qquad z'' \;=\; z \,+\, \beta\,\phi''\,, \tag{7.A.32a,b}$$

where we have now identified

$$\alpha'' \equiv \alpha' \,+\, e^{i\phi'}\,\alpha, \qquad \phi'' \equiv \phi' \,+\, \phi\,. \tag{7.A.33a,b}$$

as the new subgroup parameters for the net transformation that results from the composition of two individual transformations.

Computing the inverse transformation requires that we arrive back at the original point and that

$$\alpha'' = 0 = \alpha' + e^{i\phi'}\,\alpha\,, \qquad \phi'' = 0 = \phi' + \phi \tag{7.A.34a,b}$$

both be satisfied; which, in turn, immediately gives us

$$\alpha' = -\,e^{-i\phi}\,\alpha\,, \qquad \phi' = -\,\phi \tag{7.A.35a,b}$$

for the parameters of the inverse transformation.

Furthermore, this larger subgroup is also simply transitive (= regular), since the equations (7.A.29) again have a unique solution for the subgroup parameters

$$\phi = \frac{z' - z}{\beta}\,, \qquad \alpha = \Sigma' - \Sigma\,\exp\left[i\frac{z' - z}{\beta}\right]\,, \tag{7.A.36a,b}$$

for given (Σ', z') and (Σ, z) and values of β not equal to zero.

The Larger Subgroup's Left-invariant Differential Forms

Our next step is the computation of the left-invariant differential forms. Again, this is easiest in a matrix representation of $\mathrm{T}(\alpha, \phi; \beta)$, that is, of (7.A.29). This is given by

$$\begin{bmatrix} \Sigma' \\ z' \\ 1 \end{bmatrix} = \begin{bmatrix} e^{i\phi}\,\Sigma + \alpha \\ z + \beta\,\phi \\ 1 \end{bmatrix} = \begin{bmatrix} e^{i\phi} & 0 & \alpha \\ 0 & 1 & \beta\,\phi \\ 0 & 0 & 1 \end{bmatrix} \begin{bmatrix} \Sigma \\ z \\ 1 \end{bmatrix}. \tag{7.A.37}$$

The transformation and its inverse are given by

$$A = \begin{bmatrix} e^{i\phi} & 0 & \alpha \\ 0 & 1 & \beta\,\phi \\ 0 & 0 & 1 \end{bmatrix}, \qquad A^{-1} = \begin{bmatrix} e^{-i\phi} & 0 & -\,e^{-i\phi}\,\alpha \\ 0 & 1 & -\,\beta\,\phi \\ 0 & 0 & 1 \end{bmatrix}. \tag{7.A.38a,b}$$

The differential of the transformation matrix, with constant β, is

$$dA = \begin{bmatrix} e^{i\phi}\,i\,d\phi & 0 & d\alpha \\ 0 & 0 & \beta\,d\phi \\ 0 & 0 & 0 \end{bmatrix}. \tag{7.A.39}$$

The left-invariant differential forms are obtained from

$$A^{-1}\,dA = \begin{bmatrix} i\,d\phi & 0 & e^{-i\phi}\,d\alpha \\ 0 & 0 & \beta\,d\phi \\ 0 & 0 & 0 \end{bmatrix} \qquad (7.A.40)$$

for all points of the transformation group's manifold. Thus, we have for the left-invariant differential forms

$$\boldsymbol{\omega} \equiv e^{-i\phi}\,d\alpha\,, \qquad \omega_3 \equiv \beta\,d\phi\,. \qquad (7.A.41a,b)$$

Using (7.A.36) at the coordinate origin, where $\Sigma' = 0$ and $z' = 0$, the differential for α is

$$d\alpha = -\,e^{i\phi}\,(\,d\Sigma + i\,\Sigma\,d\phi\,) \qquad (7.A.42)$$

which, with (7.A.41a), enables us to write

$$\begin{aligned} \boldsymbol{\omega} &= -\,d\Sigma - i\,\Sigma\,d\phi \\ &= -\,d\Sigma - i\,\frac{\Sigma}{\beta}\,\omega_3\,. \end{aligned} \qquad (7.A.43)$$

It is clear that $\boldsymbol{\omega}$ is not independent of ω_3 for the larger subgroup. Expressing the metric by means of the forms (7.A.41), we get

$$\boldsymbol{\omega} = e^{iz/\beta}\,d\alpha\,, \qquad \omega_3 = -\,dz\,. \qquad (7.A.44)$$

The Metric in Left-invariant Differential Forms

We have for the line element

$$ds^2 = \boldsymbol{\omega}\,\bar{\boldsymbol{\omega}} + [\,\omega_3\,]^2 \qquad (7.A.45a)$$

$$= [\,\omega_1\,]^2 + [\,\omega_2\,]^2 + [\,\omega_3\,]^2 \qquad (7.A.45b)$$

$$= dx^2 + dy^2 + dz^2 \qquad (7.A.45c)$$

with the definitions

$$\boldsymbol{\omega} \equiv \omega_1 + i\,\omega_2\,, \qquad (7.A.46a)$$

$$\omega_1 \equiv +\cos\!\left(\frac{z}{\beta}\right)dx + \sin\!\left(\frac{z}{\beta}\right)dy\,, \qquad (7.A.46b)$$

$$\omega_2 \equiv -\sin\!\left(\frac{z}{\beta}\right)dx + \cos\!\left(\frac{z}{\beta}\right)dy\,. \qquad (7.A.46c)$$

The Connection Forms

We compute the connection forms

$$d\omega_1 - \omega_{12} \wedge \omega_2 - \omega_{13} \wedge \omega_3 = 0\,, \tag{7.A.47a}$$

$$d\omega_2 - \omega_{21} \wedge \omega_1 - \omega_{23} \wedge \omega_3 = 0\,, \tag{7.A.47b}$$

$$d\omega_3 - \omega_{31} \wedge \omega_1 - \omega_{32} \wedge \omega_2 = 0\,. \tag{7.A.47c}$$

Expanding the differentials by means of the defining equations (7.A.46b,c), we obtain

$$d\omega_1 = d\left[+\cos\left(\frac{z}{\beta}\right) dx + \sin\left(\frac{z}{\beta}\right) dy\right] \tag{7.A.48a}$$

$$= \frac{1}{\beta}\left[-\sin\left(\frac{z}{\beta}\right) dz \wedge dx + \cos\left(\frac{z}{\beta}\right) dz \wedge dy\right] \tag{7.A.48b}$$

$$= \frac{1}{\beta}\,\omega_3 \wedge \omega_2 = -\omega_{13} \wedge \omega_2. \tag{7.A.48c}$$

Thus:

$$\omega_{12} = \frac{1}{\beta}\,dz = \frac{1}{\beta}\,\omega_3\,. \tag{7.A.49}$$

We follow the same procedure for the other differential forms and get for the second

$$d\omega_2 = d\left[-\sin\left(\frac{z}{\beta}\right) dx + \cos\left(\frac{z}{\beta}\right) dy\right] \tag{7.A.50a}$$

$$= \frac{1}{\beta}\left[-\cos\left(\frac{z}{\beta}\right) dz \wedge dx - \sin\left(\frac{z}{\beta}\right) dz \wedge dy\right] \tag{7.A.50b}$$

$$= -\frac{1}{\beta}\,\omega_3 \wedge \omega_1 = -\omega_{12} \wedge \omega_1\,, \tag{7.A.50c}$$

which is what we expect from the skew-symmetry of the connection forms. The remaining form, ω_3, gives

$$d\omega_3 = 0\,. \tag{7.A.51}$$

We can guess the connection forms as

$$\omega_{12} = \frac{1}{\beta}\,\omega_3\,, \qquad \omega_{13} = \omega_{23} = 0\,. \tag{7.A.52a,b}$$

The Ricci coefficients are constants. This is the requirement set forth in our definition. The subgroup is a *homogeneous field.*

The Curvature Form

The curvature two form (11.A.4) gives

$$\boldsymbol{\Omega}_{12} = d\omega_{12} - \omega_{13} \wedge \omega_{32} = 0\,, \qquad (7.A.53a)$$

$$\boldsymbol{\Omega}_{23} = d\omega_{23} - \omega_{21} \wedge \omega_{13} = 0\,, \qquad (7.A.53b)$$

$$\boldsymbol{\Omega}_{31} = d\omega_{31} - \omega_{32} \wedge \omega_{21} = 0\,. \qquad (7.A.53c)$$

The Larger Subgroup's Structure Constants

The structure constants of the subgroup are defined by

$$d\omega^j = -\frac{1}{2}\, c^j{}_{kl}\, \omega^k \wedge \omega^l\,, \qquad (7.A.54)$$

where, for our specific space and subgroup, we have

$$c^1{}_{23} = \frac{1}{\beta}\,, \qquad c^2{}_{31} = \frac{1}{\beta} \qquad (7.A.55a,b)$$

as the only non-zero subgroup structure constants.

Discussion

We saw earlier that the Euclidean plane cannot be viewed as a homogeneous field except if we count the vacuum case where the field strength is zero. The calculations above show now that the three-dimensional case is completely different. Here, homogeneous fields exist of arbitrary field strength, given by $1/\beta$ according to equations (7.A.52). However, the notion of "field strength" has a different meaning from the one in the two-dimensional case. There, the field strength was a pseudovector which always had "sidewise orientation"; here, in three dimensions, the field strength is given by the Ricci coefficient tensor (under rotations and reflections)

$$\omega_{jk} = g_{jkl}\, \omega_l\,, \qquad (7.A.56)$$

with $j, k, l \in (1, 2, 3)$ and where those tensor components satisfy

$$g_{jkl} = -\, g_{kjl}\,, \qquad (7.A.57)$$

from the antisymmetry of the connection forms. The irreducible parts of the Ricci coefficient tensor are a pseudoscalar, a vector, and a traceless pseudotensor. We have, in our case

$$g_{jkl} = \frac{1}{\beta}\left[\delta^1{}_j\,\delta^2{}_k - \delta^1{}_k\,\delta^2{}_j\right]\delta^3{}_l\,. \tag{7.A.58}$$

The vector, formed by contracting the second and third indices of this expression for g_{jkl},

$$g_j = g_{jkk} = 0 \tag{7.A.59}$$

vanishes. The pseudoscalar is given by

$$\tilde{g} = \frac{1}{3!}\,\epsilon_{jkl}\,g_{jkl} = \frac{1}{3\,\beta}\,. \tag{7.A.60}$$

The pseudotensor is obtained as

$$\tilde{T}_{lm} = \frac{1}{2}\left[\epsilon_{jkm}\,g_{jkl} + \epsilon_{jkl}\,g_{jkm}\right] - 2\,\tilde{g}\,\delta_{lm}\,, \tag{7.A.61}$$

which, in our present case, gives

$$\tilde{T}_{lm} = \frac{2}{3\,\beta}\begin{bmatrix} -1 & 0 & 0 \\ 0 & -1 & 0 \\ 0 & 0 & +2 \end{bmatrix}. \tag{7.A.62}$$

The existence of homogeneous fields is tied to the appearance of a locally transitive group of isometries in the manifold. This group is always a subgroup of the full isometry group, also known as the congruence group, of the manifold. In the Euclidean plane there is only one transitive subgroup of motions, namely that of the translations. Their left-invariant differential forms are dx and dy, which lead to the usual Cartesian frames.

In the Euclidean $\mathbb{R}^3$, we discovered the subgroup (7.A.29) which is different from the pure translation group. The group, $\mathrm{T}(\alpha, \phi; \beta)$, combines translations perpendicular to the z-axis with screw motions of pitch $1/\beta$ about the z-axis. The group is not only locally but also globally transitive.

There was nothing special in the choice of the z-axis as the screw axis. Any other direction would have done as well. If we characterize such an arbitrary direction by the spherical polar coordinates θ and λ (colatitude and azimuth, respectively), we have found three-dimensional subgroups of Euclidean motions which depend on the three parameters β, θ, and λ.

The Lie Algebras

We get for the Lie algebra of these subgroups,

$$[X_1, X_2] = 0, \tag{7.A.63a}$$

$$[X_2, X_3] = \frac{1}{\beta} X_1, \qquad [X_3, X_1] = \frac{1}{\beta} X_2. \tag{7.A.63b,c}$$

By redefining the generator βX_3 as X_3, we obtain

$$[X_1, X_2] = 0, \tag{7.A.64a}$$

$$[X_2, X_3] = X_1, \qquad [X_3, X_1] = X_2. \tag{7.A.64b,c}$$

This is one of the Lie algebras which appear in Bianchi's classification [Bianchi, 1918]. In a revised form of the classification given by MacCallum [MacCallum, 1979], this is Bianchi Type VII_0.

No other types besides type VII_0 and type I (all structure constants zero) can occur as Lie algebras of locally transitive subgroups of the Euclidean group of motions in three dimensions. This follows from the results of **Section 11.N** where we apply **Chapter 11**'s general study of homogeneous fields in Euclidean three-dimensional spaces to the restricted case of Euclidean $\mathbb{R}^3$ (our present case).

The homogeneous field that has been discussed in this section is the only non-trivial one in the Euclidean $\mathbb{R}^3$.

B. Homogeneous Fields on 3-Dimensional Minkowski Spacetime

For studies in the Special and General Theories of Relativity, we are primarily concerned with the indefinite metrics of their respective spacetimes. The extension of the theory of homogeneous fields to these higher dimensional spaces naturally includes three-dimensional Minkowski spacetime:

$$ds^2 = dt^2 - dx^2 - dy^2. \tag{7.B.1}$$

The structural similarity to the positive definite line element allows us to proceed with an analysis as in **Section A** of this chapter. We again define a complex coordinate and its conjugate

$$\Sigma \equiv x + i\,y, \qquad \bar{\Sigma} \equiv x - i\,y \tag{7.B.2a,b}$$

with the derived one-forms

$$d\Sigma = dx + i\,dy, \qquad d\bar{\Sigma} = dx - i\,dy. \tag{7.B.3a,b}$$

The line element becomes

$$ds^2 = dt^2 - d\Sigma\, d\bar{\Sigma}. \tag{7.B.4}$$

Another Simply Transitive Subgroup

Again, there is a subgroup which we can now exhibit by the following coordinate transformations, $M(\alpha, \phi; \beta)$,

$$t' = t + \beta\,\phi\,, \qquad \beta = constant \neq 0\,, \quad \beta \in \mathbb{R} \tag{7.B.5a}$$

$$\Sigma' = e^{i\phi}\,\Sigma + \alpha\,, \qquad \alpha \in \mathbb{C} \tag{7.B.5b}$$

where ϕ is real, α is complex, and β is a fixed non-zero real constant. The three subgroup parameters are α (2 real parameters) and ϕ (1 real parameter). The line element is invariant under these transformations, that is,

$$ds^2 = dt^2 - d\Sigma\, d\bar{\Sigma} = dt'^2 - d\Sigma'\, d\bar{\Sigma}' \tag{7.B.6}$$

where the differentials are

$$dt' = dt, \qquad d\Sigma' = e^{i\phi}\, d\Sigma\,. \tag{7.B.7a,b}$$

The parameters of the transformation between any two points, (t, Σ) and (t', Σ'), of the manifold can be determined uniquely for $\beta \neq 0$:

$$\phi = \frac{t' - t}{\beta}\,, \qquad \alpha = \Sigma' - \Sigma\, \exp\left[i\,\frac{t' - t}{\beta}\right]. \tag{7.B.8a,b}$$

Therefore, these transformations form a simply transitive subgroup.

The Subgroup's Left-invariant Differential Forms

Our next step is the computation of the left-invariant differential forms. The differentials of the coordinate transformation functions are

$$dt' = dt + \beta\, d\phi, \tag{7.B.9a}$$

$$d\Sigma' = e^{i\phi}\,[\,d\Sigma + i\,\Sigma\, d\phi\,] + d\alpha\,. \tag{7.B.9b}$$

The matrix representation for the coordinate transformation given by equations (7.B.5) is

$$A = \begin{bmatrix} 1 & 0 & \beta\,\phi \\ 0 & e^{i\phi} & \alpha \\ 0 & 0 & 1 \end{bmatrix}. \tag{7.B.10}$$

Its inverse is

$$A^{-1} = \begin{bmatrix} 1 & 0 & -\beta\,\phi \\ 0 & e^{-i\phi} & -\alpha\, e^{-i\phi} \\ 0 & 0 & 1 \end{bmatrix}. \tag{7.B.11}$$

The left-invariant differential forms are obtained by forming

$$A^{-1}\,dA = \begin{bmatrix} 0 & 0 & \beta\,d\phi \\ 0 & i\,d\phi & e^{-i\phi}\,d\alpha \\ 0 & 0 & 0 \end{bmatrix}. \tag{7.B.12}$$

In more condensed form, this is just

$$\boldsymbol{\omega}^0 \equiv d\phi\,, \qquad \boldsymbol{\omega} \equiv e^{-i\phi}\,d\alpha \tag{7.B.13a,b}$$

for the left-invariant differential forms. The points that the transformation equations (7.B.5) move into the origin are given by solving

$$t' = 0\,, \qquad \Sigma' = 0 \tag{7.B.14a,b}$$

and we obtain the hitherto undetermined subgroup parameters appearing in the coordinate transformation implicitly

$$t = -\,\beta\,\phi\,, \qquad \Sigma = -\,e^{-i\phi}\,\alpha\,, \tag{7.B.15a,b}$$

or, solving for them, we get their explicit forms

$$\phi = -\,\frac{t}{\beta}\,, \qquad \alpha = -\,\Sigma\,e^{-it/\beta} \tag{7.B.16a,b}$$

with invariant differential forms (7.B.13) expressed (from 7.B.9) as

$$\boldsymbol{\omega}_0 = d\phi = -\,\frac{1}{\beta}\,dt\,, \qquad \boldsymbol{\omega} = e^{it/\beta}d\alpha = -\,d\Sigma + i\,\frac{1}{\beta}\,\Sigma\,dt\,. \tag{7.B.17a,b}$$

The Metric in Differential Forms

The line element becomes

$$ds^2 = dt^2 - \mu\,\bar{\mu} \qquad \text{where} \qquad \mu \equiv e^{it/\beta}\,d\Sigma \tag{7.B.18}$$

and, reiterating equations (7.B.3), we have

$$d\Sigma = dx + i\,dy\,, \qquad d\bar{\Sigma} = dx - i\,dy\,. \tag{7.B.19a,b}$$

This is just 3-dimensional Minkowski spacetime

$$ds^2 = M_{ab}\,\boldsymbol{\omega}^a\,\boldsymbol{\omega}^b \tag{7.B.20}$$

with $a, b \in \{0, 1, 2\}$ and where

$$M_{ab} = \begin{bmatrix} 1 & 0 & 0 \\ 0 & -1 & 0 \\ 0 & 0 & -1 \end{bmatrix} \tag{7.B.21}$$

is just the three-dimensional Minkowski metric. We define a set of differential one forms with the definitions:

$$\omega^0 = +\omega_0 = dt\,, \tag{7.B.22a}$$

$$\omega^1 = -\omega_1 = \cos\left(\frac{t}{\beta}\right) dx - \sin\left(\frac{t}{\beta}\right) dy\,, \tag{7.B.22b}$$

$$\omega^2 = -\omega_2 = \sin\left(\frac{t}{\beta}\right) dx + \cos\left(\frac{t}{\beta}\right) dy\,. \tag{7.B.22c}$$

The Connection Forms

We compute the connection forms

$$d\omega_0 - \omega_{01} \wedge \omega^1 - \omega_{02} \wedge \omega^2 = 0\,, \tag{7.B.23a}$$

$$d\omega_1 - \omega_{10} \wedge \omega^0 - \omega_{12} \wedge \omega^2 = 0\,, \tag{7.B.23b}$$

$$d\omega_2 - \omega_{20} \wedge \omega^0 - \omega_{21} \wedge \omega^1 = 0\,. \tag{7.B.23c}$$

Since $d\omega_0 = ddt = 0$, we may suppose that

$$\omega_{01} = \omega_{02} = 0\,, \tag{7.B.24}$$

and then we get from the first Cartan structure equations

$$d\omega_1 = \omega_{12} \wedge \omega^2\,, \qquad d\omega_2 = \omega_{21} \wedge \omega^1\,. \tag{7.B.25a,b}$$

Expanding the differentials by means of equations (7.B.22), we obtain

$$d\omega_1 = d\left[-\cos\left(\frac{t}{\beta}\right) dx + \sin\left(\frac{t}{\beta}\right) dy\right] \tag{7.B.26a}$$

$$= \frac{1}{\beta} + \sin\left(\frac{t}{\beta}\right) dt \wedge dx + \cos\left(\frac{t}{\beta}\right) dt \wedge dy \tag{7.B.26b}$$

$$= \frac{1}{\beta}\, dt \wedge \omega^2 = \omega_{12} \wedge \omega^2\,. \tag{7.B.26c}$$

Thus the only independent connection form is

$$\boldsymbol{\omega}_{12} = \frac{1}{\beta}\, dt = \frac{1}{\beta}\, \boldsymbol{\omega}_0 \,. \tag{7.B.27}$$

We follow the same procedure for the other differential forms and get for the second

$$d\boldsymbol{\omega}_2 = \left[-\sin\left(\frac{t}{\beta}\right) dx - \cos\left(\frac{t}{\beta}\right) dy \right] \tag{7.B.28a}$$

$$= \frac{1}{\beta} - \cos\left(\frac{t}{\beta}\right) dt \wedge dx + \sin\left(\frac{t}{\beta}\right) dt \wedge dy \tag{7.B.28b}$$

$$= -\frac{1}{\beta}\, dt \wedge \boldsymbol{\omega}^1 \tag{7.B.28c}$$

$$= -\boldsymbol{\omega}_{12} \wedge \boldsymbol{\omega}^1 \,. \tag{7.B.28d}$$

In summary, we have connection forms

$$\boldsymbol{\omega}_{01} = 0\,, \qquad \boldsymbol{\omega}_{02} = 0\,, \qquad \boldsymbol{\omega}_{12} = \frac{1}{\beta}\, \boldsymbol{\omega}_0 \,. \tag{7.B.29a,b,c}$$

The Connection Coefficients

The connection coefficients may be written

$$\boldsymbol{\omega}_{jk} = g_{jk}{}^{l}\, \boldsymbol{\omega}_l \tag{7.B.30}$$

with

$$g_{12}{}^{0} = \frac{1}{\beta} \tag{7.B.31}$$

as the only non-zero term.

The Curvature Form

The curvature form (11.A.4) gives

$$\boldsymbol{\Omega}_{01} = d\boldsymbol{\omega}_{01} - \boldsymbol{\omega}_{02} \wedge \boldsymbol{\omega}^2{}_1 = 0\,, \tag{7.B.32a}$$

$$\boldsymbol{\Omega}_{02} = d\boldsymbol{\omega}_{02} - \boldsymbol{\omega}_{01} \wedge \boldsymbol{\omega}^1{}_2 = 0\,, \tag{7.B.32b}$$

$$\boldsymbol{\Omega}_{12} = d\boldsymbol{\omega}_{12} - \boldsymbol{\omega}_{10} \wedge \boldsymbol{\omega}^0{}_2 = 0\,. \tag{7.B.32c}$$

Interpretation

We can interpret our results by observing the line element in the coordinates introduced earlier; this gives us

$$ds^2 = dt^2 - dx^2 - dy^2 = dt^2 - d\Sigma\, d\bar{\Sigma} \tag{7.B.33a}$$

$$= \left[1 - \frac{1}{\beta^2}\Sigma\,\bar{\Sigma}\right] dt^2 - d\Sigma' d\bar{\Sigma}' \tag{7.B.33b}$$

where

$$\Sigma = e^{it/\beta}\,\Sigma' \tag{7.B.34}$$

and, upon deriving,

$$d\Sigma = \exp\left[i\,\frac{t}{\beta}\right]\left[i\,\frac{1}{\beta}\Sigma'\,dt + d\Sigma'\right], \tag{7.B.35}$$

which tells us that we can make the interpretation

$$\frac{1}{\beta} = \textit{angular velocity of the frame.} \tag{7.B.36}$$

CHAPTER 8

PROPER LORENTZ TRANSFORMATIONS

X. Overview

Having demonstrated the presence of homogeneous fields in positive definite and indefinite three-dimensional spaces, we develop the apparatus of vector calculus for our three-dimensional cases with particular attention to proper Lorentz transformations. Their unique eigenvector permits their characterization and we pay extra attention to the null rotations of Wigner's "little group". Then, we look at some consequences for astronomy with the results applied to the study of redshifts.

A. Three-Dimensional Minkowski Spacetime

Before we study the null rotations of three dimensional Minkowski space, that is, the null rotations of **SO(2,1)**, we need to examine our usual theoretical apparatus to see where it continues to serve us reliably and where it must be modified or where caution must be taken.

For this discussion indices will be members of the set $\{0, 1, 2\}$. A vector is given in terms of basis vectors by

$$\boldsymbol{X} = X^j \mathbf{e}_j \,. \tag{8.A.1}$$

Typically we will map time and space coordinates according to the rule:

$$X^0 \equiv t\,, \qquad X^1 \equiv x\,, \qquad X^2 \equiv y\,. \tag{8.A.2}$$

The scalar products of the basis vectors are given for an orthonormal basis by

$$\mathbf{e}_j \cdot \mathbf{e}_k = M_{jk}\,, \qquad \text{where} \qquad M_{jk} = \begin{bmatrix} 1 & 0 & 0 \\ 0 & -1 & 0 \\ 0 & 0 & -1 \end{bmatrix}, \tag{8.A.3}$$

and where we further define

$$M^{mj} = \begin{bmatrix} 1 & 0 & 0 \\ 0 & -1 & 0 \\ 0 & 0 & -1 \end{bmatrix} \tag{8.A.4}$$

as the inverse of the matrix defined in (8.A.3), by means of

$$M^{mj} M_{jk} = \delta^{m}{}_{k} \,. \tag{8.A.5}$$

This will assist us in raising and lowering indices.

We restrict the frame of arbitrary basis vectors so that the first basis vector is always directed toward the future, that is,

$$\mathbf{e}_0 \cdot \mathbf{e}_0 > 0 \,, \tag{8.A.6}$$

and the other two basis vectors, $\mathbf{e}_1$ and $\mathbf{e}_2$, have positive orientation in the ($\mathbf{e}_1$- $\mathbf{e}_2$)-plane. The scalar product of two vectors is given by

$$\boldsymbol{X} \cdot \boldsymbol{X}' = X^j \, \mathbf{e}_j \cdot X'^k \, \mathbf{e}_k \tag{8.A.7a}$$

$$= X^j \, M_{jk} \, X'^k \tag{8.A.7b}$$

$$= t\,t' - x\,x' - y\,y' \,. \tag{8.A.7c}$$

Let ϵ_{jkl} be completely skew-symmetric with

$$\epsilon_{012} = +1 \,. \tag{8.A.8}$$

We also have

$$\epsilon^{0}{}_{12} = +1 \,, \quad \epsilon^{1}{}_{20} = -1 \,, \quad \epsilon^{2}{}_{01} = -1 \,, \tag{8.A.9}$$

when we use M^{jk} to raise the indices.

We carry out a Lorentz transformation, that is, a linear mapping, $\boldsymbol{\Lambda}$, of the vector space that preserves the scalar product:

$$\boldsymbol{\Lambda}[\boldsymbol{X}] \cdot \boldsymbol{\Lambda}[\boldsymbol{Y}] = \boldsymbol{X} \cdot \boldsymbol{Y} \,. \tag{8.A.10}$$

Notice that the basis vectors are kept fixed once and for all. We obtain a matrix representation for the linear map, $\boldsymbol{\Lambda}$, by observing its operation on the basis vectors:

$$\boldsymbol{\Lambda}[\mathbf{e}_j] = \mathbf{e}_k \, \alpha^{k}{}_{j} \,. \tag{8.A.11}$$

The vector $\boldsymbol{X}$ is transformed into the vector $\boldsymbol{X}'$ by application of $\boldsymbol{\Lambda}$:

$$\boldsymbol{X}' \equiv \boldsymbol{\Lambda}[\boldsymbol{X}] = \boldsymbol{\Lambda}[X^j \, \mathbf{e}_j] = \mathbf{e}_k \, \alpha^k{}_j \, X^j \equiv X'^k \, \mathbf{e}_k \,, \tag{8.A.12}$$

or, by comparing components,

$$X'^k = \alpha^k{}_j \, X^j \,. \tag{8.A.13}$$

The determinant of the linear map's matrix components is defined by

$$\det [\, \alpha^p{}_q \,] \equiv \epsilon_{jkl} \, \alpha^j{}_0 \, \alpha^k{}_1 \, \alpha^l{}_2 \,. \tag{8.A.14}$$

We then have

$$\epsilon_{jkl} \, \alpha^j{}_{j'} \, \alpha^k{}_{k'} \, \alpha^l{}_{l'} = \epsilon_{jkl} \, \det [\, \alpha^m{}_{m'} \,] \,. \tag{8.A.15}$$

Proof: The left hand side of equation (8.A.15) is skew-symmetric in j', k', l' because a determinant changes sign under interchange of columns; the right hand side is skew-symmetric by definition. Therefore, (8.A.15) is true if it holds for one choice of the indices j', k', l', for example, $j' = 0, k' = 1, l' = 2$. Then, in that case we will have

$$\epsilon_{jkl} \, \alpha^j{}_0 \, \alpha^k{}_1 \, \alpha^l{}_2 = \epsilon_{012} \, \det [\, \alpha^m{}_{m'} \,] \,. \tag{8.A.16}$$

This is true because of (8.A.8) and (8.A.14). If we restrict ourselves to proper Lorentz transformations, that is, those with determinant equal to $+1$, then (8.A.15) shows that ϵ_{jkl} transforms just as a covariant tensor of rank three. ■

Covectors

If $\boldsymbol{X}$ and $\boldsymbol{Y}$ are two vectors with components X^k and Y^l, we obtain a covector $\boldsymbol{Z}$ with components Z_j given by

$$Z_j = \epsilon_{jkl} \, X^k \, Y^l \,. \tag{8.A.17}$$

We can now define a vector

$$\boldsymbol{Z} \equiv Z^m \, \mathbf{e}_m \,, \tag{8.A.18a}$$

where

$$Z^m = M^{mj} \, Z_j = M^{mj} \, \epsilon_{jkl} \, X^k \, Y^l \,, \tag{8.A.18b}$$

or

$$Z^m = \epsilon^m{}_{kl} \, X^k \, Y^l \, , \qquad \text{where} \qquad \epsilon^m{}_{kl} = M^{mj} \, \epsilon_{jkl} \, . \tag{8.A.19}$$

We write

$$\boldsymbol{Z} = \boldsymbol{X} \times \boldsymbol{Y} \tag{8.A.20}$$

for the cross-product and have from (8.A.17)

$$\mathbf{e}_1 \times \mathbf{e}_2 = \mathbf{e}_0 \, , \qquad \mathbf{e}_0 \times \mathbf{e}_1 = -\,\mathbf{e}_2 \, , \tag{8.A.21a}$$

$$\mathbf{e}_2 \times \mathbf{e}_0 = -\,\mathbf{e}_1 \, . \tag{8.A.21b}$$

This shows that the "right hand rule" for the Euclidean vector product no longer applies in general. But we shall see that it is simply restricted to spacelike vectors with timelike cross-product.

The Lagrange Identity

We want to verify that the main working tool of Euclidean vector algebra, the Lagrange identity, applies unchanged in Minkowski three-dimensional spacetime. When we raise indices, we have

$$\epsilon^{j'k'l'} = M^{j'j} \, M^{k'k} \, M^{l'l} \, \epsilon_{jkl} \, . \tag{8.A.22}$$

Thus

$$\epsilon^{012} = \det \left[\, M^{m'm} \, \right] = +1 \, . \tag{8.A.23}$$

Therefore, by the same argument following (8.A.15), we conclude that

$$\epsilon^{jkl} = \epsilon_{jkl} \, , \tag{8.A.24}$$

which is distinct from the four-dimensional case where we have

$$\epsilon^{jkl} = -\,\epsilon_{jkl} \, , \tag{8.A.25}$$

with $j, k, l \in \{0, 1, 2\}$. Then

$$\epsilon^{jkl} \, \epsilon_{j'k'l'} = \det \begin{bmatrix} \delta^j{}_{j'} & \delta^j{}_{k'} & \delta^j{}_{l'} \\ \delta^k{}_{j'} & \delta^k{}_{k'} & \delta^k{}_{l'} \\ \delta^l{}_{j'} & \delta^l{}_{k'} & \delta^l{}_{l'} \end{bmatrix} . \tag{8.A.26}$$

This can be proven by the same argument following (8.A.15) with respect to the indices j, k, l and j', k', l'. All we need is to see that

$$\epsilon^{012} \, \epsilon_{012} = \det \begin{bmatrix} 1 & 0 & 0 \\ 0 & 1 & 0 \\ 0 & 0 & 1 \end{bmatrix} = +1 \, . \tag{8.A.27}$$

From (8.A.26) follows by developing with respect to the first row of the determinant

$$\begin{aligned}\epsilon^{jkl}\,\epsilon_{jk'l'} &= +\,3\left[\delta^k{}_{k'}\,\delta^l{}_{l'} - \delta^k{}_{l'}\,\delta^l{}_{k'}\right]\\ &\quad -\,\delta^j{}_{k'}\left[\delta^k{}_j\,\delta^l{}_{l'} - \delta^k{}_{l'}\,\delta^l{}_j\right]\\ &\quad +\,\delta^j{}_{l'}\left[\delta^k{}_j\,\delta^l{}_{k'} - \delta^k{}_{k'}\,\delta^l{}_j\right]\end{aligned} \tag{8.A.28a}$$

$$\begin{aligned}\epsilon^{jkl}\,\epsilon_{jk'l'} &= +\,3\left[\delta^k{}_{k'}\,\delta^l{}_{l'} - \delta^k{}_{l'}\,\delta^l{}_{k'}\right]\\ &\quad -\left[\delta^k{}_{k'}\,\delta^l{}_{l'} - \delta^k{}_{l'}\,\delta^l{}_{k'}\right]\\ &\quad +\left[\delta^k{}_{l'}\,\delta^l{}_{k'} - \delta^k{}_{k'}\,\delta^l{}_{l'}\right]\end{aligned} \tag{8.A.28b}$$

$$\epsilon^{jkl}\,\epsilon_{jk'l'} = \left[\delta^k{}_{k'}\,\delta^l{}_{l'} - \delta^k{}_{l'}\,\delta^l{}_{k'}\right]. \tag{8.A.28c}$$

Thus by lowering indices k and l, we have

$$\epsilon^j{}_{kl}\,\epsilon_{jk'l'} = \left[M_{kk'}\,M_{ll'} - M_{kl'}\,M_{lk'}\right]. \tag{8.A.29}$$

Therefore with vectors $\boldsymbol{X}, \boldsymbol{Y}, \boldsymbol{U}$, and $\boldsymbol{V}$, we have the Lagrange identity:

$$[\boldsymbol{X}\times\boldsymbol{Y}]\cdot[\boldsymbol{U}\times\boldsymbol{V}] = (\boldsymbol{X}\cdot\boldsymbol{U})\ (\boldsymbol{Y}\cdot\boldsymbol{V}) - (\boldsymbol{X}\cdot\boldsymbol{V})\ (\boldsymbol{Y}\cdot\boldsymbol{U})\,. \tag{8.A.30}$$

The other vector identities follow similarly. If $\boldsymbol{X}, \boldsymbol{Y}, \boldsymbol{U}$ have components X^j, Y^k, and U^l, respectively, then

$$[\boldsymbol{X}\times\boldsymbol{Y}]\cdot\boldsymbol{U} = \epsilon_{jkl}\,X^j\,Y^k\,U^l\,, \tag{8.A.31}$$

which immediately implies

$$[\boldsymbol{X}\times\boldsymbol{Y}]\cdot\boldsymbol{U} = \boldsymbol{X}\cdot[\boldsymbol{Y}\times\boldsymbol{U}] = [\boldsymbol{Y}\times\boldsymbol{U}]\cdot\boldsymbol{X}\,, \tag{8.A.32}$$

with similar expressions holding for permutations of the vectors. We may consolidate all such equations into

$$[[\boldsymbol{X}\times\boldsymbol{Y}]\times\boldsymbol{U}] = \epsilon_{jkl}\,X^k\,Y^l\,\epsilon^m{}_{jn}\,U^n\,\mathbf{e}_m \tag{8.A.33a}$$

$$= -\left[\delta^m{}_k\,M_{ln} - \delta^m{}_l M_{kn}\right]X^k\,Y^l\,U^n\,\mathbf{e}_m \tag{8.A.33b}$$

$$[[\boldsymbol{X}\times\boldsymbol{Y}]\times\boldsymbol{U}] = (\boldsymbol{X}\cdot\boldsymbol{U})\,\boldsymbol{Y} - (\boldsymbol{Y}\cdot\boldsymbol{U})\,\boldsymbol{X}\,. \tag{8.A.33c}$$

B. Spacetime Interpretation of the Scalar Product

A vector is termed timelike, spacelike, or null as its scalar product is greater than zero, less than zero, or zero, respectively. That is:

$$\boldsymbol{A}\cdot\boldsymbol{A} > 0 \quad \textbf{timelike}\,, \tag{8.B.1a}$$

$$\boldsymbol{A}\cdot\boldsymbol{A} < 0 \quad \textbf{spacelike}\,, \tag{8.B.1b}$$

$$\boldsymbol{A}\cdot\boldsymbol{A} = 0 \quad \textbf{null (lightlike)}\,. \tag{8.B.1c}$$

A vector $\boldsymbol{A}$ is said to "future directed" when $A^0 > 0$.

For a timelike or spacelike vector, the scalar product is interpreted as the length of that vector and is equal to

$$|\boldsymbol{A}| = [\pm\boldsymbol{A}\cdot\boldsymbol{A}]^{1/2}\,, \tag{8.B.2}$$

where the plus sign is chosen for timelike vectors and the minus sign for spacelike vectors. The length of a null vector is undefined.

The Rapidity or "Spacetime Angle" of the Scalar Product

The "angle" or "rapidity", χ, is defined between two timelike future directed vectors $\boldsymbol{A}$ and $\boldsymbol{B}$, that is, if $\boldsymbol{A}\cdot\boldsymbol{A} > 0$, $\boldsymbol{B}\cdot\boldsymbol{B} > 0$, and $\boldsymbol{A}\cdot\boldsymbol{B} > 0$, then

$$\cosh(\chi) = \frac{\boldsymbol{A}\cdot\boldsymbol{B}}{|\boldsymbol{A}|\,|\boldsymbol{B}|}\,. \tag{8.B.3}$$

Notice: An angle is not defined if the two vectors point into two different time directions (one into the future and the other into the past). Further, an angle between a timelike vector and any other vector is only defined if the second vector is also timelike.

An angle between a null vector and any other vector is not defined.

An angle between two spacelike vectors is only defined if the cross-product is timelike or the zero vector. Let $\boldsymbol{A}\cdot\boldsymbol{A} < 0$ and $\boldsymbol{B}\cdot\boldsymbol{B} < 0$ and

$$[\boldsymbol{A}\times\boldsymbol{B}]\cdot[\boldsymbol{A}\times\boldsymbol{B}] \geq 0\,. \tag{8.B.4}$$

Then from the Lagrange identity (8.A.30)

$$[\boldsymbol{A}\times\boldsymbol{B}]\cdot[\boldsymbol{A}\times\boldsymbol{B}] = (\boldsymbol{A}\cdot\boldsymbol{A})\,(\boldsymbol{B}\cdot\boldsymbol{B}) - (\boldsymbol{A}\cdot\boldsymbol{B})^2 \geq 0\,. \tag{8.B.5}$$

Even if the vectors are proportional, that is, if $\boldsymbol{A} = \lambda\boldsymbol{B}$, the cross product is still zero. Then, for both cases, when the cross product is a timelike or zero vector, we have

$$\cosh(\phi) = -\frac{\boldsymbol{A}\cdot\boldsymbol{B}}{|\boldsymbol{A}|\,|\boldsymbol{B}|}\,. \tag{8.B.6}$$

One also has

$$|\boldsymbol{A}\times\boldsymbol{B}| = |\boldsymbol{A}|\,|\boldsymbol{B}|\sin(\phi)\,. \tag{8.B.7}$$

The Orientation of Spacetime Vectors

The orientation of three vectors is given by the sign of

$$[\boldsymbol{A} \times \boldsymbol{B}] \cdot \boldsymbol{C} . \tag{8.B.8}$$

If our third vector is the cross-product of the first two, then, by (8.B.5),

$$\Delta = [\,\boldsymbol{A} \times \boldsymbol{B}\,] \cdot [\,\boldsymbol{A} \times \boldsymbol{B}\,] \tag{8.B.9a}$$

$$= (\boldsymbol{A} \cdot \boldsymbol{A})\,(\boldsymbol{B} \cdot \boldsymbol{B}) \;-\; (\boldsymbol{A} \cdot \boldsymbol{B})^2 . \tag{8.B.9b}$$

Assume that timelike vectors $\boldsymbol{A}$ or $\boldsymbol{B}$ are directed into the future (that is, $A^0 > 0$ and $B^0 > 0$). Then Δ positive can only occur for $\boldsymbol{A}$ and $\boldsymbol{B}$ both spacelike. If, in this case, $\Delta > 0$, then (8.B.6) obtains. For two timelike future directed vectors one has, from (8.B.2),

$$|\boldsymbol{A} \times \boldsymbol{B}| \;=\; |\boldsymbol{A}|\; |\boldsymbol{B}|\; \sinh(\chi) , \tag{8.B.10}$$

with negative orientation.

C. Lorentz Transformations in Three-Dimensional Spacetime

A Lorentz transformation defines a map

$$\mathbf{X}' \,=\, \boldsymbol{\Lambda}[\mathbf{X}] , \qquad \mathbf{Y}' \,=\, \boldsymbol{\Lambda}[\mathbf{Y}] \tag{8.C.1}$$

such that the scalar product is preserved, that is,

$$\mathbf{X}' \cdot \mathbf{Y}' = \mathbf{X} \cdot \mathbf{Y} . \tag{8.C.2}$$

With respect to a basis we have:

$$\mathbf{e}'_i \;=\; \boldsymbol{\Lambda}\,[\mathbf{e}_i] . \tag{8.C.3}$$

The transformation is called "proper" if

$$\det\,[\,\Lambda^j{}_i\,] \;>\; 0 , \tag{8.C.4}$$

where $[\Lambda^j{}_i]$ is the matrix representing the transformation (8.C.1). Let

$$g_{ik} \;=\; \mathbf{e}'_i \cdot \mathbf{e}'_k \;=\; \mathbf{e}_j\,\Lambda^j{}_i \cdot \mathbf{e}_l\,\Lambda^l{}_k \;=\; \Lambda^j{}_i\, g_{jl}\,\Lambda^l{}_k . \tag{8.C.5}$$

We write this equation in matrix form

$$\mathbf{G} = \mathbf{\Lambda}^t \, \mathbf{G} \, \mathbf{\Lambda} \,, \qquad \mathbf{G} = \mathbf{G}^t \,, \qquad g \equiv \det[\,\mathbf{G}\,] > 0 \,. \tag{8.C.6}$$

Eigenvector

We want to show now that a proper Lorentz transformation always has a non-zero eigenvector with eigenvalue equal to unity:

$$\mathbf{\Lambda}[\,\boldsymbol{\xi}\,] = +1\,\boldsymbol{\xi} \quad \Longleftrightarrow \quad (\,\mathbf{\Lambda} - \mathbf{I}\,)\,\boldsymbol{\xi} = 0\,, \quad \boldsymbol{\xi} \neq 0\,. \tag{8.C.7}$$

A non-vanishing $\boldsymbol{\xi}$ exists if and only if

$$\det[\,\mathbf{\Lambda} - \mathbf{I}\,] = 0\,. \tag{8.C.8}$$

We have

$$g \det[\,\mathbf{\Lambda} - \mathbf{I}\,] = \det[\,\mathbf{G}\,\mathbf{\Lambda} - \mathbf{G}\,] \tag{8.C.9a}$$
$$= \det[\,\mathbf{G}\,\mathbf{\Lambda} - \mathbf{\Lambda}^t\,\mathbf{G}\,\mathbf{\Lambda}\,] \tag{8.C.9b}$$
$$= \det[\,\mathbf{G} - \mathbf{\Lambda}^t\,\mathbf{G}\,] \det[\,\mathbf{\Lambda}\,] \tag{8.C.9c}$$
$$= \det[\,\mathbf{G}^t - \mathbf{G}^t\,\mathbf{\Lambda}\,] \det[\,\mathbf{\Lambda}\,] \tag{8.C.9d}$$
$$= g \det[\,\mathbf{\Lambda}\,] \det[\,\mathbf{I} - \mathbf{\Lambda}\,] \tag{8.C.9e}$$
$$= g \det[\,\mathbf{\Lambda}\,]\,(-1)^3 \det[\,\mathbf{\Lambda} - \mathbf{I}\,] \tag{8.C.9f}$$
$$= -\,g \det[\,\mathbf{\Lambda}\,] \det[\,\mathbf{\Lambda} - \mathbf{I}\,]\,. \tag{8.C.9g}$$

Thus, since, by definition, the determinant of a proper Lorentz transformation is greater than zero, we must have

$$\det[\,\mathbf{\Lambda} - \mathbf{I}\,] = 0\,. \tag{8.C.10}$$

Uniqueness

Our next observation is that the eigenvector $\boldsymbol{\xi}$ is uniquely determined for a proper Lorentz transformation that is not the identity.

Suppose there exists another vector $\boldsymbol{\eta}$ which is an eigenvector belonging to eigenvalue $+1$ and which is also linearly independent of $\boldsymbol{\xi}$. We chose then a vector $\boldsymbol{\zeta}$ independent of $\boldsymbol{\xi}$ and $\boldsymbol{\eta}$ and have with respect to the basis formed by $\boldsymbol{\xi}$, $\boldsymbol{\eta}$, and $\boldsymbol{\zeta}$

$$\boldsymbol{\xi}' = \boldsymbol{\xi}\,, \quad \boldsymbol{\eta}' = \boldsymbol{\eta}\,, \quad \boldsymbol{\zeta}' = a\,\boldsymbol{\xi} + b\,\boldsymbol{\eta} + c\,\boldsymbol{\zeta}\,. \tag{8.C.11}$$

From (8.C.5) it follows that the determinant of a proper Lorentz transformation is unity. Therefore the coefficient c in (8.C.11) is 1. From

$$\boldsymbol{\xi}'\cdot\boldsymbol{\zeta}' = \boldsymbol{\xi}\cdot\boldsymbol{\zeta} \quad \longrightarrow \quad (\boldsymbol{\zeta}' - \boldsymbol{\zeta})\cdot\boldsymbol{\xi} = 0\,, \tag{8.C.12a}$$

$$\boldsymbol{\eta}'\cdot\boldsymbol{\zeta}' = \boldsymbol{\eta}\cdot\boldsymbol{\zeta} \quad \longrightarrow \quad (\boldsymbol{\zeta}' - \boldsymbol{\zeta})\cdot\boldsymbol{\eta} = 0\,, \tag{8.C.12b}$$

$$\boldsymbol{\zeta}'\cdot\boldsymbol{\zeta}' = \boldsymbol{\zeta}\cdot\boldsymbol{\zeta} \tag{8.C.12ca}$$

$$\longrightarrow \quad (\boldsymbol{\zeta}' - \boldsymbol{\zeta})\cdot(\boldsymbol{\zeta}' + \boldsymbol{\zeta}) = 0 \tag{8.C.12cb}$$

$$\longrightarrow \quad (\boldsymbol{\zeta}' - \boldsymbol{\zeta})\cdot(a\,\boldsymbol{\xi} + b\,\boldsymbol{\eta} + 2\,\boldsymbol{\zeta}) = 0 \tag{8.C.12cc}$$

$$\longrightarrow \quad 2\,(\boldsymbol{\zeta}' - \boldsymbol{\zeta})\cdot\boldsymbol{\zeta} = 0\,. \tag{8.C.12cd}$$

Thus

$$(\boldsymbol{\zeta}' - \boldsymbol{\zeta})^j\, g_{jk} = 0\,. \tag{8.C.13}$$

Since the determinant of $\mathbf{G}$ is not equal to zero, $\boldsymbol{\zeta}' - \boldsymbol{\zeta} = 0$ and the transformation is the identity.

D. Lorentz Transformations Characterized in Three-Dimensional Spacetime

We can now characterize the proper Lorentz transformations in terms of their eigenvector $\boldsymbol{\xi}$. There are three cases that we must now examine individually. They are defined by the scalar product of the unit eigenvector with itself, that is, whether it is +1, −1, or 0.

Case I: $\boldsymbol{\xi}\cdot\boldsymbol{\xi} = +1$ (Timelike)

The plane $\boldsymbol{x}\cdot\boldsymbol{\xi} = 0$ goes over into itself. Since the transformation is proper, it includes a rotation in this plane, (but *not* a reflection). We have for an arbitrary vector $\boldsymbol{x}$:

$$\boldsymbol{x} = (\boldsymbol{x}\cdot\boldsymbol{\xi})\,\boldsymbol{\xi} + \boldsymbol{x}_\perp\,, \quad \boldsymbol{x}_\perp = \boldsymbol{x} - (\boldsymbol{x}\cdot\boldsymbol{\xi})\,\boldsymbol{\xi}\,, \tag{8.D.1a}$$

$$\boldsymbol{x}' = (\boldsymbol{x}\cdot\boldsymbol{\xi})\,\boldsymbol{\xi} + \boldsymbol{x}'_\perp\,, \quad \boldsymbol{x}'_\perp = \boldsymbol{x}' - (\boldsymbol{x}\cdot\boldsymbol{\xi})\,\boldsymbol{\xi}'\,. \tag{8.D.1b}$$

If φ is the angle of rotation counted positive in the sense that turns $\boldsymbol{\xi}_1$ into $\boldsymbol{\xi}_2$ (with $\left[\boldsymbol{\xi}_0 \times \boldsymbol{\xi}_1\right]\cdot\boldsymbol{\xi}_2 = +1$). We have

$$\boldsymbol{x}'_\perp = \boldsymbol{x}' - (\boldsymbol{x}\cdot\boldsymbol{\xi})\,\boldsymbol{\xi}' \tag{8.D.2a}$$

$$= \boldsymbol{x}_\perp \cos(\varphi) + \left[\boldsymbol{\xi} \times \boldsymbol{x}_\perp\right] \sin(\varphi) \tag{8.D.2b}$$

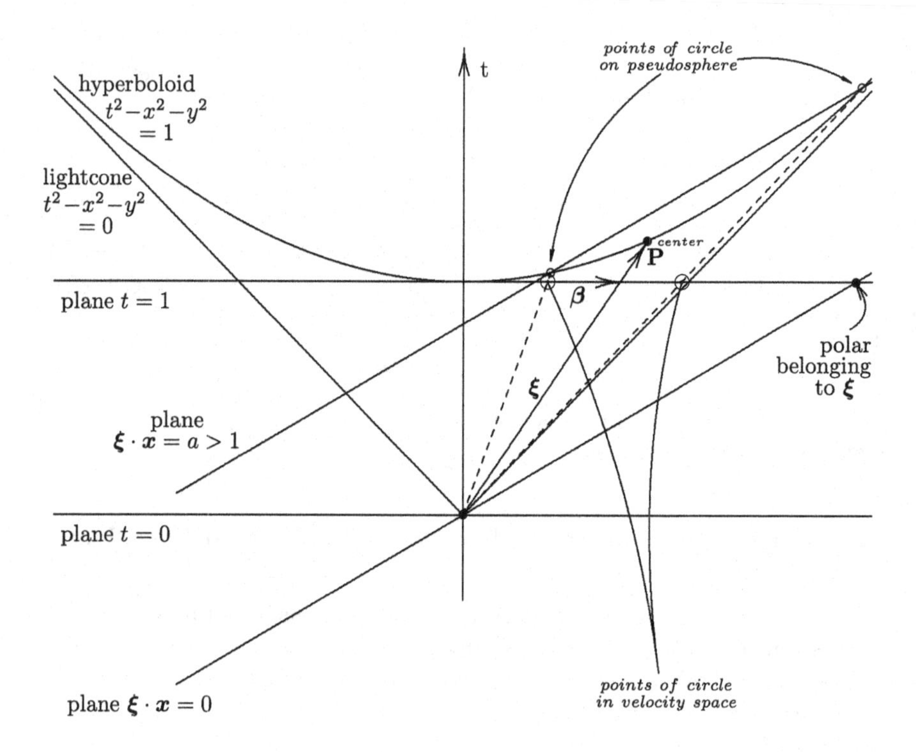

Figure 8.D.1 Cross-section through the pseudosphere containing the t-axis and $\boldsymbol{\xi}$ The polar through $\boldsymbol{\xi}$ goes into the plane of the paper. $a = \cosh(\chi_0)$ where χ_0 is the rapidity radius of the circle. The points of the circle all have rapidity χ_0 with respect to the center, **P**. The dashed line intersections with the plane $t = 1$ are points of a circle in velocity space. The points where the dashed lines end at the hyperboloid are points of a circle on the pseudosphere.

or

$$\boldsymbol{x}' = \cos(\varphi)\,\boldsymbol{x} + (1 - \cos(\varphi))\,(\boldsymbol{x}\cdot\boldsymbol{\xi})\,\boldsymbol{\xi} + \sin(\varphi)\,[\,\boldsymbol{\xi}\times\boldsymbol{x}\,]\,, \tag{8.D.3}$$

exactly as in the Euclidean case.

All planes $\boldsymbol{x}\cdot\boldsymbol{\xi} = \mathit{constant}$ are rotated into themselves. They are all spacelike and intersect the pseudosphere

$$\boldsymbol{x}\cdot\boldsymbol{x} = R, \qquad R > 0 \tag{8.D.4}$$

in circles if

$$\boldsymbol{x}\cdot\boldsymbol{\xi} = A \geq R\,. \tag{8.D.5}$$

These circles are the orbits (trajectories) of the rotations about $\boldsymbol{\xi}$. Every fixed φ corresponds to a plane through the vector $\boldsymbol{\xi}$ giving a geodesic on the pseudosphere.

In the Klein-Beltrami map of the velocity space $(t = 1)$ these planes through $\boldsymbol{\xi}$ give straight lines through the point corresponding to $\boldsymbol{\xi}$.

This point is now characterized in terms of a two-vector $\boldsymbol{\beta}$ as

$$\boldsymbol{\xi} = \frac{1}{\sqrt{1-\beta^2}}\,(1\,,\,\beta_x\,,\,\beta_y)\,. \tag{8.D.6}$$

The circles appear in velocity space as ellipses about $\boldsymbol{\beta}$ that degenerate into circles in the Klein-Beltrami map if $\boldsymbol{\beta}$ vanishes.

The plane $\boldsymbol{x}\cdot\boldsymbol{\xi} = 0$ intersects $t = 1$ in a straight line outside of the circle $\beta^2 = 1$. This line is the **polar** belonging to all these circles and it stays fixed under all rotations for all values of φ.

Case II: $\boldsymbol{\xi}\cdot\boldsymbol{\xi} = -1$ (Spacelike)

The plane $\boldsymbol{x}\cdot\boldsymbol{\xi} = 0$ goes over into itself. Since the transformation is proper, the mapping of this plane into itself has determinant of unity. The plane is like the $(t\text{-}x)$-plane and there are now two types of transformation: those that do not change the direction of time (orthochronous) and those which do. We restrict the discussion to orthochronous transformations. The Lorentz transformation induces in the plane $\boldsymbol{x}\cdot\boldsymbol{\xi} = 0$ a boost with rapidity χ. We have for arbitrary vector $\boldsymbol{x}$

$$\boldsymbol{x} = -(\boldsymbol{x}\cdot\boldsymbol{\xi})\,\boldsymbol{\xi} + \boldsymbol{x}_\perp\,, \quad \boldsymbol{x}_\perp = \boldsymbol{x} + (\boldsymbol{x}\cdot\boldsymbol{\xi})\,\boldsymbol{\xi}\,, \tag{8.D.7a}$$

$$\boldsymbol{x}' = -(\boldsymbol{x}\cdot\boldsymbol{\xi})\,\boldsymbol{\xi} + \boldsymbol{x}'_\perp\,, \quad \boldsymbol{x}'_\perp = \boldsymbol{x}' + (\boldsymbol{x}\cdot\boldsymbol{\xi})\,\boldsymbol{\xi}'\,. \tag{8.D.7b}$$

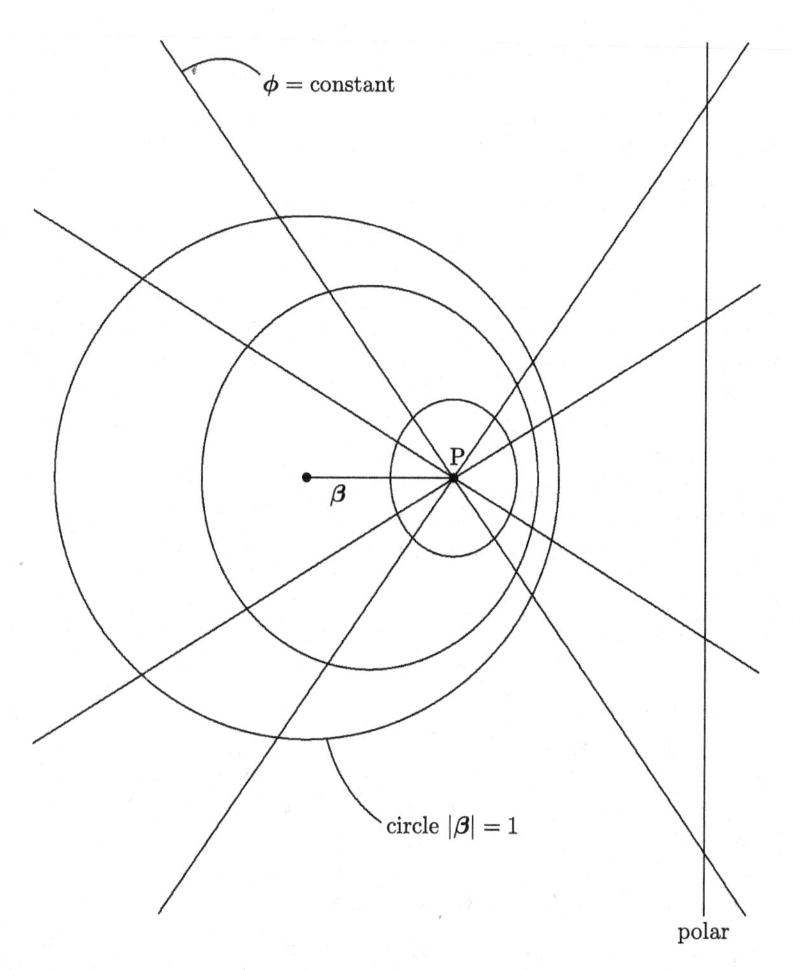

Figure 8.D.2 Two circles in the velocity space Rotations about **P** leave **P** fixed and transform circles into themselves and polar into polar. In this velocity space diagram a pair of circles appear as ellipses inside the circle $|\boldsymbol{\beta}| = 1$.

Notice the changed signs compared with (8.D.1) since $\boldsymbol{\xi}$ is now spacelike. Let $\boldsymbol{\zeta}$ be a timelike unit vector in the plane $\boldsymbol{x}\cdot\boldsymbol{\xi} = 0$ directed into the future.

$$\boldsymbol{\zeta}\cdot\boldsymbol{\zeta} = +1\,, \qquad \boldsymbol{\zeta}\cdot\boldsymbol{\xi} = 0\,. \tag{8.D.8}$$

We define then

$$\boldsymbol{\eta} \equiv \boldsymbol{\xi}\times\boldsymbol{\zeta}\,. \tag{8.D.9}$$

The vector $\boldsymbol{\eta}$ is a unit vector since

$$\boldsymbol{\eta}\cdot\boldsymbol{\eta} = -1 \tag{8.D.10}$$

and

$$\left[\,\boldsymbol{\xi}\times\boldsymbol{\zeta}\,\right]\cdot\boldsymbol{\eta} = -1 \tag{8.D.11}$$

or

$$\left[\,\boldsymbol{\xi}\times\boldsymbol{\eta}\,\right]\cdot\boldsymbol{\zeta} = +1 \tag{8.D.12}$$

which means that $\boldsymbol{\xi}$, $\boldsymbol{\eta}$, $\boldsymbol{\zeta}$ are positively oriented. A positive sense for χ is then implied if $\boldsymbol{\xi}'\cdot\boldsymbol{\eta}\,(=\sinh(\chi))$ is positive. We have then

$$\boldsymbol{x}'_\perp = \boldsymbol{x}' + (\,\boldsymbol{x}\cdot\boldsymbol{\xi}\,)\,\boldsymbol{\xi} \tag{8.D.13a}$$

$$= \cosh(\chi)\,\boldsymbol{x}_\perp + \sinh(\chi)\left[\,\boldsymbol{\xi}\times\boldsymbol{x}_\perp\,\right] \tag{8.D.13b}$$

or, rewriting the last two expressions using (8.D.7a,b), we have

$$\boldsymbol{x}' = \cosh(\chi)\,\boldsymbol{x} + \left[\,\cosh(\chi) - 1\,\right](\,\boldsymbol{x}\cdot\boldsymbol{\xi}\,)\,\boldsymbol{\xi} + \sinh(\chi)\left[\,\boldsymbol{\xi}\times\boldsymbol{x}\,\right]. \tag{8.D.14}$$

All planes $\boldsymbol{x}\cdot\boldsymbol{\xi} = \mathit{constant}$ are boosted into themselves. They are all timelike and intersect the pseudosphere

$$\boldsymbol{x}\cdot\boldsymbol{x} = R^2\,, \qquad R > 0 \tag{8.D.15}$$

in equidistant curves to the geodesics cut out by the plane $\boldsymbol{x}\cdot\boldsymbol{\xi} = 0$. The equidistant curves are the orbits (trajectories) of the boosts about $\boldsymbol{\xi}$. Every fixed χ determines a plane through χ which cuts the pseudosphere in a geodesic. Just as in Case I, the circles about a point **P** and the geodesics through **P** form a system of orthogonal lines like latitude circles about the north pole **P** of a sphere and the meridians through **P**. We have in Case II also an analogy with the sphere. The geodesic cut out by the plane $\boldsymbol{x}\cdot\boldsymbol{\xi} = 0$ corresponds to the equator of the sphere, the equidistant lines to

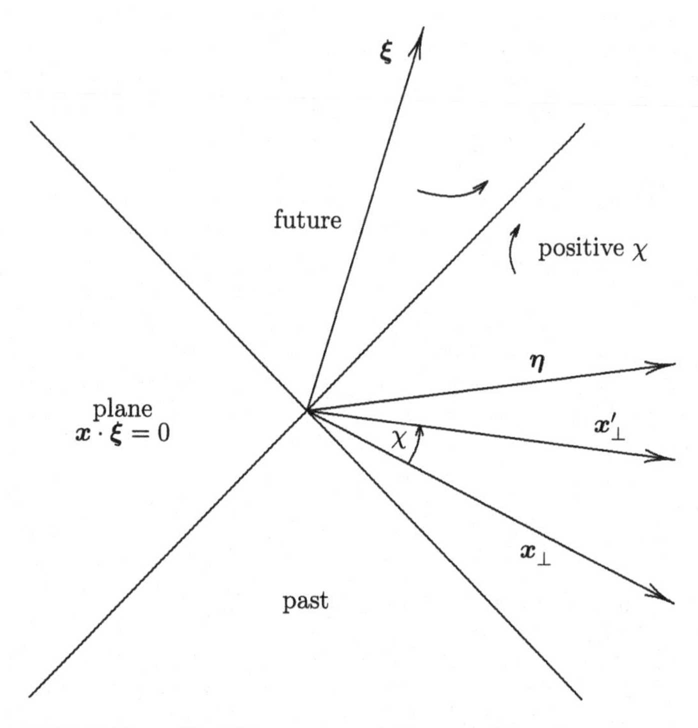

Figure 8.D.3 Case II: The vector $\boldsymbol{\xi}$ is spacelike The vector $\boldsymbol{\xi}$ comes out of the plane of the paper.

the latitude circles and lines $\chi = constant$ correspond to the meridians on the sphere that intersect the equator and all latitude circles perpendicularly.

In the Klein-Beltrami map of the velocity space, where $t = 1$, the plane $\boldsymbol{x} \cdot \boldsymbol{\xi} = 0$, the polar of $\boldsymbol{\xi}$ is now a straight line through the unit circle which it intersects in two points, $\mathbf{P}_1$ and $\mathbf{P}_2$. Two generators of the light cone go through those points. The planes through these two generators and eigenvector $\boldsymbol{\xi}$ give two planes that touch the light cone. They correspond to $\boldsymbol{x} = -\infty$ and $\boldsymbol{x} = +\infty$. In the Klein-Beltrami map these planes appear as two tangents tangent to the unit circle. They intersect in the point where the extension of the vector $\boldsymbol{\xi}$ intersects the plane $t = 1$. This point is the pole.

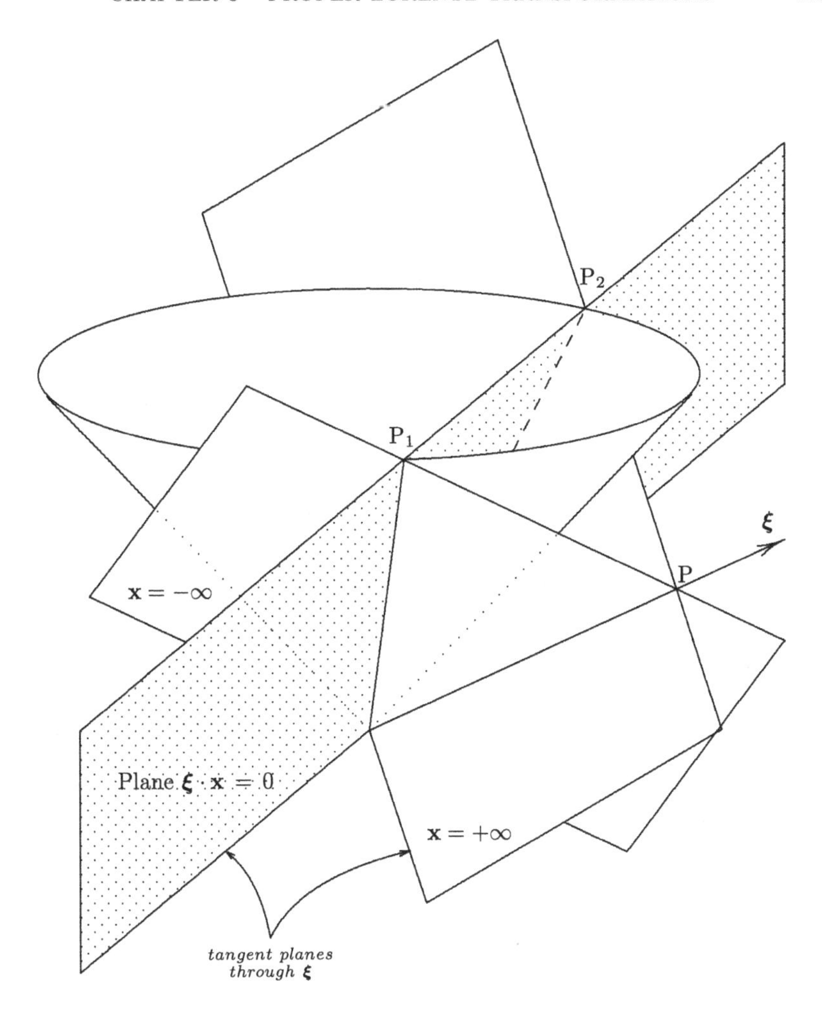

Figure 8.D.4 Case II: Klein-Beltrami Map I: The two tangents are not changed under the boosts about $\boldsymbol{\xi}$ The equidistant curves become ellipses all going through $\mathbf{P}_1$ and $\mathbf{P}_2$. They touch the unit circle in these points from the interior.

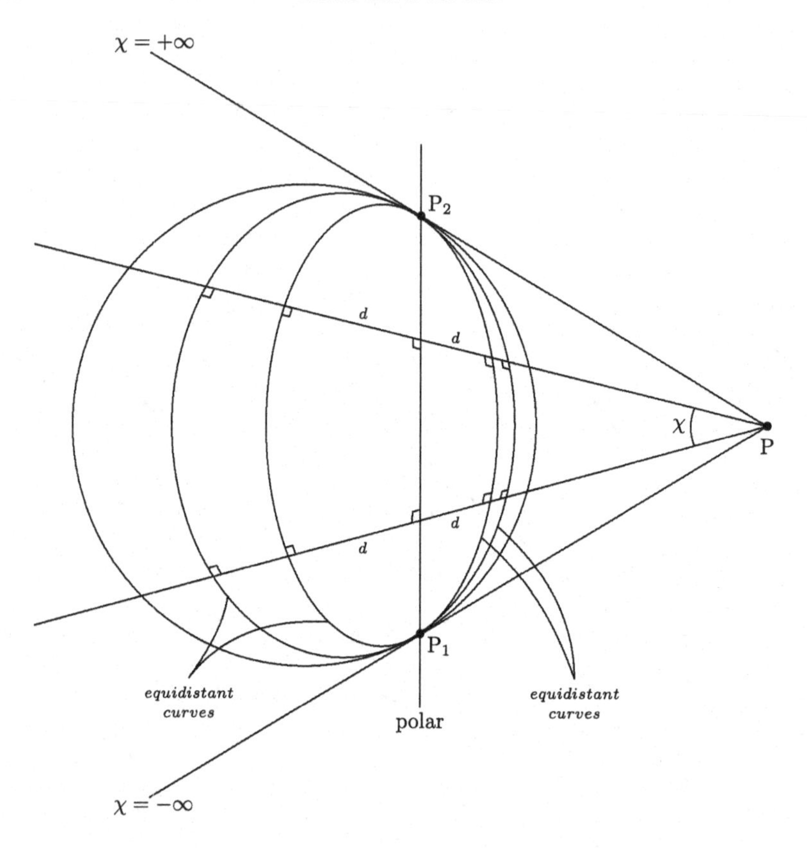

Figure 8.D.5 Case II: Klein-Beltrami Map II: The two tangents are not changed under the boosts about $\boldsymbol{\xi}$ The equidistant curves become ellipses all going through $\mathbf{P}_1$ and $\mathbf{P}_2$. They touch the unit circle in these points from the interior.

Case III: $\boldsymbol{\xi}\cdot\boldsymbol{\xi} = 0$ (Null)

The plane $\boldsymbol{x}\cdot\boldsymbol{\xi} = 0$ goes over into itself. This plane also contains the eigenvector since null vectors are self-orthogonal, that is, $\boldsymbol{\xi}\cdot\boldsymbol{\xi} = 0$. The two planes tangent to the null cone in Case II have now coalesced into one plane that touches the null cone along a generator in the direction of $\boldsymbol{\xi}$.

We can no longer decompose vectors into components parallel and orthogonal to the eigenvector. We proceed, instead, along a different route from the one followed in the previous cases. Now we introduce a future directed ($\zeta^0 > 0$) null vector, $\boldsymbol{\zeta}$, that is linearly independent of our eigenvector, $\boldsymbol{\xi}$, and has a scalar product with $\boldsymbol{\xi}$ that is equal to unity:

$$\boldsymbol{\zeta}\cdot\boldsymbol{\zeta} = 0\,, \qquad \boldsymbol{\zeta}\cdot\boldsymbol{\xi} = +1\,. \tag{8.D.16}$$

We define a spacelike unit vector by the cross-product

$$\boldsymbol{\eta} = \boldsymbol{\xi}\times\boldsymbol{\zeta}\,, \tag{8.D.17}$$

that is orthogonal to the eigenvector $\boldsymbol{\xi}$ ($\boldsymbol{\eta}\cdot\boldsymbol{\xi} = 0$) and lies in the plane $\boldsymbol{x}\cdot\boldsymbol{\xi} = 0$ and is clearly linearly independent of $\boldsymbol{\xi}$. The vectors $\boldsymbol{\xi}$ and $\boldsymbol{\eta}$ span the plane $\boldsymbol{x}\cdot\boldsymbol{\xi} = 0$. Using (8.B.9), we have:

$$\boldsymbol{\eta}\cdot\boldsymbol{\eta} = [\boldsymbol{\xi}\times\boldsymbol{\zeta}]\cdot[\boldsymbol{\xi}\times\boldsymbol{\zeta}] = -\boldsymbol{\xi}\cdot\boldsymbol{\zeta} = -1\,, \tag{8.D.18a}$$

$$\boldsymbol{\xi}\cdot\boldsymbol{\eta} = \boldsymbol{\xi}\cdot[\boldsymbol{\xi}\times\boldsymbol{\zeta}] = 0\,. \tag{8.D.18b}$$

The three vectors $\boldsymbol{\xi}$, $\boldsymbol{\eta}$, $\boldsymbol{\zeta}$ are positively oriented since

$$[\boldsymbol{\xi}\times\boldsymbol{\eta}]\cdot\boldsymbol{\zeta} = [\boldsymbol{\zeta}\times\boldsymbol{\xi}]\cdot\boldsymbol{\eta} \tag{8.D.19a}$$

$$= [\boldsymbol{\zeta}\times\boldsymbol{\xi}]\cdot[\boldsymbol{\xi}\times\boldsymbol{\zeta}] = +1\,. \tag{8.D.19b}$$

This also proves that the three vectors form a basis whose scalar products are

$$\boldsymbol{\xi}\cdot\boldsymbol{\xi} = 0\,, \qquad \boldsymbol{\eta}\cdot\boldsymbol{\eta} = -1\,, \qquad \boldsymbol{\zeta}\cdot\boldsymbol{\zeta} = 0\,, \tag{8.D.20a}$$

$$\boldsymbol{\xi}\cdot\boldsymbol{\eta} = 0\,, \qquad \boldsymbol{\xi}\cdot\boldsymbol{\zeta} = +1\,, \qquad \boldsymbol{\eta}\cdot\boldsymbol{\zeta} = 0\,. \tag{8.D.20b}$$

E. The Null Rotations of SO(2,1)

We now want to find those proper, orthochronous Lorentz transformations that leave the direction of $\boldsymbol{\xi}$ unchanged. They are called the null rotations. We require that

$$\boldsymbol{\xi}' = a\,\boldsymbol{\xi} \tag{8.E.1}$$

with $a > 0$ to maintain orthochronicity. We put

$$\boldsymbol{\zeta}' = \frac{1}{a}\left[d\,\boldsymbol{\xi} + b\,\boldsymbol{\eta} + c\,\boldsymbol{\zeta}\right], \tag{8.E.2a}$$

$$\boldsymbol{\eta}' = \boldsymbol{\xi}' \times \boldsymbol{\zeta}' = \boldsymbol{\xi} \times \left[b\,\boldsymbol{\eta} + c\,\boldsymbol{\zeta}\right], \tag{8.E.2b}$$

where b, c, and d are constants to be determined. We have

$$\begin{aligned} \boldsymbol{\eta}' &= b\left[\boldsymbol{\xi} \times \boldsymbol{\eta}\right] + c\,\boldsymbol{\eta} & (8.E.3a)\\ &= b\,\boldsymbol{\xi} \times \left[\boldsymbol{\xi} \times \boldsymbol{\zeta}\right] + c\,\boldsymbol{\eta} & (8.E.3b)\\ &= b\,\boldsymbol{\xi} + c\,\boldsymbol{\eta}\,. & (8.E.3c) \end{aligned}$$

This last expression, together with our initial requirement (8.E.1), puts into evidence the mapping of the ($\boldsymbol{\xi}$-$\boldsymbol{\eta}$)-plane into itself. We have

$$\boldsymbol{\xi}' \cdot \boldsymbol{\zeta}' = a\,\boldsymbol{\xi} \cdot \frac{1}{a}\left[d\,\boldsymbol{\xi} + b\,\boldsymbol{\eta} + c\,\boldsymbol{\zeta}\right] \tag{8.E.4a}$$

$$= c = +1\,, \tag{8.E.4b}$$

and

$$\boldsymbol{\zeta}' \cdot \boldsymbol{\zeta}' = \frac{1}{a^2}\left[d\,\boldsymbol{\xi} + b\,\boldsymbol{\eta} + c\,\boldsymbol{\zeta}\right] \cdot \left[d\,\boldsymbol{\xi} + b\,\boldsymbol{\eta} + c\,\boldsymbol{\zeta}\right] \tag{8.E.5a}$$

$$= \frac{1}{a^2}\left[-b^2 + 2\,d\right]. \tag{8.E.5b}$$

Thus

$$d = \frac{1}{2}\,b^2\,. \tag{8.E.6}$$

Collecting our results, we obtain

$$\boldsymbol{\xi}' = a\,\boldsymbol{\xi}\,, \qquad \boldsymbol{\eta}' = b\,\boldsymbol{\xi} + \boldsymbol{\eta}\,, \tag{8.E.7a}$$

$$\boldsymbol{\zeta}' = \frac{1}{a}\left[\frac{1}{2}\,b^2\,\boldsymbol{\xi} + b\,\boldsymbol{\eta} + \boldsymbol{\zeta}\right]. \tag{8.E.7b}$$

This transformation has determinant $+1$ and leaves the scalar products of the basis vectors unchanged. A matrix representation is given by

$$\mathbf{\Lambda}^j{}_i = \begin{bmatrix} a & b & b/a \\ 0 & 1 & 0 \\ 0 & 0 & 1/a \end{bmatrix}, \qquad a > 0\,. \tag{8.E.8}$$

The group of null rotations, also known as the "little group" of a rest mass zero particle [Wigner, 1939], has two parameters a and b. [It has four parameters in the four-dimensional spacetime.] And as we intended to show, it transforms the ($\boldsymbol{\xi}$-$\boldsymbol{\eta}$)-plane into itself. If one refers the group to orthonormal frames defined by

$$\mathbf{e}_0 = \frac{1}{\sqrt{2}}[\,\boldsymbol{\xi} + \boldsymbol{\zeta}\,], \qquad \mathbf{e}_1 = \frac{1}{\sqrt{2}}[\,\boldsymbol{\xi} - \boldsymbol{\zeta}\,], \qquad \mathbf{e}_2 = \boldsymbol{\eta}, \tag{8.E.9a}$$

$$\mathbf{e}'_0 = \frac{1}{\sqrt{2}}[\,\boldsymbol{\xi} + \boldsymbol{\zeta}'\,], \qquad \mathbf{e}'_1 = \frac{1}{\sqrt{2}}[\,\boldsymbol{\xi}' - \boldsymbol{\zeta}'\,], \qquad \mathbf{e}'_2 = \boldsymbol{\eta}', \tag{8.E.9b}$$

it looks somewhat more complicated. One gets

$$\mathbf{e}'_0 = \left[\frac{b^2}{4a} + \frac{1}{a} + \frac{1}{2a}\right]\mathbf{e}_0 + \left[\frac{b^2}{4a} + \frac{1}{a} - \frac{1}{2a}\right]\mathbf{e}_1 + \frac{b}{a\sqrt{2}}\,\mathbf{e}_2\,, \tag{8.E.10a}$$

$$\mathbf{e}'_1 = \left[-\frac{b^2}{4a} + \frac{1}{a} - \frac{1}{2a}\right]\mathbf{e}_0 + \left[-\frac{b^2}{4a} + \frac{1}{a} + \frac{1}{2a}\right]\mathbf{e}_1 - \frac{b}{a\sqrt{2}}\,\mathbf{e}_2\,, \tag{8.E.10b}$$

$$\mathbf{e}'_2 = \frac{b}{a\sqrt{2}}\,\mathbf{e}_0 + \frac{b}{a\sqrt{2}}\,\mathbf{e}_1 + \mathbf{e}_2\,. \tag{8.E.10c}$$

One subgroup of the null rotations, given by $b = 0$ with $a > 0$, consists of the boosts in the 1-direction:

$$\mathbf{e}'_0 = \cosh(\chi)\,\mathbf{e}_0 + \sinh(\chi)\,\mathbf{e}_1\,, \tag{8.E.11a}$$

$$\mathbf{e}'_1 = \sinh(\chi)\,\mathbf{e}_0 + \cosh(\chi)\,\mathbf{e}_1\,, \tag{8.E.11b}$$

$$\mathbf{e}'_2 = \mathbf{e}_2\,, \tag{8.E.11c}$$

where $a = e^{\chi}$. The general null rotation can be obtained through an arbitrary boost followed by a rotation that compensates for the aberration caused by the boost. If the boost moves $\boldsymbol{\xi}$ into $\boldsymbol{\xi}''$, then a rotation will move

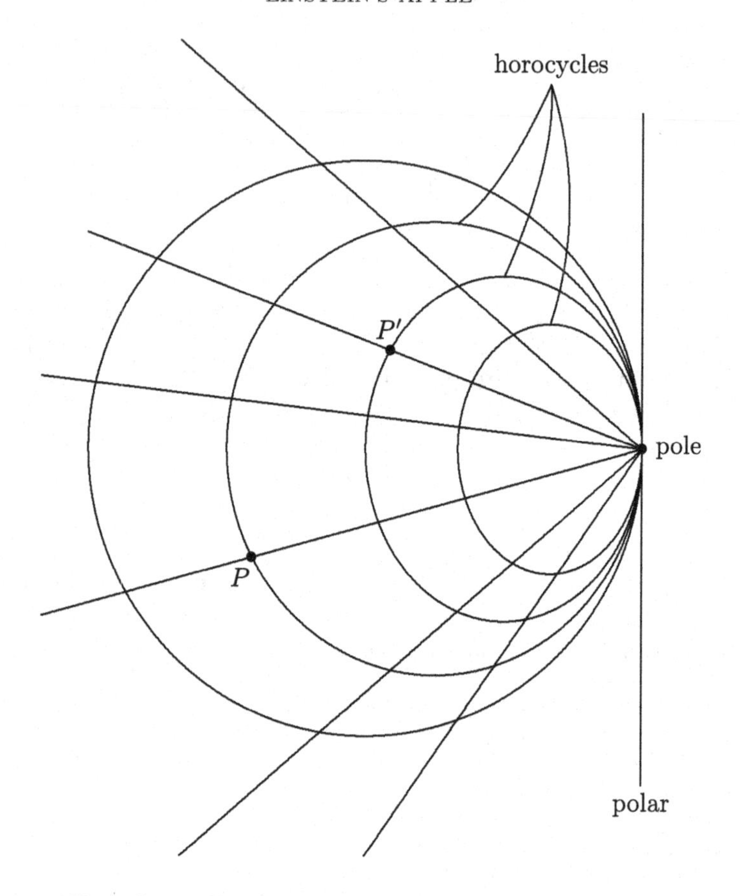

Figure 8.E.1 The null rotations are those Lorentz transformations that leave the polar fixed They transform all horocycles through the pole into each other and all geodesics into each other. For arbitrary points **P** and **P**′ there is exactly one null rotation that moves **P** into **P**′ (both are assumed in the interior of the circle) for a given pole of the null rotations.

$\boldsymbol{\xi}''$ into $\boldsymbol{\xi}' = a\,\boldsymbol{\xi}$ and thus keep the null direction fixed. Such a representation of the null rotations does not bring out their intrinsic simplicity.

All the planes $\boldsymbol{x}\cdot\boldsymbol{\xi} = \textit{constant}$ are null and are transformed into each other under null rotations. The plane $\boldsymbol{x}\cdot\boldsymbol{\xi} = 0$ is the only one that is transformed into itself. Only the subgroup given by $a = 1$ of the null rotations leaves the planes individually fixed. A plane $\boldsymbol{x}\cdot\boldsymbol{\xi} = c > 0$ intersects the pseudosphere in a horocycle. These horocycles are orbits of the subgroup given by $a = 1$.

The planes through the vector $\boldsymbol{\xi}$ intersect the pseudosphere in geodesics that are transformed into each other by the null rotations.

To represent null rotations one can choose convenient coordinates defined by

$$\lambda \equiv \boldsymbol{x} \cdot \boldsymbol{\xi} , \qquad \mu \equiv - \boldsymbol{x} \cdot \boldsymbol{\eta} , \tag{8.E.12a}$$

$$\lambda' \equiv \boldsymbol{x}' \cdot \boldsymbol{\xi}, \qquad \mu' \equiv - \boldsymbol{x}' \cdot \boldsymbol{\eta} . \tag{8.E.12b}$$

But we have

$$\lambda = \boldsymbol{x}' \cdot \boldsymbol{\xi}' = a\, \boldsymbol{x}' \cdot \boldsymbol{\xi} = a\, \lambda' \tag{8.E.13a}$$

and

$$\mu = - \boldsymbol{x}' \cdot \boldsymbol{\eta}' = - b\, \boldsymbol{x}' \cdot \boldsymbol{\xi} - \boldsymbol{x}' \cdot \boldsymbol{\eta} = - b\, \lambda' + \mu' . \tag{8.E.13b}$$

Thus

$$\lambda = a\, \lambda' , \qquad \mu = \mu' - b\, \lambda' , \tag{8.E.14a}$$

$$\lambda' = \frac{1}{a}\, \lambda , \qquad \mu' = \mu + \frac{b}{a}\, \lambda . \tag{8.E.14b}$$

The meaning of these coordinates will be discussed soon.

The null rotations are those Lorentz transformations that leave the pole fixed. They transform all horocycles through the pole into each other and all geodesics into each other. For arbitrary $\mathbf{P}$ and $\mathbf{P}'$ there is exactly one null rotation that moves $\mathbf{P}$ into $\mathbf{P}'$ (both assumed to be in the interior of the circle) for given pole of the null rotation.

F. The Redshift and Superluminal Velocities

Let $\mathbf{x}$ and $\mathbf{x}'$ be timelike unit vectors

$$\mathbf{x} = \frac{1}{\sqrt{1 - \beta^2}} \left(1 , \beta_x , \beta_y \right) , \tag{8.F.1a}$$

$$\mathbf{x}' = \frac{1}{\sqrt{1 - \beta'^2}} \left(1 , \beta'_x , \beta'_y \right) . \tag{8.F.1b}$$

We choose, for example,

$$\boldsymbol{\xi} = (1, 1, 0) , \qquad \boldsymbol{\eta} = (0, 0, 1) . \tag{8.F.2}$$

Then

$$\lambda = \frac{1}{\sqrt{1 - \beta^2}} \, (1 - \beta_x), \tag{8.F.3a}$$

$$\mu = \frac{1}{\sqrt{1 - \beta^2}} \, \beta_y, \tag{8.F.3b}$$

$$\frac{\mu}{\lambda} = \frac{1}{1 - \beta_x} \, \beta_y, \tag{8.F.3c}$$

where λ is the redshift factor $1/(1+z)$ while μ/λ is the "apparent transverse velocity",

$$\lambda' = \frac{1}{a} \lambda, \qquad \frac{\mu'}{\lambda'} = a \frac{\mu}{\lambda} + b. \tag{8.F.4}$$

Calling the redshift $\zeta \equiv (1+z)$ and also giving the apparent transverse velocity by $\sigma \equiv \mu/\lambda$, the null rotation then gives the "observable coordinates" ζ and σ as

$$\zeta' = a\zeta, \qquad \sigma' = a\sigma + b. \tag{8.F.5}$$

The notion of an "observable transverse velocity" [**See Figure 8.F.1.**] can easily be justified: Assume an observer at a large distance, L, from a radiating object on the x-axis and stationary at $x = L$. The object has velocity components in the x and y directions. At time $t = 0$ it emits a light ray from position $x = 0$, $y = 0$ which reaches the observer at time L (as $c = 1$). At $t = 1$ the object emits a second light ray which reaches the observer at time $L + 1 - \beta_x$, where β_x is the x-component of the velocity. The time difference between between the arrival of the first and second signals at the observer is $1 - \beta_x$ seconds. During this time the object moved transverse to the line of sight the distance β_y, or apparently, that is, per unit time of the observer

$$\frac{\beta_y}{1 - \beta_x}. \tag{8.F.6}$$

This makes superluminal velocities appear possible in observations.

A null vector $\mathbf{k}$ can be expressed as

$$\mathbf{k} = [\alpha, \, \alpha\beta_x, \, \alpha\beta_y] \tag{8.F.7}$$

where

$$\alpha > 0 \quad \text{and} \quad [\beta_x]^2 + [\beta_y]^2 = 1. \tag{8.F.8}$$

Then

$$\sigma = \frac{\mu}{\lambda} = \frac{1}{1 - \beta_x} \beta_y \tag{8.F.9}$$

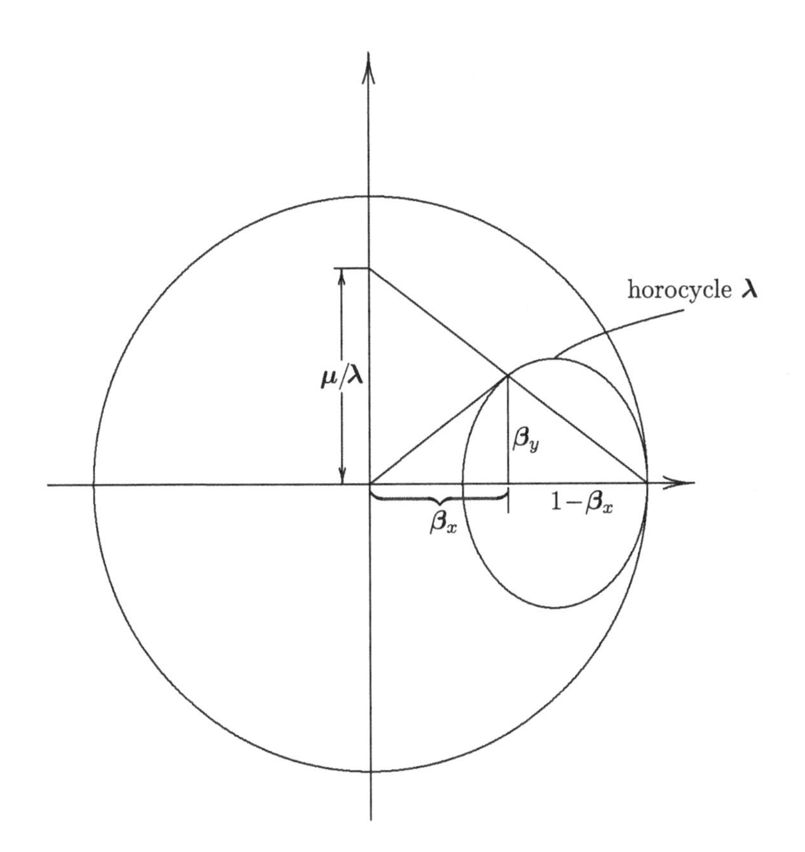

Figure 8.F.1 Observable Transverse Velocity

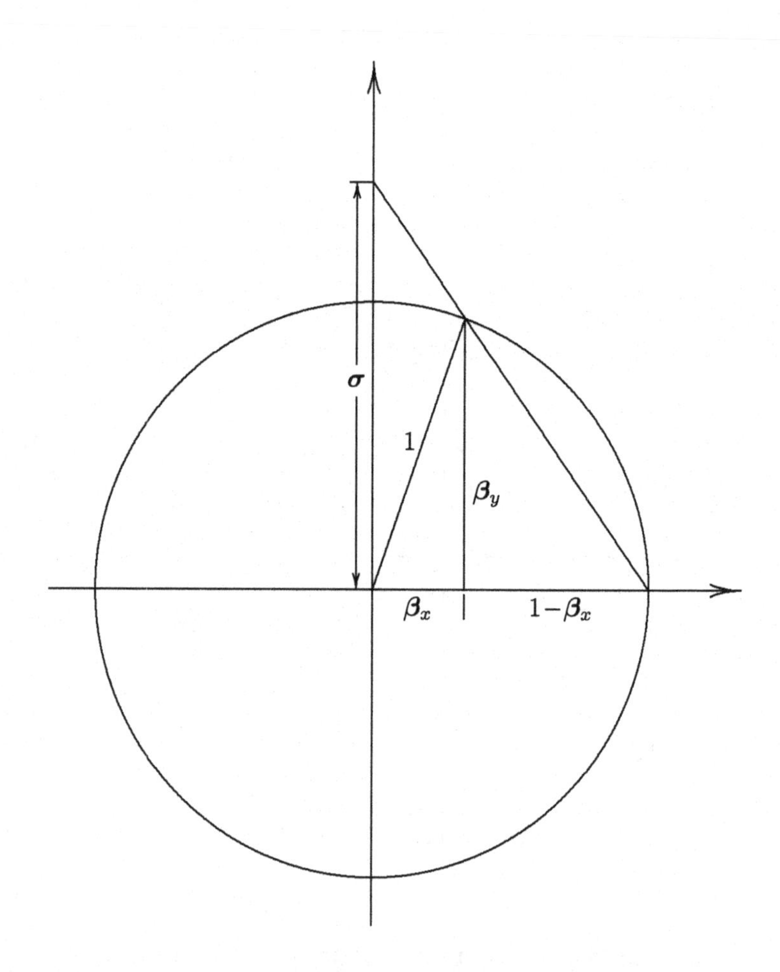

Figure 8.F.2 Apparent Transverse Velocity This makes the apparent superluminal velocities possible.

and, therefore, we have under null rotation

$$\sigma' = a\,\sigma + b\,. \tag{8.F.10}$$

The new components are obtained through stereographic projection as

$$\beta_x = \frac{\sigma^2 - 1}{\sigma^2 + 1}, \qquad \beta_y = \frac{2\,\sigma}{\sigma^2 + 1}, \tag{8.F.11a}$$

$$\beta_{x'} = \frac{\sigma'^2 - 1}{\sigma'^2 + 1}, \qquad \beta_{y'} = \frac{2\,\sigma'}{\sigma'^2 + 1}, \tag{8.F.11b}$$

which leads to a very simple transformation. [**See Figure 8.F.2.**]

CHAPTER 9

LIMITS OF SPACETIMES

A. Introduction

When we try to analyze the limit behavior of solutions of Einstein's gravitational field equations, we are dealing with a situation that has been described by this comment [Stephani, 1980.1st]:

> "There is no general recipe for this procedure. The examples given in the literature have been found intuitively."

The authors of that statement were referring to limits found that transform one solution into another whose surface appearance may not indicate any obvious relationship to the source solution. Finding the literature is also a act of intuition, in that, quite typically, results are buried in papers that do not mention that a limit will be shown, such as Taub does [Taub, 1980]. Abstracts do not provide the space to detail the content of papers and authors may easily omit any mention of an interesting limit derivation. In the time interval since our quotation appeared, the situation is somewhat better as noted in the second edition [Stephani, 2003.2nd]. However, there is no exhaustive method for finding all limits at this time.

The singular limit process of "contractions" was originally described by Wigner and Inonu [Wigner, 1953] in group theoretical terms and generalized by Saletan [Saletan, 1961]. An extremely general line element was obtained by Plebanski and Demianski [Plebanski, 1976] for Petrov type D solutions. As part of their paper, they obtain some additional solutions via "contractions". Plebanski and Demianski use it to refer to coordinate transformations that are taken to a singular limit. A more restricted version of that limit was given by Geroch [Geroch, 1969]. The coordinate transformations are not described in terms of their possible physical or geometrical significance, but instead are handled as if they are only the parameterized stretching of coordinates. The physical interpretation was left as an exercise to the reader.

A. H. Taub briefly mentions the intent to systematically study the possible limits of solutions in a paper [Taub, 1980] that also gives a suggestive example. The program that he sketches, in a single paragraph, includes the determination of the spacetime metrics that can be obtained from those known metrics that contain elements with the symmetry properties of spheres, cylinders, and lines. This plan has never been followed through by Taub [Taub, 1986]. To his list, we could also add the desirability of including the ring solutions that have a toroidal symmetry. The development of these areas of research has not been pursued here, but even the simple calculations of homogeneous fields in **Chapter 5** led directly to manifolds of these identical symmetries.

A primary characteristic of the type of solutions that Taub wanted to examine is their high degree of symmetry. MacCallum [MacCallum, 1985] reviews and expands on some of the work describing highly symmetric solutions of Einstein's field equations, but does not indicate connections between solutions via limitations. Many of these highly symmetric solutions are typified by the presence of 2-surfaces of symmetry, either spheres, planes or cylinders. We would also expect to find toroidal surfaces in a solution for the gravitational field of a ring with low mass and charge per unit length. It has been seen that the theory of homogeneous fields leads directly to the conclusion that their two-dimensional solutions include the topological torus among the vacua, that is, homogeneous fields of zero strength.

We will now apply the theory to highly specific solutions of Einstein's field equations to see how closely they approximate homogeneous fields.

B. About Ricci Coefficients

It is desirable to have an interpretation of the Ricci coefficients g_{jkl} that is especially useful in the four-dimensional case and can be applied in all dimensions.

We proceed as follows: Any continuous function can be decomposed into it symmetric and antisymmetric parts. In particular, the Ricci coefficients can be decomposed as

$$g_{ijk} = \frac{1}{2}\left(g_{ijk} + g_{ikj} \right) + \frac{1}{2}\left(g_{ijk} - g_{ikj} \right) . \tag{9.B.0}$$

We will deal with the first expression on the right momentarily. The second expression will be identified with the structure constants of a local Lie group. [**See Appendix A.**] Much of the rest of this book will be working out the consequences being able use the group structure to study homogeneous fields. Such a study was proposed by Cartan [Cartan, 1927].

We define

$$2\,f_{ijk} \equiv g_{ijk} + g_{ikj}\,. \tag{9.B.1}$$

The f_{ijk} represent that part of the Ricci coefficients that is symmetric in the last two indices. By writing down the equations for these components three times with cyclic permutation of the indices, we obtain

$$\begin{aligned} 2\,f_{ijk} &= g_{ijk} + g_{ikj}\,, \\ 2\,f_{jki} &= g_{jki} + g_{jik}\,, \\ 2\,f_{kij} &= g_{kij} + g_{kji}\,. \end{aligned} \tag{9.B.2}$$

Adding all three and dividing by two, we get

$$f_{ijk} + f_{jki} + f_{kij} = 0\,. \tag{9.B.3}$$

The components are subject to these side conditions.

We define further

$$h_{ijk} \equiv g_{ijk} + g_{jki} + g_{kij}\,. \tag{9.B.4}$$

This tensor is totally skew-symmetric. From f_{ijk} and h_{ijk} we can reconstitute g_{ijk}. We have

$$\begin{aligned} 2\,(f_{ikj} - f_{jki}) - h_{ikj} &= +\,g_{ikj} + g_{ijk} - g_{jki} - g_{jik} \\ &\quad -\,g_{ikj} - g_{kji} - g_{jik} \\ &= 3\,g_{ijk}\,. \end{aligned} \tag{9.B.5}$$

We call h_{ijk} the *spirality tensor* and f_{ijk} the *force components* of the field strengths. In two dimensions the spirality of a frame always vanishes. Generally, the field strengths are obtained as a combination of force components and spirality. The spirality can also be described by means of the three-form

$$\begin{aligned} \Theta &= \omega^i \wedge d\omega_i \\ &= \omega^i \wedge \omega_{ij} \wedge \omega^j \\ &= -\,g_{ijk}\,\omega^i \wedge \omega^j \wedge \omega^k \\ &= -\frac{1}{3}\,h_{ijk}\,\omega^i \wedge \omega^j \wedge \omega^k\,. \end{aligned} \tag{9.B.6}$$

Let u_a be the frame component of a geodesic. We have then from the equation of a geodesic

$$\dot{u}_a = g_{abc}\,u^b\,u^c = f_{abc}\,u^b\,u^c, \qquad |\,u_a\,u^a\,| = 1\,. \tag{9.B.7}$$

One is thus able to define the force components from the free-fall behavior of particles. The side conditions for the force components imply that

$$u^a \, \dot{u}_a = 0 \,. \tag{9.B.8}$$

However, the spirality does not affect the acceleration of particles as measured from a non-inertial frame.

The spirality shows up as rotation (that is, precession) of a vector which is parallelly transported along a line. If we chose a triplet of unit vectors that are linearly independent and tangent to geodesics, we have, from the equation for the parallel transport of a vector,

$$\bar{u}_a = g_{abc} \, u^b \, w^c \,, \qquad | \, u_a \, u^a \, | = 1 \,, \tag{9.B.9a}$$

$$\widetilde{w}_a = g_{abc} \, w^b \, v^c \,, \qquad | \, w_a \, w^a \, | = 1 \,, \tag{9.B.9b}$$

$$\breve{v}_a = g_{abc} \, v^b \, u^c \,, \qquad | \, v_a \, v^a \, | = 1 \,. \tag{9.B.9c}$$

These are the respective derivatives of the vector components u_a, w_a, and v_a in directions w^c, v^c, and u^c when undergoing parallel transport in these directions. We have for the sum of three terms

$$\begin{aligned} \bar{u}_a v^a + \breve{v}_a w^a + \widetilde{w}_a u^a &= g_{abc} \left[u^b \, w^c \, v^a + v^b \, u^c \, w^a + w^b \, v^c \, u^a \right] \\ &= g_{abc} \left[u^b \, v^a \, w^c + u^c \, v^b \, w^a + u^a \, v^c \, w^b \right] \\ &= h_{abc} \, u^a \, v^b \, w^c \,. \end{aligned} \tag{9.B.10}$$

By using linearly independent vector triplets $\mathbf{u}, \mathbf{v}, \mathbf{w}$, the spirality tensor can be determined. In a frame that is geodesic at a point, force components and spirality vanish at this point.

It might be worth while to find the conditions for the Riemann tensor that determine the local existence of frames with spirality zero.

In this chapter we have not assumed the existence of homogeneous fields. To demonstrate the formalism and to understand its physical meaning, we now discuss a few simple examples.

C. Levi-Civita's Constant Gravitational Field

This field is constant in time and constant in direction, but not constant in strength. We set $c = 1$ and $\omega_4 = i\,\omega_0$ and will write all indices downstairs with Greek lower case indices now running from one to three.

The frame is usually most easily obtained by normalization of the tangent vectors to an orthogonal coordinate system We take

$$ds^2 = dx_\alpha\, dx_\alpha - dt^2\,(1 + k_\beta\, x_\beta)^2 \tag{9.C.1}$$

where the k_β are three constants. We have then

$$\omega_\alpha = dx_\alpha\,, \qquad \omega_4 = i\,(1 + k_\beta\, x_\beta)\, dt\,. \tag{9.C.2}$$

This gives

$$d\omega_\alpha = \omega_{\alpha\beta}\wedge\omega_\beta + \omega_{\alpha 4}\wedge\omega_4 = 0\,, \tag{9.C.3a}$$

$$d\omega_4 = \omega_{4\beta}\wedge\omega_\beta = i\,k_\beta\wedge dt\,. \tag{9.C.3b}$$

We can write the last expression as

$$d\omega_4 = -\frac{k_\beta}{(1 + k_\beta\, x_\beta)}\,\omega_4\wedge\omega_\beta \tag{9.C.4}$$

and thus see that

$$\omega_{\alpha\beta} = 0, \qquad \omega_{\alpha 4} = -\,\omega_{4\alpha} = \frac{k_\alpha}{(1 + k_\beta\, x_\beta)}\,\omega_4\,, \tag{9.C.5}$$

solves (9.C.3). We have thus found the connection forms and now get for the field strengths

$$g_{\alpha\beta\kappa} = 0\,, \qquad g_{\alpha 4\kappa} = -\,g_{4\alpha\kappa} = \frac{k_\alpha}{(1 + k_\beta\, x_\beta)}\,\delta_{\kappa 4}\,. \tag{9.C.6}$$

The spirality is zero and the field strengths coincide with the force components. The force components are, clearly, not constant. The equations of motions for a free particle are

$$\begin{aligned}\dot{u}_\alpha &= g_{\alpha j\kappa}\, u_j\, u_\kappa = \frac{k_\alpha}{(1 + k_\beta\, x_\beta)}\,[\,u_4\,]^2 \\ &= -\,\frac{k_\alpha}{(1 + k_\beta\, x_\beta)}\,\frac{1}{1 - \beta^2}\end{aligned} \tag{9.C.7}$$

where $\beta = v/c$ is the velocity. In vector notation

$$\frac{d}{ds}\frac{\beta}{\sqrt{1-\beta^2}} = -\frac{\mathbf{k}}{(1+\mathbf{k}\cdot\mathbf{x})}\frac{1}{1-\beta^2}\,. \tag{9.C.8}$$

For a particle of mass m and energy $E = m\,(1-\beta^2)^{1/2}$, we have for the inertial (gravitational) force

$$\mathbf{F} = -\frac{\mathbf{k}}{(1+\mathbf{k}\cdot\mathbf{x})}\frac{E}{\sqrt{1-\beta^2}}\,. \tag{9.C.9}$$

If we write

$$\mathbf{f} \equiv \frac{ds}{dt}\mathbf{F} = \frac{d\mathbf{P}}{dt}\,, \tag{9.C.10}$$

we get

$$\mathbf{f} = -\frac{\mathbf{k}}{(1+\mathbf{k}\cdot\mathbf{x})}E = \frac{d\mathbf{P}}{dt} \tag{9.C.11}$$

and this should also apply to particles that move with the speed of light.

D. Minkowski Spacetime as an Infinite Mass Limit of Schwarzschild Spacetime

To invariantly define a homogeneous field limit of the Schwarzschild solution, let us start with the positive mass Schwarzschild line element and apply Israel's coordinate transformation. The line element is

$$ds^2 = \left[1-\frac{2M}{r}\right]dt^2 - \left[1-\frac{2M}{r}\right]^{-1}dr^2 - r^2\left[d\theta^2 + \sin^2\theta\, d\phi^2\right]\,. \tag{9.D.1}$$

To this we apply Israel's transformation [Israel, 1967] to the timelike and radial coordinates

$$r = 2M + \frac{u\,z}{4M}, \qquad t = r - 2M\ln\left[\frac{u}{2\,z}\right]\,, \tag{9.D.2}$$

and their differentials

$$dr = \frac{1}{4M}\left[z\,du + u\,dz\right]\,, \tag{9.D.3a}$$

$$dt = \left[\frac{z}{4M} - \frac{2M}{u}\right]du + \left[\frac{u}{4M} - \frac{2M}{z}\right]dz\,. \tag{9.D.3b}$$

We can now rewrite the line element as

$$ds^2 = -2\,du\,dz - \frac{z^2}{2Mr}\,du^2 - r^2\left[d\theta^2 + \sin^2\theta\,d\phi^2\right]. \tag{9.D.4}$$

The last expression for the line element does not conveniently express an orthonormal set of one-forms. However, it can be rewritten as

$$\begin{aligned} ds^2 = &+\left[\frac{1}{\sqrt{2}}\left\{\left[1 - \frac{z^2}{4Mr}\right]du + dz\right\}\right]^2 \\ &-\left[\frac{1}{\sqrt{2}}\left\{\left[1 + \frac{z^2}{4Mr}\right]du - dz\right\}\right]^2 \\ &- r^2\left[d\theta^2 + \sin^2\theta\,d\phi^2\right]. \end{aligned} \tag{9.D.5}$$

This displays the set of differential one-forms that we will use to obtain a limit:

$$\omega^0 = \frac{1}{\sqrt{2}}\left\{\left[1 - \frac{z^2}{4Mr}\right]du + dz\right\}, \tag{9.D.6a}$$

$$\omega^1 = \frac{1}{\sqrt{2}}\left\{\left[1 + \frac{z^2}{4Mr}\right]du - dz\right\}, \tag{9.D.6b}$$

$$\omega^2 = r\,d\theta, \tag{9.D.6c}$$

$$\omega^3 = r\sin\theta\;d\phi. \tag{9.D.6d}$$

The first two of these one-forms are obviously well behaved if we attempt to take the limit $M \to \infty$ for values of $r > 2M$. The other pair is troublesome without some further work. Our initial Israel transformation introduced the divergence by scaling the radial coordinate with the leading term that dominates in the infinite mass limit. To impose finite behavior on them we can utilize the freedom of transforming the angular coordinates. With that in mind and to simplify understanding of the limit, we stereographically project a point of the topological sphere onto an imaginary plane in contact with the "north pole" of the sphere. The line of projection originates at the "south pole", runs through the point being projected and terminates at the projection plane. The angle between the axis and the line of projection is related to the colatitude angle, θ, that appears in the line element by

$$x + i\,y = 2\,\Gamma\,\tan\left[\frac{\theta}{2}\right]e^{i\phi}, \tag{9.D.7}$$

where Γ is as the radius of the sphere. This gives the substitutions

$$\sin^2\theta = \left[\frac{x^2 + y^2}{\Gamma^2}\right]\left[1 + \frac{x^2 + y^2}{4\,\Gamma^2}\right]^{-2}, \tag{9.D.8a}$$

$$\cos^2\theta = \left[1 - \frac{x^2 + y^2}{4\,\Gamma^2}\right]^2\left[1 + \frac{x^2 + y^2}{4\,\Gamma^2}\right]^{-2}, \tag{9.D.8b}$$

$$d\theta^2 = \left[dx^2 + dy^2 - \left(x^2 + y^2\right)\left[\frac{x\,dy - y\,dx}{x^2 + y^2}\right]^2\right] \times \frac{1}{\Gamma^2}\left[1 + \frac{x^2 + y^2}{4\,\Gamma^2}\right]^{-2}, \tag{9.D.8c}$$

$$d\phi^2 = \left[\frac{x\,dy - y\,dx}{x^2 + y^2}\right]^2. \tag{9.D.8d}$$

This projection, centered on the "north pole" of the topological sphere that appears in the structure of the line element, permits us to see the effect of the limit on the angular coordinates and the "contraction" of the group. We can now substitute any radius larger than the Schwarzschild event horizon radius for Γ, that is,

$$r = \Gamma > 2M\,. \tag{9.D.9}$$

With the reparameterization and allowing M, and therefore also Γ, to go to infinite values, we will then have the approximate limit forms

$$\lim_{\Gamma\to\infty} \sin^2\theta \approx \left[\frac{x^2 + y^2}{\Gamma^2}\right], \tag{9.D.10a}$$

$$\lim_{\Gamma\to\infty} \cos^2\theta \approx 1\,, \tag{9.D.10b}$$

$$\lim_{\Gamma\to\infty} d\theta^2 \approx \frac{1}{\Gamma^2}\left[dx^2 + dy^2 - \left(x^2 + y^2\right)\left[\frac{x\,dy - y\,dx}{x^2 + y^2}\right]^2\right]. \tag{9.D.10c}$$

The expression for $d\phi$, (9.D.8d), remains unchanged.

From the expressions (9.D.10) we conclude that since both $\sin\theta$ and $d\theta$ have limit behavior on the order of $1/\Gamma$, the one-forms (9.D.6c) and (9.D.6d) that appeared to have singular behavior, actually remain finite with their behavior given by

$$\omega^2 = r\left[\frac{1}{r}\right]\left[dx^2 + dy^2 - \left(x^2 + y^2\right)\left[\frac{x\,dy - y\,dx}{x^2 + y^2}\right]^2\right]^{\frac{1}{2}}, \tag{9.D.11a}$$

$$\omega^3 = r\left[\frac{x^2 + y^2}{r^2}\right]^{\frac{1}{2}}\left[\frac{x\,dy - y\,dx}{x^2 + y^2}\right]. \tag{9.D.11b}$$

These expressions are independent of r (though it has been retained to show the relationship to previous expressions). In the infinite mass limit, they reduce to the flat Minkowski line element in the form

$$ds^2 = 2\,du\,dz - dx^2 - dy^2\,. \tag{9.D.12}$$

We still must show that the Ricci coefficients also have the appropriate limit. As the physical components of the gravitational field, their limit behavior is crucial. We proceed by calculating the differentials of the one-forms (9.D.6) and obtain

$$d\omega^0 = +\frac{1}{\sqrt{2}}\frac{z}{4Mr}\left[-2 + \frac{u\,z}{4Mr}\right]\ \omega^0\wedge\omega^1\,, \tag{9.D.13a}$$

$$d\omega^1 = -\frac{1}{\sqrt{2}}\frac{z}{4Mr}\left[-2 + \frac{u\,z}{4Mr}\right]\ \omega^0\wedge\omega^1\,, \tag{9.D.13b}$$

$$d\omega^2 = +\frac{1}{\sqrt{2}}\frac{z}{4Mr}\left\{\,\omega^0\wedge\omega^1 + \omega^1\wedge\omega^2\,\right\}, \tag{9.D.13c}$$

$$\begin{aligned} d\omega^3 = &+\frac{1}{\sqrt{2}}\left[\frac{z}{4Mr} + \frac{u}{4Mr}\left[1 + \frac{z^2}{4Mr}\right]\right]\omega^0\wedge\omega^3 \\ &+\left[\frac{z}{4Mr} - \frac{u}{4Mr}\left[1 - \frac{z^2}{4Mr}\right]\right]\omega^1\wedge\omega^3 \\ &+\frac{1}{r}\frac{\cos\theta}{\sin\theta}\ \omega^2\wedge\omega^3\,. \end{aligned} \tag{9.D.13d}$$

From these expressions and by applying the first Cartan structure equation and the Maurer-Cartan equations, we obtain the non-zero Ricci coefficients for the Schwarzschild metric expressed in Israel coordinates:

$$g^0{}_{10} = +\frac{1}{\sqrt{2}}\frac{z}{4Mr}\left[-2 + \frac{u\,z}{4Mr}\right], \tag{9.D.14a}$$

$$g^0{}_{11} = +\frac{1}{\sqrt{2}}\frac{z}{4Mr}\left[-2 + \frac{uz}{4Mr}\right], \tag{9.D.14b}$$

$$g^0{}_{22} = -\frac{1}{\sqrt{2}}\frac{z}{4Mr}, \tag{9.D.14c}$$

$$g^{0}{}_{33} = -\frac{1}{\sqrt{2}}\frac{z}{4Mr}\left[z + u\left[1 + \frac{z^2}{4Mr}\right]\right], \tag{9.D.14d}$$

$$g^{1}{}_{00} = +\frac{1}{\sqrt{2}}\frac{z}{4Mr}\left[-2 + \frac{uz}{4Mr}\right], \tag{9.D.14e}$$

$$g^{1}{}_{01} = +\frac{1}{\sqrt{2}}\frac{z}{4Mr}\left[-2 + \frac{u\,z}{4Mr}\right], \tag{9.D.14f}$$

$$g^{1}{}_{22} = +\frac{1}{\sqrt{2}}\frac{z}{4Mr}, \tag{9.D.14g}$$

$$g^{1}{}_{33} = +\frac{1}{\sqrt{2}}\frac{z}{4Mr}\left[z - u\left[1 - \frac{z^2}{4Mr}\right]\right], \tag{9.D.14h}$$

$$g^{2}{}_{02} = -\frac{1}{\sqrt{2}}\frac{z}{4Mr}, \tag{9.D.14i}$$

$$g^{2}{}_{12} = -\frac{1}{\sqrt{2}}\frac{z}{4Mr}, \tag{9.D.14j}$$

$$g^{2}{}_{33} = +\frac{1}{r}\frac{\cos\theta}{\sin\theta}, \tag{9.D.14k}$$

$$g^{3}{}_{03} = -\frac{1}{\sqrt{2}}\frac{z}{4Mr}\left[z + u\left[1 + \frac{z^2}{4Mr}\right]\right], \tag{9.D.14l}$$

$$g^{3}{}_{13} = -\frac{1}{\sqrt{2}}\frac{z}{4Mr}\left[z - u\left[1 - \frac{z^2}{4Mr}\right]\right], \tag{9.D.14m}$$

$$g^{3}{}_{23} = -\frac{1}{r}\frac{\cos\theta}{\sin\theta}. \tag{9.D.14n}$$

The expressions for the circular functions and their differentials have not been substituted. Our work earlier in this section has shown that they are well behaved. An examination of the Ricci coefficients shows that they all have the same limit as Minkowski spacetime when the equivalent calculation is made for it. The limit is taken by increasing the mass, M, in expressions (9.D.14), towards infinity while also requiring that $r > 2M$.

As a further expression of this conclusion, we also see that the connection coefficients have the similar limiting behavior. The connection coefficients are given by

$$\boldsymbol{\omega}^{0}{}_{1} = +\frac{z}{4M}\left[-2 + \frac{u\,z}{4Mr}\right]du, \tag{9.D.15a}$$

$$\boldsymbol{\omega}^{0}{}_{2} = -\frac{z}{4M\sqrt{2}}\,d\theta, \tag{9.D.15b}$$

$$\boldsymbol{\omega}^{0}{}_{3} = +\frac{1}{4M\sqrt{2}}\left[z + u\left[1 + \frac{z^2}{4Mr}\right]\right]\sin\theta\,d\phi, \tag{9.D.15c}$$

$$\omega^1{}_2 = + \frac{z}{4M\sqrt{2}}\, d\theta\,, \tag{9.D.15d}$$

$$\omega^1{}_3 = + \frac{1}{4M\sqrt{2}} \left[z - u \left[1 - \frac{z^2}{4Mr} \right] \right] \sin\theta\, d\phi\,, \tag{9.D.15e}$$

$$\omega^2{}_3 = + \cos\theta\, d\phi\,. \tag{9.D.15f}$$

With the exception of (9.D.15f), all of these go to zero. The lone non-zero limit (9.D.15f) is a consequence of the choice of coordinates; the equivalent calculation for Minkowski spacetime in spherical polar coordinates has the same non-zero connection coefficient. The weakness of this choice of coordinates can be ameliorated by using isotropic coordinates [Fock, 1964] which treat spatial coordinates symmetrically. This will be our choice in the sequel.

E. The Constant Electric or Magnetic Field as a Limit of the Spherically Symmetric Field of a Point Charge

The constant electric or magnetic field in Einstein's theory of gravitation was given by T. Levi-Civita [Levi-Civita, 1917] [Schücking, 1985B]. The metric can be written as

$$ds^2 = a^2 \left[\cosh^2 z\, d\tau^2 - dz^2 - \left(d\theta^2 + \sin^2\theta\, d\phi^2 \right) \right]\,. \tag{9.E.1}$$

In the next section, we will show this line element to be a limit of the Reissner-Nordstrom solution.

The field lies in the z-direction and its strength is given by

$$\mathbf{F} = \mathbf{E} + i\,\mathbf{B}\,, \quad F_x = F_y = 0\,, \quad F_z = F e^{i\,a}\,, \tag{9.E.2}$$

where a is real and with

$$F = \frac{1}{a} \frac{c^2}{\sqrt{G}}\,, \tag{9.E.3}$$

where G is Newton's gravitational constant. The total flux of the field is given by

$$4\,\pi\, a^2\, F = 4\,\pi\, a\, \frac{c^2}{\sqrt{G}} = 4\,\pi\, e\,, \tag{9.E.4}$$

and, therefore, the charge, e (electric or magnetic or combined), is related to the radius of curvature, a, of the metric (9.E.1) through

$$a = \frac{\sqrt{G}}{c^2}\, e\,. \tag{9.E.5}$$

F. The Reissner-Nordstrom Limit Procedure Near m = e

We try to obtain the Levi-Civita field as the limiting case of a point charge of strength e. The spherically symmetric field of such a charge was given by Reissner [Reissner, 1916] and Nordstrom. It can be written

$$ds^2 = a^2 \left\{ \left[1 - \frac{2(1-\varepsilon)}{r} + \frac{1}{r^2}\right] dt^2 - \left[1 - \frac{2(1-\varepsilon)}{r} + \frac{1}{r^2}\right]^{-1} dr^2 - r^2 \left(d\theta^2 + \sin^2\theta\, d\phi^2\right) \right\}, \tag{9.F.1}$$

where the mass, m, of the metric is given by

$$m = (1 - \varepsilon)\, a\,. \tag{9.F.2}$$

The field, F, lies in the r-direction and is given by

$$F = \frac{1}{a} \frac{c^2}{\sqrt{G}} \frac{1}{r^2}\,. \tag{9.F.3}$$

The field strength of the Levi-Civita field is obtained for $r = 1$.

We now study the Reissner-Nordstrom solution in the neighborhood of $r = 1$. The geometry is determined by the function

$$\begin{aligned} f(r) &\equiv 1 - \frac{2\,(1-\varepsilon)}{r} + \frac{1}{r^2} \\ &= \frac{1}{r^2}\Big[\left[r - (1-\varepsilon)\right]^2 + \varepsilon\,(2-\varepsilon)\Big]\,. \end{aligned} \tag{9.F.4}$$

The function f has a double root at $r = 1$ for $\varepsilon = 0$. The hypersurface $r = 1$ then becomes null. Quite generally, we have for $r = 1$

$$f(1) = 2\varepsilon\,. \tag{9.F.5}$$

If we want to identify hypersurfaces $r = constant$ near $r = 1$ in the Reissner-Nordstrom metric with the timelike surfaces $z = constant$ of the Levi-Civita metric, we must have $\varepsilon > 0$.

We now transform the Reissner-Nordstrom solution by introducing new coordinates t and z which are defined by

$$t \equiv \frac{\tau}{\sqrt{\varepsilon\,(2-\varepsilon)}}\,, \qquad z \equiv \int_{1-\varepsilon}^{r} \frac{d\lambda}{\sqrt{\lambda^2 - 2(1-\varepsilon)\lambda + 1}}\,. \tag{9.F.6}$$

The coordinate z measures the radial distance along the r-direction from the hypersurface $r = 1 - \varepsilon$. We have for $r > 1 - \varepsilon$

$$z = \left\{ [r - (1 - \varepsilon)]^2 + \varepsilon\,(2 - \varepsilon) \right\}^{1/2} - \{\varepsilon\,(2 - \varepsilon)\}^{1/2} + (1 - \varepsilon)\sinh^{-1}\left(\frac{r - (1 - \varepsilon)}{\sqrt{\varepsilon\,(2 - \varepsilon)}}\right) \tag{9.F.7}$$

and for $r < 1 - \varepsilon$

$$z = -\left\{ [r - (1 - \varepsilon)]^2 + \varepsilon\,(2 - \varepsilon) \right\}^{1/2} + \{\varepsilon\,(2 - \varepsilon)\}^{1/2} + (1 - \varepsilon)\sinh^{-1}\left(\frac{r - (1 - \varepsilon)}{\sqrt{\varepsilon\,(2 - \varepsilon)}}\right). \tag{9.F.8}$$

If we call R the radial distance to the singularity at $r = 0$, we have from (9.F.8)

$$R = 1 - \sqrt{\varepsilon\,(2 - \varepsilon)} + \frac{1 - \varepsilon}{2}\ln(2 - \varepsilon) - \frac{1 - \varepsilon}{2}\ln(\varepsilon) \tag{9.F.9}$$

which means that for $\varepsilon \to 0$

$$R = \frac{1}{2}\ln\left(\frac{1}{\varepsilon}\right). \tag{9.F.10}$$

This shows that for $\varepsilon \to 0$ the charge at $r = 0$ in the Reissner-Nordstrom solution becomes infinitely distant from the hypersurface $r = 1 - \varepsilon$ along the r-direction.

Introducing the abbreviation

$$\rho \equiv \frac{r - (1 - \varepsilon)}{\sqrt{\varepsilon\,(2 - \varepsilon)}}, \tag{9.F.11}$$

we can write (9.F.7) and (9.F.8) for $\rho \geq 0$ as:

$$z = \sqrt{\varepsilon\,(2 - \varepsilon)}\left\{\sqrt{1 + \rho^2} - 1\right\} + (1 - \varepsilon)\sinh^{-1}\rho, \tag{9.F.12}$$

and, for $\rho \leq 0$ as:

$$z = \sqrt{\varepsilon\,(2 - \varepsilon)}\left\{1 - \sqrt{1 + \rho^2}\right\} + (1 - \varepsilon)\sinh^{-1}\rho. \tag{9.F.13}$$

If we take a slice of the Reissner-Nordstrom metric of thickness 2δ in the r-coordinate centered on $r = 1 - \varepsilon$, with δ fixed but arbitrarily small and greater than zero, we see the following: for $\varepsilon \to 0$ one learns from (9.F.11), (9.F.12), and (9.F.13) that this slice acquires a thickness L given approximately by

$$L \approx 2\,a \sinh^{-1}\left(\frac{\delta}{\sqrt{2\varepsilon}}\right) \tag{9.F.14}$$

which becomes infinite as ε tends to zero. The t = *constant* sections of this slice form a three-dimensional cylinder given by the direct product of a 2-sphere of radius $\approx A$ with a segment of length L.

The Reissner-Nordstrom solution with charge a and mass m, slightly smaller than a, develops near $r = a$ an infinitely long "throat" where the radii of spheres r = *constant* no longer diminish in the radial direction.

G. The Levi-Civita Limit

We shall show now that this infinitely long throat is the Levi-Civita metric (9.E.1). It follows from (9.F.12) and (9.F.13) that

$$\frac{z}{d\rho} = \frac{\sqrt{\varepsilon\,(2-\varepsilon)}\,|\,\rho\,| + (1-\varepsilon)}{\sqrt{1+\rho^2}} \geq 0\,. \tag{9.G.1}$$

Therefore, the inverse function $\rho = \rho(z, \varepsilon)$ exists. We have

$$\lim_{\varepsilon \to 0} \rho(z, \varepsilon) = \sinh(z)\,. \tag{9.G.2}$$

Introducing τ and z into the Reissner-Nordstrom solution, we get

$$\begin{aligned} ds^2 = a^2 \Bigg\{ &- \frac{1+\rho^2}{\left[(1-\varepsilon) + \sqrt{\varepsilon\,(2-\varepsilon)}\,\rho\right]^2}\, d\tau^2 + dz^2 \\ &+ \left[(1-\varepsilon) + \sqrt{\varepsilon\,(2-\varepsilon)}\,\rho\right]^2 \left(d\theta^2 + \sin^2\theta\, d\phi^2\right) \Bigg\}. \end{aligned} \tag{9.G.3}$$

In the limit $\varepsilon \to 0$, from the Reissner-Nordstrom solution together with (9.G.2), we get

$$ds^2 = a^2 \Big\{ -\cosh^2 z\, d\tau^2 + dz^2 + \left(d\theta^2 + \sin^2\theta\, d\phi^2\right) \Big\}. \tag{9.G.4}$$

We have, thus, shown that the constant field can be obtained as a limit of the spherically symmetric field of a point charge.

We have, also, the result that the Levi-Civita metric can be assigned a mass

$$M = \frac{a\,c^2}{G}\,, \tag{9.G.5}$$

although the integral over the field energy clearly diverges.

One should also be able to recover the Levi-Civita metric directly from the $m = e$ case of the Reissner-Nordstrom metric by a coordinate transformation of a slice in r near $r = m$. The geometry of the Reissner-Nordstrom solution has been discussed by Carter for the $m = e$ case; see especially Carter [Carter, 1972], where he calls the Levi-Civita solution the "Robinson-Bertotti solution".

H. Isotropic Reissner-Nordstrom Solution

It is a simple exercise to derive the isotropic form of the Reissner-Nordstrom line element, using the same method that is available in an introductory textbook [Adler, 1975]. That calculation is provided in **Appendix F**, since it is not readily found in the literature.

In the derived isotropic coordinates, the Reissner-Nordstrom line element becomes

$$\begin{aligned} ds^2 &= \left[\left(1+\frac{m}{2r}\right)^2 - \frac{e^2}{4r^2}\right]^2 \left[\,dx_1{}^2 + dx_2{}^2 + dx_3{}^2\right] \\ &\quad + \left[\frac{(\,2r\,+\,m\,)\,(\,2r\,-\,m\,)\,+\,e^2}{(\,2r\,+\,m\,+\,e\,)\,(\,2r\,+\,m\,-\,e\,)}\right]^2 dx_4{}^2\,. \end{aligned} \tag{9.H.1}$$

The differential forms adapted to the geometry of the solution are just read off from the line element as

$$\omega_A = \left[\left(1+\frac{m}{2r}\right)^2 - \frac{e^2}{4r^2}\right] dx_A\,, \tag{9.H.2a}$$

$$\omega_4 = \left[\frac{(\,2r\,+\,m\,)\,(\,2r\,-\,m\,)\,+\,e^2}{(\,2r\,+\,m\,+\,e\,)\,(\,2r\,+\,m\,-\,e\,)}\right] dx_4\,. \tag{9.H.2b}$$

The analysis now proceeds identically as if it had been restricted to the simpler Schwarzschild case whose isotropic form is more available in the literature. We obtain for their differentials

$$d\boldsymbol{\omega}_A = \left[-m\left(1+\frac{m}{2r}\right)+\frac{e^2}{2r}\right]\frac{x_B}{r^3}\,dx_B\wedge dx_A\,, \tag{9.H.3a}$$

$$d\boldsymbol{\omega}_4 = 4\left[\frac{m\,(2r+m)^2-(2r+m)\,e^2-2r\,e^2}{(2r+m+e)^2\,(2r+m-e)^2}\right] \times \frac{x_B}{r}\,dx_B\wedge dx_4\,, \tag{9.H.3b}$$

which give us, by means of the same method of assuming a form for the connection forms, the factors

$$\lambda(r) = 4\frac{1}{r}\left[\left(1+\frac{m}{2r}\right)^2-\frac{e^2}{4r^2}\right]^{-1} \times\left[\frac{m\,(2r+m)^2-(2r+m)\,e^2-2r\,e^2}{(2r+m+e)^2\,(2r+m-e)^2}\right], \tag{9.H.4a}$$

$$\Lambda(r) = \frac{1}{r^3}\left[\left(1+\frac{m}{2r}\right)^2-\frac{e^2}{4r^2}\right]^{-1}\left[-m\left(1+\frac{m}{2r}\right)+\frac{e^2}{2r}\right], \tag{9.H.4b}$$

as well as the explicit expressions

$$\begin{aligned}\boldsymbol{\omega}_{4B} &= \lambda(r)\,x_B\,dx_4\\ &= 4\,\frac{x_B}{r}\,(2r)^2\left[\frac{1}{(2r+m)^2-e^2}\right]^2\\ &\quad\times\left[\frac{m\,(2r+m)^2-2\,(2r+m)\,e^2+m\,e^2}{(2r+m)(2r-m)+e^2}\right]\boldsymbol{\omega}_4\,,\end{aligned} \tag{9.H.5a}$$

$$\begin{aligned}\boldsymbol{\omega}_{AB} &= \Lambda(r)\,[x_B\,dx_A-x_A\,dx_B]\\ &= \frac{1}{r^3}\,(2r)^3\left[\frac{1}{(2r+m)^2-e^2}\right]^2\\ &\quad\times\left[-m\,(2r+m)+e^2\right]\left[x_B\,\boldsymbol{\omega}_A-x_A\,\boldsymbol{\omega}_B\right].\end{aligned} \tag{9.H.5b}$$

I. Reissner-Nordstrom Force Field - Rest Case

A test particle's motion must satisfy the geodesic equation

$$\dot{u}_\mu + \frac{\omega_{\mu\nu}}{ds} u_\nu = 0 . \tag{9.I.1}$$

This equation describes the change in the four-velocity as a consequence of motion relative to a frame pointwise adapted to the local geometry. The consequent geodesic motion reflects the symmetries of the solution of Einstein's gravitational field equations. We will select a frame adapted to motion in the field and solve the geodesic equation which supplies the four-acceleration. We first consider the rest case, where $u_A = 0$ and $u_4 = i$.

By inserting the various quantities that we have obtained, we can decompose the changes in the four-velocity into various terms that are identifiable as the classical Newtonian gravitational force as well others that originate in the pseudo-Riemannian geometry of the Schwarzschild solution. The geodesic equation for the spatial components of the four-acceleration

$$\dot{u}_A + \frac{\omega_{AB}}{ds} u_B + \frac{\omega_{A4}}{ds} u_4 = 0 , \tag{9.I.2}$$

reduces to

$$\dot{u}_A + \frac{\omega_{A4}}{ds} u_4 = 0 \tag{9.I.3}$$

since $u_A = 0$. The geodesic equation for the time component of the four-acceleration is

$$\dot{u}_4 + \frac{\omega_{4A}}{ds} u_A = 0 , \tag{9.I.4}$$

with the time component making no contribution to the force law for this static case since we have $\dot{u}_4 = 0$. We do not have gravitational or electromagnetic radiation. On the other hand, we have for the spatial components of the acceleration

$$\begin{aligned} \dot{u}_A &= + \frac{\omega_{4A}}{ds} u_4 \\ &= -4 \frac{x_A}{r} (2r)^2 \left[\frac{1}{(2r+m)^2 - e^2} \right]^2 \\ &\quad \times \left[\frac{m(2r+m)^2 - 2(2r+m)e^2 + me^2}{(2r+m)(2r-m) + e^2} \right] . \end{aligned} \tag{9.I.5}$$

From expression (9.I.5), we can immediately observe that for $r >> m$ and $r >> e$, we have the limit form

$$\dot{u}_A \to -m \frac{x_A}{r^3} \left[1 - \frac{e}{r} \left(\frac{e}{m} \right) \right] \tag{9.I.6}$$

which gives us the first order correction to the Newtonian force of gravity for a body possessing both mass and charge. The physically significant charge to mass ratio shows itself as the natural expansion factor to express the corrections. Since this limit was taken at zero spatial velocity, it can also be considered as the appropriate limit for very low velocities.

J. Reissner-Nordstrom Force Field - General Case

For "relativistic" velocities we have consider the full geodesic equation at non-zero spatial components of the four-velocity:

$$\begin{aligned}
\dot{u}_A &= -\frac{\boldsymbol{\omega}_{AB}}{ds} u_B - \frac{\boldsymbol{\omega}_{A4}}{ds} u_4 \\
&= -\frac{1}{r^3} (2r)^3 \left[\frac{1}{(2r+m)^2 - e^2} \right]^2 \left[-m(2r+m) + e^2 \right] \\
&\quad \times \left[x_B \frac{\boldsymbol{\omega}_A}{ds} - x_A \frac{\boldsymbol{\omega}_B}{ds} \right] u_B \\
&\quad + 4 \frac{x_A}{r} (2r)^2 \left[\frac{1}{(2r+m)^2 - e^2} \right]^2 \\
&\quad \times \left[\frac{m(2r+m)^2 - 2(2r+m)e^2 + me^2}{(2r+m)(2r-m) + e^2} \right] \frac{\boldsymbol{\omega}_4}{ds} u_4 \, .
\end{aligned} \tag{9.J.1}$$

To display the last result in a compact form, we define

$$A = A(r; m, e) = \frac{2r}{(2r+m)^2 - e^2}, \tag{9.J.2a}$$

$$B = B(r; m, e) = \frac{m(2r+m) - e^2}{2r}, \tag{9.J.2b}$$

$$C = C(r; m, e) = \frac{m(2r+m)^2 - 2(2r+m)e^2 + me^2}{(2r+m)(2r-m) + e^2}. \tag{9.J.2c}$$

With these definitions and using $u_4 \equiv \omega_4/ds$, we have for the spatial components of the four-velocity,

$$\begin{aligned}
\dot{u}_A &= +\,4\,[A]^2\,[B]\left[\frac{x_B}{r}\,\frac{\omega_A}{ds} - \frac{x_A}{r}\,\frac{\omega_B}{ds}\right] u_B \\
&\quad +\,4\,[A]^2\,[C]\,\frac{x_A}{r}\,u_4\,u_4 \qquad (9.J.3a) \\
&= +\,4\,[A]^2\,[B]\left[\frac{x_B}{r}\,u_A - \frac{x_A}{r}\,u_B\right] u_B \\
&\quad +\,4\,[A]^2\,[C]\,\frac{x_A}{r}\,u_4\,u_4 \qquad (9.J.3b) \\
&= 4\,\frac{1}{r}\left[\frac{[A]^2}{1-v^2}\right]\Big\{[B]\left[\mathbf{v}\,(\mathbf{x}\cdot\mathbf{v}) - v^2\,\mathbf{x}\right] - [C]\,\mathbf{x}\Big\}. \qquad (9.J.3c)
\end{aligned}$$

For the energy loss, we obtain from equation (9.I.4)

$$\begin{aligned}
\frac{dE}{ds} &= -\,4\,\frac{M}{r}\,[A]^2\,[C]\left[\frac{1}{\sqrt{1-v^2}}\right]\frac{(\mathbf{x}\cdot\mathbf{v})}{\sqrt{1-v^2}} \\
&= -\,4\,\frac{1}{r}\,E\,[A]^2\,[C]\,\frac{(\mathbf{x}\cdot\mathbf{v})}{\sqrt{1-v^2}}\,. \qquad (9.J.4)
\end{aligned}$$

K. Schwarzschild's Field in Isotropic Coordinates

Now that we have the acceleration field for the Reissner-Nordstrom solution, we are able to restrict the equations to the case $e = 0$, which gives us the acceleration field for the Schwarzschild spacetime metric. The resulting acceleration field will allow us to gain insight into some of the published "force fields" for the Schwarzschild solution.

We obtain, from (9.J.2a,b,c) respectively,

$$A = A\,(r; m, e = 0) = \frac{2r}{(\,2r + m\,)^2}\,, \qquad (9.K.1a)$$

$$B = B\,(r; m, e = 0) = \frac{m\,(\,2r + m\,)}{2r}\,, \qquad (9.K.1b)$$

$$C = C\,(r; m, e = 0) = \frac{m\,(\,2r + m\,)^2}{(\,2r + m\,)\,(\,2r - m\,)}\,. \qquad (9.K.1c)$$

Inserting these values for A, B, and C into (9.J.3c) provides the acceleration field for the spatial components of the four-velocity:

$$\dot{u}_A = 4\,\frac{1}{r}\left[\frac{1}{1-v^2}\right]\left[\frac{2r}{(2r+m)^2}\right]^2 \times\left\{\left[\frac{m\,(2r+m)}{2r}\right]\left[\mathbf{v}\,(\mathbf{x}\cdot\mathbf{v}) - v^2\,\mathbf{x}\right] - \left[\frac{m\,(2r+m)^2}{(2r+m)\,(2r-m)}\right]\mathbf{x}\right\}. \qquad (9.K.2)$$

To obtain this last expression in a somewhat more familiar form we only need to rewrite it in terms of $m/2$, which gives us:

$$\frac{d\mathbf{u}}{ds} = -\frac{m}{\left(r+\frac{m}{2}\right)^3}\left[\frac{1}{1-v^2}\right] \times\left\{\mathbf{x}\left[\left(1+v^2\right)+\left(\frac{m/2}{r-\frac{m}{2}}\right)\right] - \mathbf{v}\,(\mathbf{x}\cdot\mathbf{v})\right\}. \qquad (9.K.3)$$

L. Schwarzschild Spatial Acceleration

The last expression reduces, for the case where $r >> m$, to one similar to the one given by Okun [Okun, 1991; equation 13]:

$$\mathbf{a} = \frac{\mathbf{F} - (\mathbf{F}\cdot\boldsymbol{\beta})\,\boldsymbol{\beta}}{m\,\beta}. \qquad (9.L.1)$$

This equation omits the term of (9.K.3) depends on v^2 and also the one that is of fourth order in $1/r$. Okun attributes his expression to Bowler [Bowler, 1976].

Bowler obtains the following formula by way of a semi-empirical construction:

$$\frac{d\mathbf{p}}{dt} = -m\left(1+v^2\right)\boldsymbol{\nabla}\phi - m\,\mathbf{v}\,(\mathbf{v}\cdot\boldsymbol{\nabla}\phi)\,, \qquad (9.L.2)$$

which is his equation [Bowler, 1976: 5.4.6a]. For a Newtonian potential $\phi = 1/r$ this becomes

$$\frac{d\mathbf{p}}{dt} = -\frac{m}{r^3}\left[\mathbf{x}\left(1+v^2\right) - \mathbf{v}\,(\mathbf{x}\cdot\mathbf{v})\right]. \qquad (9.L.3)$$

M. Schwarzschild Energy Loss

Inserting the $e = 0$ values for A and C into (9.J.4) gives the rate of energy loss for Schwarzschild's metric:

$$\frac{dE}{ds} = -4\frac{1}{r}E\left[\frac{2r}{(2r+m)^2}\right]^2 \times\left[\frac{m\,(2r+m)^2}{(2r+m)\,(2r-m)}\right]\frac{(\mathbf{x}\cdot\mathbf{v})}{\sqrt{1-v^2}}. \tag{9.M.1}$$

With a bit of elementary manipulation of the various elements of the last equation, we obtain

$$\frac{dE}{ds} = -E\,\frac{m}{r^3}\left(\frac{1}{1+\frac{m}{2r}}\right)^3\left(\frac{1}{1-\frac{m}{2r}}\right)\frac{(\mathbf{x}\cdot\mathbf{v})}{\sqrt{1-v^2}}. \tag{9.M.2}$$

N. The de Sitter Isotropic Form

The de Sitter solution to Einstein's gravitational field equations is

$$ds^2 = \left[1-\frac{\Lambda}{3}r^2\right]dt^2 - \left[1-\frac{\Lambda}{3}r^2\right]^{-1}dr^2 - r^2\left(d\theta^2 + \sin^2\theta\,d\phi^2\right). \tag{9.N.1}$$

When we transform to isotropic coordinates, we are finding coordinates that treat all the spatial coordinates on an equal basis. The final result will give the line element in coordinates that resemble rectangular coordinates with an overall factor scaling the axes. We will start, however, by taking the spatial coordinates in de Sitter's solution and converting them so that they all share the same factor. This brings the line element into the form

$$ds^2 = \left[1-\frac{\Lambda}{3}r^2\right]dt^2 - \lambda^2(\rho)\left[d\rho^2 + \rho^2\left(d\theta^2+\sin^2\theta\,d\phi^2\right)\right] \tag{9.N.2}$$

where we have defined

$$r^2 \equiv \lambda^2\rho^2 \tag{9.N.3}$$

and assumed that λ will need to depend only on a radial coordinate. This assumption is based on the recognizing that the original line element (9.N.1) is independent of the angular coordinates. We need to determine λ as it is incompletely specified by (9.N.3) which gives it in terms of both sets of coordinates. To eliminate r from the expression for λ, we must obtain an expression for the radial coordinate r as a function of ρ. Since (9.N.3) contains three mutually dependent variables, we need another relationship to eliminate one of them, in this instance, λ. The invariant ds^2 provides

$$\lambda^2\, d\rho^2 \;=\; \left(\frac{r}{\rho}\right)^2 d\rho^2 \;=\; \frac{dr^2}{1-\frac{\Lambda}{3}r^2} \tag{9.N.4}$$

by equating $ds^2(t,r,\theta,\phi) = ds^2(t,\rho,\theta,\phi)$ at fixed time and angular position. Integration of the square root of (9.N.4) provides the required function of $r\,(\rho)$. We have

$$\int\frac{d\rho}{\rho} \;=\; \pm\int\frac{1}{r}\,\frac{dr}{\sqrt{1-\frac{\Lambda}{3}r^2}}\,, \tag{9.N.5}$$

which integrates readily after we make the trigonometric substitution

$$\sqrt{\frac{\Lambda}{3}}\,r = \sin\,\theta\,. \tag{9.N.6}$$

Our integral becomes

$$\ln\rho + \ln C \;=\; \pm\int \csc\,\theta\, d\theta \;=\; \pm\ln\left[\csc\,\theta - \cot\,\theta\right] \tag{9.N.7}$$

and we can revert to our original coordinates

$$\ln\rho + \ln C \;=\; \pm\ln\left[1-\sqrt{1-\frac{\Lambda}{3}\,r^2}\right] \mp \ln\left[\sqrt{\frac{\Lambda}{3}}\,r\right]. \tag{9.N.8}$$

Dropping out of logarithmic form, we have

$$C\,\rho \;=\; \pm\,\frac{1-\sqrt{1-\frac{\Lambda}{3}\,r^2}}{\sqrt{\frac{\Lambda}{3}}\,r}\,, \tag{9.N.9}$$

where we must determine the integration constant C from boundary conditions presumed appropriate to our problem. We obtain the value of C by observing that for $r \to \sqrt{3/\Lambda}$ we will have

$$\rho\, C \;\approx\; \pm 1\,. \tag{9.N.10}$$

This immediately yields

$$\rho = \frac{1 - \sqrt{1 - \frac{\Lambda}{3}\, r^2}}{\sqrt{\frac{\Lambda}{3}}\, r}\,, \tag{9.N.11}$$

which leads us to observe that ρ is a dimensionless quantity and that the actual coordinate in the isotropic form is $\lambda\rho$. We can now solve for $r(\rho)$. Since $\lambda = r/\rho$, we have

$$\lambda = \frac{3}{\Lambda}\left(\frac{2}{\rho^2+1}\right)^2 . \tag{9.N.12}$$

With these forms for $r(\rho)$ and $\lambda(\rho)$, we can rewrite (9.N.2) in the required isotropic form. The coefficient of the squared "time" differential dt^2 is given, in terms of the isotropic coordinate, as

$$1 - \left(\frac{2\,\rho}{\rho^2+1}\right)^2 . \tag{9.N.13}$$

In the derived isotropic coordinates, the line element becomes:

$$\begin{aligned} ds^2 &= \left[1 - \left(\frac{2\,\rho}{\rho^2+1}\right)^2\right] dt^2 \\ &\quad - \left[\frac{3}{\Lambda}\left(\frac{2}{\rho^2+1}\right)^2\right] \left[d\rho^2 + \rho^2\left(d\theta^2 + \sin^2\theta\, d\phi^2\right)\right] . \end{aligned} \tag{9.N.14}$$

We now convert to rectilinear coordinates by rewriting

$$d\rho^2 + \rho^2\left(d\theta^2 + \sin^2\theta\, d\phi^2\right) = \left[dx^1\right]^2 + \left[dx^2\right]^2 + \left[dx^3\right]^2 , \tag{9.N.15}$$

where we use rectilinear coordinates to express the complete equivalence of all spatial directions, and by scaling the "time" coordinate

$$(\Lambda/3)\, t \to 2\,\tau\, . \tag{9.N.16}$$

The de Sitter metric's squared line element now takes on its complete isotropic form

$$ds^2 = 4\,\frac{3}{\Lambda}\left\{\left[\frac{1-\rho^2}{1+\rho^2}\right]^2 d\tau^2 - \left[\frac{1}{1+\rho^2}\right]^2 \left[dx^k\, dx^k\right]\right\} . \tag{9.N.17}$$

O. De Sitter Analysis

Calculations on a solution to Einstein's field equations are carried out in a system of frames that are concretely specified in terms of a coordinate system most suited to the symmetries of the problem. For the de Sitter solution it would be tempting to use spherical polar coordinates. These however do not treat all points on spherical surfaces equally since they have singular behavior at the poles. Instead we will use a system that most resembles rectangular coordinates; this allows for a maximal exhibition of the de Sitter solution's symmetry. In this system the metric components are only dependent on the radius as measured from the origin.

We use the line element formed by the potentials $g_{\mu\nu}$ in isotropic coordinates:

$$ds^2 = A^2 \left\{ \left[\frac{1 - \rho^2}{1 + \rho^2}\right]^2 [dx_0]^2 - \left[\frac{1}{1 + \rho^2}\right]^2 [dx_k \, dx_k] \right\}, \qquad (9.\text{O}.1)$$

where $\rho^2 \equiv x_k x_k$ $(k, l, m = 1, 2, 3)$ and $dx_0 = d\tau$ and where we have used $A \equiv 12/\Lambda$. The isotropic form for the de Sitter line element puts all the spatial coordinates on an equal footing; by implication, the tangent and dual spaces will also reflect the isotropy. We now restate the line element in terms of differential forms adapted to the geometry of the solution

$$ds^2 = \omega_\mu \, \omega_\mu = \omega_0 \, \omega_0 - \omega_k \, \omega_k \,, \qquad (9.\text{O}.2)$$

where we define the covariant moving frame by

$$\omega_0 = A \left[\frac{1 - \rho^2}{1 + \rho^2}\right] dx_0 \,, \qquad (9.\text{O}.3\text{a})$$

$$\omega_k = A \left[\frac{1}{1 + \rho^2}\right] dx_k \,, \qquad (9.\text{O}.3\text{b})$$

for the differential forms. We can obtain the differentials of ω_μ directly:

$$d\omega_0 = -A \left[\frac{2}{1 + \rho^2}\right]^2 x_k \; dx_k \wedge dx_0 \,, \qquad (9.\text{O}.4\text{a})$$

$$d\omega_l = -\frac{1}{2} A \left[\frac{2}{1 + \rho^2}\right]^2 x_k \; dx_k \wedge dx_l \,. \qquad (9.\text{O}.4\text{b})$$

The differential forms ω_{0k} can be obtained, indirectly, by finding the unique solution of the torsion-free Cartan structure equations

$$d\omega_0 + \omega_{0k} \wedge \omega_k = 0 \,, \qquad (9.\text{O}.5\text{a})$$

$$d\omega_j + \omega_{jk} \wedge \omega_k + \omega_{j0} \wedge \omega_0 = 0 \,. \qquad (9.\text{O}.5\text{b})$$

By assuming that the solution for ω_{0k} will have the form

$$\omega_{0k} = \delta(\rho)\, x_k\, dx_0\,, \tag{9.O.6}$$

where $\delta(\rho)$ is an as yet unknown function, we first attempt to go directly to the unique solution of the simpler structure equation (9.O.5a). We have

$$- A\left[\frac{2}{1+\rho^2}\right]^2 x_k\; dx_k \wedge dx_0$$

$$+\ \delta(\rho)\, x_k\, dx_0 \wedge A\left[\frac{1}{1+\rho^2}\right] dx_k = 0\,, \tag{9.O.7}$$

which immediately provides

$$\delta(\rho) = -\frac{4}{1+\rho^2} \tag{9.O.8}$$

as our needed explicit expression for $\delta(\rho)$ and, directly from the postulated form for ω_{0k}, we obtain:

$$\omega_{0k} = -\frac{4}{1+\rho^2}\, x_k\, dx_0 \tag{9.O.9a}$$

$$= -\frac{4}{A}\,\frac{x_k}{1-\rho^2}\,\omega_0\,. \tag{9.O.9b}$$

Using results (9.O.9a,b), we can now produce an expression involving ω_{jk} from the second structure equation (9.O.5b) all the while taking into account that $\omega_{k0} \wedge \omega_0 = 0$. We have:

$$-\frac{1}{2}A\left[\frac{2}{1+\rho^2}\right]^2 x_k\; dx_k \wedge dx_j + \omega_{jk} \wedge A\left[\frac{1}{1+\rho^2}\right] dx_k = 0\,. \tag{9.O.10}$$

We can now solve this expression by making another guess at the form of ω_{jk}, again involving a function which must be determined by satisfying (9.O.10):

$$\omega_{jk} = \Delta(\rho)\,\{\, x_k\, dx_j - x_j\, dx_k \,\}\,. \tag{9.O.11}$$

Inserting (9.O.11) into (9.O.10), provides

$$0 = -\frac{1}{2}\left[\frac{2}{1+\rho^2}\right]^2 x_k\; dx_k \wedge dx_j$$

$$+\ \Delta(\rho)\,\{\, x_k\, dx_j - x_j\, dx_k \,\} \wedge \left[\frac{1}{1+\rho^2}\right] dx_k\,. \tag{9.O.12}$$

It immediately follows that

$$\Delta(\rho) = -\frac{2}{1 + \rho^2} \tag{9.O.13}$$

and we complete our determination of the connection forms by obtaining

$$\omega_{jk} = -\frac{2}{1 + \rho^2}\{x_k\, dx_j - x_j\, dx_k\}. \tag{9.O.14}$$

P. Geodesic Equation

As indicated earlier, the four-velocity in our chosen frame must also satisfy the geodesic equation. By inserting the various quantities that we have obtained, we can decompose the changes in the four-velocity into various terms that are identifiable as the classical Newtonian gravitational force as well as others that originate in the pseudo-Riemannian geometry of the de Sitter solution.

For the spatial components of the four-acceleration, the geodesic equation expands into

$$\dot{u}_k + \frac{\omega_{kj}}{ds} u_j + \frac{\omega_{k0}}{ds} u_0 = 0, \tag{9.P.1}$$

which, upon inserting the appropriate values for the connection forms, gives

$$\dot{u}_k - \frac{2}{1 + \rho^2}\{x_j\, u_k - x_k\, u_j\} u_j + \frac{4}{1 + \rho^2} x_k\, [u_0]^2 = 0. \tag{9.P.2}$$

The geodesic equation for the time component of the four-acceleration is given by

$$\dot{u}_0 + \frac{\omega_{0k}}{ds} u_k = 0 \tag{9.P.3}$$

and becomes

$$\dot{u}_0 - 4\frac{1}{1 + \rho^2} x_k\, u_0\, u_k = 0. \tag{9.P.4}$$

CHAPTER 10

HOMOGENEOUS FIELDS IN MINKOWSKI SPACETIMES

A. Subalgebras for the Lie Algebra of the Poincaré Group

We now produce a set of subalgebras for the Lie Algebra of the Poincaré Group. It will turn out to have the same number of parameters as the general homogeneous Lorentz transformation.

We take the metric of M^4 as

$$ds^2 = dt^2 - dx^2 - dy^2 - dz^2 \tag{10.A.1}$$

and rewrite it in the proper time form as

$$d\tau^2 = -ds^2 = 2\,du\,dv + dw\,d\bar{w}\,, \tag{10.A.2}$$

where u and v are real coordinates and w is a complex coordinate; they are given by

$$u \equiv z + t\,, \qquad w \equiv x + i\,y\,, \tag{10.A.3a}$$

$$v \equiv z - t\,, \qquad \bar{w} \equiv x - i\,y\,. \tag{10.A.3b}$$

A bar denotes the complex conjugate. We combine u, v, and w into a 2×2 matrix with the aid of the Pauli spin matrices

$$C = \begin{pmatrix} \sqrt{2}\,u & w \\ \bar{w} & -\sqrt{2}\,v \end{pmatrix}. \tag{10.A.4}$$

Clearly, matrices of this form can represent any point of the manifold.

To express the subgroup of transformations of coordinates that is the focus of our attention, we provide matrices that will, momentarily, help give us the equivalent point under the transformation of the subgroup in the form of a matrix such as (10.A.4). We introduce the matrix A by putting

$$A = \begin{pmatrix} e^{+\sigma/2} & -\gamma e^{-\sigma/2}/\sqrt{2} \\ 0 & e^{-\sigma/2} \end{pmatrix} \tag{10.A.5}$$

with σ and γ the complex parameters given by

$$\sigma \equiv p + i q, \qquad \gamma \equiv r + i s. \tag{10.A.6}$$

We further introduce the matrix B by

$$B = \begin{pmatrix} \sqrt{2}\, t_u & t_x + i\, t_y \\ t_x - i\, t_y & -\sqrt{2}\, t_v \end{pmatrix} \tag{10.A.7}$$

with four real parameters t_x, t_y, t_u, t_v. Upon writing another point of the manifold as the matrix

$$C' = \begin{pmatrix} \sqrt{2}\, u' & w' \\ \bar{w}' & -\sqrt{2}\, v' \end{pmatrix}, \tag{10.A.8}$$

we obtain an eight-dimensional subgroup of the Poincaré group by putting

$$C' = A\, C\, A^{\dagger} + B, \tag{10.A.9}$$

where the cross, †, denotes the adjoint.

We define a basis for the infinitesimal transformations of this subgroup by the operators (where we use "∂" to indicate the partial derivative)

$$X \equiv \partial_p, \qquad F \equiv \partial_q, \qquad D_u \equiv e^p\, \partial_{t_u}, \tag{10.A.10a}$$

$$T_1 \equiv e^p\, (+\cos q\, \partial_r + \sin q\, \partial_s) \tag{10.A.10b}$$

$$T_2 \equiv e^p\, (-\sin q\, \partial_r + \cos q\, \partial_s) \tag{10.A.10c}$$

$$D_x \equiv \left(+\cos q\, \partial_{t_x} + \sin q\, \partial_{t_y}\right) - (r \cos q + s \sin q)\, \partial_{t_u}, \tag{10.A.10d}$$

$$D_y \equiv \left(-\sin q\, \partial_{t_x} + \cos q\, \partial_{t_y}\right) + (r \sin q - s \cos q)\, \partial_{t_u}, \tag{10.A.10e}$$

$$D_v \equiv e^{-p} \left[\partial_{t_v} - \frac{1}{2}\, (r^2 + s^2)\, \partial_{t_u} + r\, \partial_{t_x} + s\, \partial_{t_y} \right]. \tag{10.A.10f}$$

We find that X commutes with F and the displacements D_x, D_y, D_u, and D_v commute among each other. The only non-vanishing commutators are

$$[X, T_1] = +T_1, \qquad [X, T_2] = +T_2, \tag{10.A.11a}$$

$$[X, D_u] = +D_u, \qquad [X, D_v] = -D_v, \tag{10.A.11b}$$

$$[F, T_1] = +T_2, \qquad [F, T_2] = -T_1, \tag{10.A.11c}$$

$$[F, D_x] = +D_y, \qquad [F, D_y] = -D_x, \tag{10.A.11d}$$

$$[T_1, D_x] = -D_u, \qquad [T_1, D_v] = +D_x, \tag{10.A.11e}$$

$$[T_2, D_y] = -D_u, \qquad [T_2, D_v] = -D_y. \tag{10.A.11f}$$

Defining a new operator

$$Y \equiv D_v + a\,X + b\,F + c\,T_1 + d\,T_2 \tag{10.A.12}$$

with arbitrary coefficients a, b, c, and d, we find that D_x, D_y, D_u, and Y form the basis of a Lie algebra with the following non-vanishing commutators

$$[Y, D_u] = a\,D_u, \tag{10.A.13a}$$

$$[Y, D_x] = b\,D_y - c\,D_u, \tag{10.A.13b}$$

$$[Y, D_y] = -b\,D_x - d\,D_u. \tag{10.A.13c}$$

The four-dimensional Lie groups will act transitively since they contain all translations. If a, b, c, d vanish, we have the subgroup of translations.

The eight-dimensional subgroup of the Poincaré group (10.A.9) leaves the set of null hyperplanes $v = constant$ invariant. Any other set of parallel null hyperplanes can be obtained by a rotation depending on two parameters (θ, ϕ).

We have thus obtained a set of four-dimensional transitive Lie subgroups of the Poincaré group which depends on six parameters, namely, a, b, c, d, θ, ϕ. This is same the number of parameters as in the general homogeneous Lorentz transformation.

One may expect, therefore, that all homogeneous fields for four-dimensional Minkowski space have been found.

B. Null Rotations and Translations in the (2+N)–Dimensional Minkowski Space M^{2+N}

We take the metric in M^{2+N} as

$$ds^2 = 2\,du\,dv + dx_\alpha\,dx_\alpha\,, \tag{10.B.1}$$

where Greek indices run from 1 to N. We study the transformations of the Poincaré group acting on M^{2+N} which leave the N–direction fixed. We represent these transformations by means of $(N+3)\times(N+3)$ matrices of the form

$$\begin{pmatrix} u' \\ x'_\alpha \\ v' \\ 1 \end{pmatrix} = \begin{pmatrix} e^{\psi} & -\gamma_\mu A_{\mu\beta} & -\frac{1}{2}e^{-\psi}\gamma_\mu\gamma_\mu & t_u \\ 0 & A_{\alpha\beta} & e^{-\psi}\gamma_\alpha & t_\alpha \\ 0 & 0 & e^{-\psi} & t_v \\ 0 & 0 & 0 & 1 \end{pmatrix} \begin{pmatrix} u \\ x_\beta \\ v \\ 1 \end{pmatrix}, \tag{10.B.2}$$

where γ_α and t_α are N–vectors, t_u and t_v real numbers representing null translations in the u– and v–directions, respectively, and $A_{\alpha\beta}$ is an orthogonal $N\times N$ matrix with

$$A_{\alpha\beta}\,A_{\alpha\rho} = \delta_{\beta\rho}\,. \tag{10.B.3}$$

We write equation (10.B.2) as

$$\boldsymbol{\xi}' = \mathbf{M}\,\boldsymbol{\xi} \tag{10.B.4}$$

in terms of column vectors $\boldsymbol{\xi}$ and $\boldsymbol{\xi}'$, denoting the matrix in (10.B.2) by $\mathbf{M}$. The matrix $\mathbf{M}$ depends on $N+2$ translations, $N(N-1)/2$ Euclidean rotation parameters, and the $N+1$ parameters ψ and γ_α. Altogether, it depends on

$$(N+2) + \frac{N(N-1)}{2} + (N+1) = 2 + \frac{(N+1)(N+2)}{2} \tag{10.B.5}$$

parameters. The full Poincaré group for M^{2+N} has N additional parameters.

Inverse Transformation

We compute the inverse to $\mathbf{M}$, denoted by $\mathbf{M}^{-1}$, as equal to

$$\begin{pmatrix} e^{-\psi} & e^{-\psi}\gamma_\alpha & -\frac{1}{2}e^{-\psi}\gamma_\mu\gamma_\mu & -e^{-\psi}[t_u+\gamma_\mu t_\mu - \frac{1}{2}\gamma_\mu\gamma_\mu t_v] \\ 0 & (A^{-1})_{\rho\alpha} & -(A^{-1})_{\rho\mu}\gamma_\mu & (A^{-1})_{\rho\mu}(\gamma_\mu t_v - t_\mu) \\ 0 & 0 & e^{\psi} & -e^{\psi}t_v \\ 0 & 0 & 0 & 1 \end{pmatrix} \tag{10.B.6}$$

This matrix can be written as

$$\mathbf{M}^{-1} = \begin{pmatrix} e^{\overline{\psi}} & -\bar{\gamma}_\mu A_{\mu\alpha} & -\frac{1}{2}e^{-\overline{\psi}}\bar{\gamma}_\mu\bar{\gamma}_\mu & \bar{t}_u \\ 0 & \overline{A}_{\rho\alpha} & e^{-\overline{\psi}}\bar{\gamma}_\rho & \bar{t}_\rho \\ 0 & 0 & e^{-\overline{\psi}} & \bar{t}_v \\ 0 & 0 & 0 & 1 \end{pmatrix} \tag{10.B.7}$$

if one makes the following identifications

$$\overline{\psi} = -\psi\,, \tag{10.B.8a}$$

$$\overline{A}_{\rho\alpha} = (A^{-1})_{\rho\alpha}\,, \tag{10.B.8b}$$

$$\bar{\gamma}_\rho = -e^{-\psi}(A^{-1})_{\rho\mu}\gamma_\mu\,, \tag{10.B.8c}$$

$$\bar{t}_\rho = (A^{-1})_{\rho\mu}(\gamma_\mu t_v - t_\mu)\,, \tag{10.B.8d}$$

$$\bar{t}_v = -e^{\psi}t_v\,, \tag{10.B.8e}$$

$$\bar{t}_u = -e^{-\psi}\left[t_u + \gamma_\mu t_\mu - \frac{1}{2}\gamma_\mu\gamma_\mu t_v\right]. \tag{10.B.8f}$$

Then one has

$$-\bar{\gamma}_\mu\overline{A}_{\mu\alpha} = e^{-\psi}(A^{-1})_{\mu\nu}\gamma_\nu(A^{-1})_{\mu\alpha} = e^{-\psi}\gamma_\alpha \tag{10.B.9}$$

and

$$\begin{aligned} -\frac{1}{2}e^{-\overline{\psi}}\bar{\gamma}_\mu\bar{\gamma}_\mu &= -\frac{1}{2}e^{-\psi}(A^{-1})_{\mu\nu}\gamma_\nu(A^{-1})_{\mu\sigma}\gamma_\sigma \\ &= -\frac{1}{2}e^{-\psi}\gamma_\sigma\gamma_\sigma\,. \end{aligned} \tag{10.B.10}$$

All transformations have an inverse belonging to the set of matrices $\mathbf{M}$ and the product of two matrices of type $\mathbf{M}$ again belongs to the set. We shall not work this out explicitly now. We will see closure in the Lie algebra which we proceed to compute.

Left Invariant Differential Forms

To this end, we first compute the left invariant differential forms by computing the matrix $\mathbf{M}^{-1}d\mathbf{M}$. We have for $d\mathbf{M}$

$$\begin{pmatrix} e^{\psi}\,d\psi & -d\gamma_\mu A_{\mu\beta} - \gamma_\mu dA_{\mu\beta} & -\frac{1}{2}e^{-\psi}(2\gamma_\mu d\gamma_\mu - \gamma_\mu d\gamma_\mu d\psi) & dt_u \\ 0 & dA_{\alpha\beta} & e^{-\psi}(d\gamma_\alpha - \gamma_\alpha\, d\psi) & dt_\alpha \\ 0 & 0 & -\,e^{-\psi}\,d\psi & dt_v \\ 0 & 0 & 0 & 1 \end{pmatrix} \tag{10.B.11}$$

and, therefore, from (10.B.6), $\mathbf{M}^{-1}d\mathbf{M}$ becomes

$$\begin{pmatrix} d\psi & e^{-\psi}\,d\gamma_\mu A_{\mu\beta} & 0 & e^{-\psi}\left[dt_u + \gamma_\mu dt_\mu - \frac{1}{2}\gamma_\mu\gamma_\mu dt_v\right] \\ 0 & (A^{-1})_{\rho\alpha}\,dA_{\alpha\beta} & e^{-\psi}\,(A^{-1})_{\rho\alpha}\,d\gamma_\alpha & (A^{-1})_{\rho\alpha}\left[dt_\alpha - \gamma_\alpha dt_v\right] \\ 0 & 0 & -\,d\psi & e^{\psi}\,dt_v \\ 0 & 0 & 0 & 1 \end{pmatrix}. \tag{10.B.12}$$

We define

$$\boldsymbol{\omega}_{\rho\beta} \equiv (A^{-1})_{\rho\alpha}\,dA_{\alpha\beta} \tag{10.B.13}$$

and have, because of the orthonormality condition (10.B.3),

$$\boldsymbol{\omega}_{\rho\beta} = -\,\boldsymbol{\omega}_{\beta\rho}\,. \tag{10.B.14}$$

The other invariant forms are

$$d\psi\,, \tag{10.B.15a}$$

$$\boldsymbol{\theta}_\beta \equiv e^{-\psi}\,(A^{-1})_{\beta\alpha}\,d\gamma_\alpha\,, \tag{10.B.15b}$$

$$\boldsymbol{\omega}_u \equiv e^{-\psi}\left[dt_u + \gamma_\mu dt_\mu - \frac{1}{2}\gamma_\mu\gamma_\mu dt_v\right], \tag{10.B.15c}$$

$$\boldsymbol{\omega}_\beta \equiv (A^{-1})_{\beta\alpha}\left[dt_\alpha - \gamma_\alpha dt_v\right], \tag{10.B.15d}$$

$$\boldsymbol{\omega}_v \equiv e^{\psi}\,dt_v\,. \tag{10.B.15e}$$

We call the elements of the dual basis

$$X\,,\quad T_\beta\,,\quad D_u\,,\quad D_\beta\,,\quad D_v\,,\quad \text{and}\quad R_{\rho\beta}\ \ (\rho<\beta\,) \tag{10.B.16}$$

and such that

$$1 \;=\; d\psi(X) \;=\; \boldsymbol{\omega}_u(D_u) \;=\; \boldsymbol{\omega}_v(D_v)\,, \tag{10.B.17a}$$

$$\boldsymbol{\theta}_\beta(T_\alpha) \;=\; \boldsymbol{\omega}_\beta\,(D_\alpha) \;=\; \delta_{\alpha\beta}\,, \tag{10.B.17b}$$

$$\boldsymbol{\omega}_{\rho\beta}(R_{\mu\nu}) \;=\; \delta_{\rho\mu}\delta_{\beta\nu}\,, \tag{10.B.17c}$$

where $\rho < \beta$ and $\mu < \nu$ and with all other values vanishing. We take the last equation as the definition of $R_{\mu\nu}$. We then have

$$X \equiv \frac{\partial}{\partial\psi}\,, \tag{10.B.18a}$$

$$T_\alpha \equiv e^{\psi}\,A_{\mu\alpha}\,\frac{\partial}{\partial\gamma_\mu}\,, \tag{10.B.18b}$$

$$D_u \equiv e^{\psi}\,\frac{\partial}{\partial t_u}\,, \tag{10.B.18c}$$

$$D_\lambda \equiv A_{\rho\lambda}\,\frac{\partial}{\partial t_\rho} \;-\; \gamma_\rho\,A_{\rho\lambda}\,\frac{\partial}{\partial t_u}\,, \tag{10.B.18d}$$

$$D_v \equiv e^{-\psi}\left(\frac{\partial}{\partial t_v} + \gamma_\beta\,\frac{\partial}{\partial t_\beta} - \frac{1}{2}\,\gamma_\mu\,\gamma_\mu\,\frac{\partial}{\partial t_u}\right). \tag{10.B.18e}$$

The displacements D_u, D_λ, D_v commute among themselves, and so do the null translations T_α. We have

$$[\,T_\alpha,\,T_\beta\,] \;=\; 0\,. \tag{10.B.19}$$

We further have

$$[\,X,\,T_\alpha\,] \;=\; T_\alpha\,, \tag{10.B.20a}$$

$$[\,X,\,D_u\,] \;=\; D_u\,,\qquad [\,X,\,D_v\,] \;=\; -\,D_v\,,\qquad [\,X,\,D_\lambda\,] = 0\,. \tag{10.B.20b}$$

and

$$[\,T_\alpha,\,D_u\,] \;=\; 0\,,\qquad [\,T_\alpha,\,D_\lambda\,] \;=\; -\,\delta_{\alpha\lambda}\,D_u\,,\qquad [\,T_\alpha,\,D_v\,] \;=\; D_\alpha\,. \tag{10.B.21}$$

X, D_u, and D_v commute with $R_{\mu\nu}$, but T_α and D_λ do not.

Subgroup

We now restrict ourselves to the subgroup given by

$$A_{\alpha\beta} = \delta_{\alpha\beta} \,. \tag{10.B.22}$$

We define an $N{+}2$–dimensional subalgebra depending on the $N+1$ parameters a and b_λ with the basis

$$D_u \,, \quad D_\lambda \,, \quad Y \equiv D_v + a\,X + b_\alpha\,T_\alpha \tag{10.B.23}$$

and with the commutation relations

$$[\,D_u, D_\lambda\,] = 0 \,, \quad [\,Y, D_u\,] = a\,D_u \,, \quad [\,Y, D_\lambda\,] = -\,b_\lambda\,D_u \,. \tag{10.B.24}$$

For vanishing a and b_λ this coincides with the algebra of the displacements. We find, from (10.B.18) and (10.B.22),

$$D_u = e^{\psi}\frac{\partial}{\partial t_u} \,, \tag{10.B.25a}$$

$$D_\lambda = \frac{\partial}{\partial t_\lambda} - \gamma_\lambda\,\frac{\partial}{\partial t_u} \,, \tag{10.B.25b}$$

$$Y = e^{-\psi}\left(\frac{\partial}{\partial t_v} + \gamma_\beta\,\frac{\partial}{\partial t_\beta} - \frac{1}{2}\,\gamma_\mu\,\gamma_\mu\,\frac{\partial}{\partial t_u}\right) + a\,\frac{\partial}{\partial\psi} + e^{\psi}\,b_\alpha\,\frac{\partial}{\partial\gamma_\alpha} \,. \tag{10.B.25c}$$

The dual basis to D_u, D_λ, Y, X, T_α is given by

$$\boldsymbol{\omega}_u \,, \quad \boldsymbol{\omega}_\lambda \,, \quad \boldsymbol{\omega}_v \,, \quad d\psi - a\,\boldsymbol{\omega}_v \,, \quad \boldsymbol{\vartheta}_\beta - b_\beta\,\boldsymbol{\omega}_v \,. \tag{10.B.26}$$

The subalgebra in question can be found from the equations

$$d\psi - a\,\boldsymbol{\omega}_v = 0 \,, \quad \boldsymbol{\vartheta}_\beta - b_\beta\,\boldsymbol{\omega}_v = 0 \,, \tag{10.B.27}$$

or according to (10.B.15)

$$d\psi - a\,e^{\psi}\,dt_v = 0 \,, \quad e^{-\psi}\,d\gamma_\beta - b_\beta\,e^{\psi}\,dt_v = 0 \,. \tag{10.B.28}$$

In the group space with coordinates ψ, γ_β, t_u, t_λ, t_v the equations (10.B.28) define $N+2$–dimensional subspaces which depend on the $N+1$ parameters a and b_β. In order to find the space of the subgroup (for given parameters

a and b_β), we have the condition that the manifold must go through the unit element given by

$$\psi = \gamma_\beta = t_u = t_v = t_\beta = 0\,. \tag{10.B.29}$$

The first equation (10.B.28) gives

$$e^{-\psi} + a\,t_v = \textit{constant}\,. \tag{10.B.30}$$

The condition that the surface goes through the unit element then gives

$$e^{-\psi} = 1 - a\,t_v \tag{10.B.31}$$

or

$$e^{\psi} = \frac{1}{1 - a\,t_v} \quad \Longrightarrow \quad \psi = -\ln(1 - a\,t_v)\,. \tag{10.B.32}$$

Inserting that into the second set of N equations in (10.B.28), we find

$$\gamma_\beta = b_\beta \int \frac{dt_v}{(1 - a\,t_v)^2} = \frac{b_\beta}{a}\,\frac{1}{(1 - a\,t_v)} + \textit{constant} \tag{10.B.33}$$

if $a \neq 0$. For vanishing a, we have

$$\gamma_\beta = b_\beta\,t_v\,. \tag{10.B.34}$$

By determining the constant in (10.B.33), we have

$$\gamma_\beta = \frac{b_\beta}{a}\left[\frac{1}{1 - a\,t_v} - 1\right] = \frac{b_\beta}{a}\,\frac{a\,t_v}{(1 - a\,t_v)} = \frac{b_\beta\,t_v}{1 - a\,t_v}\,. \tag{10.B.35}$$

This includes the case of vanishing a also.

The subgroups are, therefore, given by the equations

$$\psi = -\ln(1 - a\,t_v)\,, \qquad \gamma_\beta = \frac{b_\beta\,t_v}{1 - a\,t_v}\,. \tag{10.B.36}$$

From (10.B.15), we then have, for the solder forms,

$$\begin{aligned} \omega_u = {} & (1 - a\,t_v)\,\times \\ & \left[dt_u + \left(\frac{b_\mu\,t_v}{1 - a\,t_v}\right) dt_\mu - \frac{1}{2}\,\frac{b_\beta\,b_\beta\,(t_v)^2}{(1 - a\,t_v)^2}\,dt_v\right] \end{aligned} \tag{10.B.37}$$

and

$$\boldsymbol{\omega}_\beta = dt_\beta - \left(\frac{b_\beta\, t_v}{1 - a\, t_v}\right) dt_v\,, \qquad \boldsymbol{\omega}_v = \left(\frac{1}{1 - a\, t_v}\right) dt_v\,. \tag{10.B.38}$$

By calling now t_u, t_v, t_α : u, v, x_α we get for the line element

$$\begin{aligned} ds^2 &= 2\, du\, dv + dx_\alpha\, dx_\alpha \\ &= 2\, dv \left[du + \frac{b_\mu\, v}{1 - a\, v}\, dx_\mu - \frac{1}{2}\, \frac{b_\beta\, b_\beta\, v^2}{(1 - a\, v)^2}\, dv\right] \\ &\quad + \left[dx_\alpha - \frac{b_\alpha\, v}{1 - a\, v}\, dv\right]\left[dx_\alpha - \frac{b_\alpha\, v}{1 - a\, v}\, dv\right]. \end{aligned} \tag{10.B.39}$$

Connection Forms

To compute the connection forms, we find

$$d\boldsymbol{\omega}_v = 0\,, \qquad d\boldsymbol{\omega}_\beta = 0\,, \qquad d\boldsymbol{\omega}_u = -\, a\; dv \wedge du + b_\mu\; dv \wedge dx_\mu\,. \tag{10.B.40}$$

We express the differentials by the forms and find

$$dv = \boldsymbol{\omega}_v\, (1 - a\, v)\,, \tag{10.B.41a}$$

$$dx_\mu = \boldsymbol{\omega}_\mu + b_\mu\, v\, \boldsymbol{\omega}_v\,, \tag{10.B.41b}$$

$$\begin{aligned} du &= \frac{1}{1 - a\, v}\, \boldsymbol{\omega}_u - \frac{b_\mu\, v}{1 - a\, v}\, dx_\mu + \frac{1}{2}\, \frac{b_\beta\, b_\beta\, v^2}{1 - a\, v}\, \boldsymbol{\omega}_v \\ &= \frac{1}{1 - a\, v}\, \boldsymbol{\omega}_u - \frac{b_\mu\, v}{1 - a\, v}\, \boldsymbol{\omega}_\mu - \frac{1}{2}\, \frac{b_\mu\, b_\mu\, v^2}{1 - a\, v}\, \boldsymbol{\omega}_v\,. \end{aligned} \tag{10.B.41c}$$

We thus obtain

$$d\boldsymbol{\omega}_v = 0\,, \qquad d\boldsymbol{\omega}_\beta = 0\,, \tag{10.B.42a}$$

$$\begin{aligned} d\boldsymbol{\omega}_u &= -\, a\; \boldsymbol{\omega}_v \wedge \boldsymbol{\omega}_u + a\, v\, b_\mu\; \boldsymbol{\omega}_v \wedge \boldsymbol{\omega}_\mu \\ &\quad + b_\mu\, (1 - a\, v)\; \boldsymbol{\omega}_v \wedge \boldsymbol{\omega}_\mu \\ &= -\, a\; \boldsymbol{\omega}_v \wedge \boldsymbol{\omega}_u + b_\mu\; \boldsymbol{\omega}_v \wedge \boldsymbol{\omega}_\mu\,. \end{aligned} \tag{10.B.42b}$$

To compute the connection forms we now write them with upper indices for consistency and have, according to (A.P.2) of **Appendix A**, the first Cartan structural equation,

$$0 = d\omega^u - \omega^u{}_u \wedge \omega^u - \omega^u{}_\alpha \wedge \omega^\alpha\,, \tag{10.B.43a}$$

$$0 = d\omega^v - \omega^v{}_v \wedge \omega^v - \omega^v{}_\alpha \wedge \omega^\alpha\,, \tag{10.B.43b}$$

$$0 = d\omega^\alpha - \omega^\alpha{}_u \wedge \omega^u - \omega^\alpha{}_v \wedge \omega^v - \omega^\alpha{}_\beta \wedge \omega^\beta\,. \tag{10.B.43c}$$

We thus find, for the connection forms,

$$\omega^u{}_u = a\,\omega^v\,, \quad \omega^u{}_\alpha = -\,b_\alpha\,\omega^v\,, \quad \omega^v{}_\alpha = 0\,, \quad \omega^\alpha{}_\beta = 0\,, \tag{10.B.44}$$

or, alternatively, we have

$$\omega_{vu} = a\,\omega^v\,, \quad \omega_{v\alpha} = -\,b_\alpha\,\omega^v\,, \quad \omega_{u\alpha} = 0\,, \quad \omega_{\alpha\beta} = 0\,. \tag{10.B.45}$$

The coefficients g_{ijk} are all constants which is the defining characteristic of a homogeneous field.

Curvature Form

As a check, we compute the components of the curvature form (11.A.4)

$$\begin{aligned}
\boldsymbol{\Omega}_{vu} &= d\omega_{vu} - \omega_{vu} \wedge \omega^u{}_u - \omega_{v\alpha} \wedge \omega^\alpha{}_u = 0\,,\\
\boldsymbol{\Omega}_{v\beta} &= d\omega_{v\beta} - \omega_{vu} \wedge \omega^u{}_\beta - \omega_{v\alpha} \wedge \omega^\alpha{}_\beta = 0\,,\\
\boldsymbol{\Omega}_{u\beta} &= d\omega_{u\beta} - \omega_{uv} \wedge \omega^v{}_\beta - \omega_{u\alpha} \wedge \omega^\alpha{}_\beta = 0\,,\\
\boldsymbol{\Omega}_{\alpha\gamma} &= d\omega_{\alpha\gamma} - \omega_{\alpha u} \wedge \omega^u{}_\gamma - \omega_{\alpha v} \wedge \omega^v{}_\gamma - \omega_{\alpha\beta} \wedge \omega^\beta{}_\gamma = 0\,.
\end{aligned} \tag{10.B.46}$$

We have, thus, found a non-vanishing homogeneous field in $N{+}2$–dimensional Minkowski spacetime.

C. Specialization to Three-Dimensional Minkowski Spacetime

It is clear that a suitable constant rotation of the frames in the N–dimensional Euclidean space can be used to put the vector b_α in the x'–direction. We then call its component b. Denoting x' by "y", we now restrict ourselves to the 3–dimensional u, v, y–spacetime. We then have for the solder forms, (10.B.37),

$$\omega^u = (1 - a\,v)\left[du + \left(\frac{b\,v}{1 - a\,v}\right)dx - \frac{1}{2}\left(\frac{b\,v}{1 - a\,v}\right)^2 dv\right], \tag{10.C.1a}$$

$$\omega^y = dy - \left(\frac{b\,v}{1 - a\,v}\right)dv\,, \tag{10.C.1b}$$

$$\omega^v = \frac{1}{1 - a\,v}dv\,. \tag{10.C.1c}$$

The connection forms are

$$\omega_{vu} = a\,\omega^{v}, \quad \omega_{vy} = -\,b\,\omega^{v}, \quad \omega_{uy} = 0. \tag{10.C.2}$$

The metric is

$$ds^2 = 2\,\omega^u\,\omega^v + (\,\omega^y\,)^2. \tag{10.C.3}$$

We now define new forms ω^x and ω^t by

$$\omega^t = \frac{1}{\sqrt{2}}(\,\omega^u - \omega^v\,), \quad \omega^x = \frac{1}{\sqrt{2}}(\,\omega^u + \omega^v\,) \tag{10.C.4}$$

and obtain

$$ds^2 = -(\,\omega^t\,)^2 + (\,\omega^x\,)^2 + (\,\omega^y\,)^2 \tag{10.C.5}$$

with

$$\omega^u = \frac{1}{\sqrt{2}}(\,\omega^x + \omega^t\,), \quad \omega^v = \frac{1}{\sqrt{2}}(\,\omega^x - \omega^t\,). \tag{10.C.6}$$

We have, for the dual bases,

$$\mathbf{e}_u = \frac{1}{\sqrt{2}}(\,\mathbf{e}_x + \mathbf{e}_t\,), \qquad \mathbf{e}_v = \frac{1}{\sqrt{2}}(\,\mathbf{e}_x - \mathbf{e}_t\,), \tag{10.C.7a,b}$$

$$\mathbf{e}_t = \frac{1}{\sqrt{2}}(\,\mathbf{e}_u - \mathbf{e}_v\,), \qquad \mathbf{e}_x = \frac{1}{\sqrt{2}}(\,\mathbf{e}_u + \mathbf{e}_v\,). \tag{10.C.7c,d}$$

For the differentials of the basis vectors, we obtain

$$\begin{aligned}
d\mathbf{e}_t &= \mathbf{e}_x\,\omega_{xt} + \mathbf{e}_y\,\omega_{yt} \\
&= \frac{1}{2}(\,d\mathbf{e}_u - d\mathbf{e}_v\,) \\
&= \frac{1}{\sqrt{2}}(\mathbf{e}_u\,\omega^u{}_u + \mathbf{e}_y\,\omega^y{}_u - \mathbf{e}_v\,\omega^v{}_v - \mathbf{e}_y\,\omega^y{}_v\,) \\
&= \frac{1}{\sqrt{2}}(\,a\,\omega^v(\,\mathbf{e}_u + \mathbf{e}_v\,) - b\,\omega^v\,\mathbf{e}_y\,) \\
&= a\,\omega^v\,\mathbf{e}_x - \frac{b}{\sqrt{2}}\,\omega^v\,\mathbf{e}_y\,.
\end{aligned} \tag{10.C.8}$$

By comparison, we find

$$\omega_{xt} = a\,\omega^v, \qquad \omega_{yt} = -\frac{b}{\sqrt{2}}\,\omega^v. \tag{10.C.9}$$

By substituting for $\boldsymbol{\omega}^v$ from (10.C.6), we find

$$\boldsymbol{\omega}_{xt} = \frac{a}{\sqrt{2}}(\boldsymbol{\omega}^x - \boldsymbol{\omega}^t), \qquad \boldsymbol{\omega}_{yt} = -\frac{b}{2}(\boldsymbol{\omega}^x - \boldsymbol{\omega}^t). \tag{10.C.10}$$

To calculate the third connection coefficient, we put

$$\begin{aligned} d\mathbf{e}_y &= -\mathbf{e}_t\,\boldsymbol{\omega}_{ty} + \mathbf{e}_x\,\boldsymbol{\omega}_{xy} \\ &= \mathbf{e}_u\,\boldsymbol{\omega}^u{}_y + \mathbf{e}_v\,\boldsymbol{\omega}^v{}_y \\ &= -b\,\boldsymbol{\omega}^v\,\mathbf{e}_u \\ &= -b\,\boldsymbol{\omega}^v\,\frac{1}{\sqrt{2}}(\mathbf{e}_t + \mathbf{e}_x). \end{aligned} \tag{10.C.11}$$

Thus

$$\boldsymbol{\omega}_{xy} = -\frac{b}{\sqrt{2}}\boldsymbol{\omega}^v = -\frac{b}{2}(\boldsymbol{\omega}^x - \boldsymbol{\omega}^t). \tag{10.C.12}$$

We have now found the connection coefficients in the t-x-y–frame and compute

$$\begin{aligned} \boldsymbol{\nabla}_{\mathbf{e}_t}\mathbf{e}_t &= \frac{D}{Ds}\mathbf{e}_t \\ &= [\mathbf{e}_x\,\boldsymbol{\omega}_{xt} + \mathbf{e}_y\,\boldsymbol{\omega}_{yt}](\mathbf{e}_t) \\ &= \mathbf{e}_x\,[\boldsymbol{\omega}_{xt}(\mathbf{e}_t)] + \mathbf{e}_y\,[\boldsymbol{\omega}_{yt}(\mathbf{e}_t)] \\ &= \mathbf{e}_x\left(-\frac{a}{\sqrt{2}}\right) + \mathbf{e}_y\left(\frac{b}{2}\right). \end{aligned} \tag{10.C.13}$$

We find thus that the t–lines are worldlines of constant geodesic acceleration of the amount

$$|g| = \frac{1}{2}\sqrt{2a^2 + b^2} = \sqrt{\frac{a^2}{2} + \frac{b^2}{4}}. \tag{10.C.14}$$

To obtain a Frenet frame, we introduce the vectors

$$\mathbf{e}_\xi = -\frac{a\sqrt{2}}{\sqrt{2a^2 + b^2}}\mathbf{e}_x + \frac{b}{\sqrt{2a^2 + b^2}}\mathbf{e}_y, \tag{10.C.15a}$$

$$\mathbf{e}_\eta = -\frac{b}{\sqrt{2a^2 + b^2}}\mathbf{e}_x - \frac{a\sqrt{2}}{\sqrt{2a^2 + b^2}}\mathbf{e}_u. \tag{10.C.15b}$$

We then have

$$\frac{D}{Ds}\,\mathbf{e}_t \;=\; g\,\mathbf{e}_\xi\,. \tag{10.C.16}$$

Now, computing

$$\begin{aligned}
\frac{D}{Ds}\,\mathbf{e}_\xi \;&=\; \boldsymbol{\nabla}_{\mathbf{e}_t}\left(-\,\frac{a\sqrt{2}}{\sqrt{2\,a^2\,+\,b^2}}\,\mathbf{e}_x\,+\,\frac{b}{\sqrt{2\,a^2\,+\,b^2}}\,\mathbf{e}_y\right)\\
&=\;\frac{1}{\sqrt{2\,a^2\,+\,b^2}}\Big\{-\,\sqrt{2}\,a\,\big[-\mathbf{e}_t\,\boldsymbol{\omega}_{tx}\,(\mathbf{e}_t)\,+\,\mathbf{e}_y\,\boldsymbol{\omega}_{yx}\,(\mathbf{e}_t)\big]\\
&\qquad\qquad +\,b\,\big[-\,\mathbf{e}_t\,\boldsymbol{\omega}_{ty}\,(\mathbf{e}_t)\,+\,\mathbf{e}_x\,\boldsymbol{\omega}_{xy}\,(\mathbf{e}_t)\big]\Big\}\\
&=\;\frac{1}{\sqrt{2\,a^2\,+\,b^2}}\left\{-\,\sqrt{2\,a}\left(-\,\frac{a}{\sqrt{2}}\,\mathbf{e}_t\,-\,\frac{b}{2}\,\mathbf{e}_y\right)\,+\,\frac{b^2}{2}\,\mathbf{e}_t\,+\,\frac{b^2}{2}\,\mathbf{e}_x\right\}\\
&=\frac{1}{2}\,\sqrt{2\,a^2\,+\,b^2}\,\mathbf{e}_t\,-\,\frac{b}{2}\left(-\,\frac{a\,\sqrt{2}}{\sqrt{2\,a^2\,+\,b^2}}\,\mathbf{e}_y\,-\,\frac{b}{\sqrt{2\,a^2\,+\,b^2}}\,\mathbf{e}_x\right)\\
&=\,g\,\mathbf{e}_t\,-\,\frac{b}{2}\,\mathbf{e}_\eta\,.
\end{aligned} \tag{10.C.17}$$

We have found that the worldlines to which $\mathbf{e}_t$ is the unit tangent vector are helices with acceleration $g\;=\;\frac{1}{2}\,\sqrt{2\,a^2\,+\,b^2}$ and first torsion $-b/2$.

D. Specialization to Circular Orbits

Do worldlines of this type occur for particles in a circular orbit with constant angular velocity, $\boldsymbol{\omega}$? Put

$$ds^2 \;=\; -\,dt^2\,+\,dx^2\,+\,dy^2\,, \tag{10.D.1}$$

where the coordinates are given, in terms of the arc length parameter, s, and the orbit's radius, R, and the angular velocity magnitude, ω, by

$$t \;=\; \frac{s}{\sqrt{1\,-\,\omega^2\,R^2}}\,, \tag{10.D.2a}$$

$$x \;=\; R\,\cos\Big(\frac{\omega\,s}{\sqrt{1\,-\,\omega^2\,R^2}}\Big)\,, \tag{10.D.2b}$$

$$y \;=\; R\,\sin\Big(\frac{\omega\,s}{\sqrt{1\,-\,\omega^2\,R^2}}\Big)\,. \tag{10.D.2c}$$

The tangent vector $\mathbf{T}$ has components

$$\mathbf{T} = \Bigg(\frac{1}{\sqrt{1 - \omega^2 R^2}}, \quad - \frac{R\omega}{\sqrt{1 - \omega^2 R^2}} \sin\big(\frac{\omega s}{\sqrt{1 - \omega^2 R^2}}\big), \quad \frac{R\omega}{\sqrt{1 - \omega^2 R^2}} \cos\big(\frac{\omega s}{\sqrt{1 - \omega^2 R^2}}\big) \Bigg). \tag{10.D.3}$$

We have

$$\frac{d\mathbf{T}}{ds} = \Bigg(0, \quad - \frac{R\omega^2}{1 - \omega^2 R^2} \cos\big(\frac{\omega s}{\sqrt{1 - \omega^2 R^2}}\big), \quad - \frac{R\omega^2}{1 - \omega^2 R^2} \sin\big(\frac{\omega s}{\sqrt{1 - \omega^2 R^2}}\big) \Bigg). \tag{10.D.4}$$

Thus the geodesic acceleration is given by

$$g = \frac{R\omega^2}{1 - \omega^2 R^2} \tag{10.D.5}$$

where the radius of the circular orbit is given by R and the magnitude of the angular velocity by ω. We write

$$\frac{d\mathbf{T}}{ds} = g\,\mathbf{N} \tag{10.D.6a}$$

where we have

$$\mathbf{N} = \Bigg(0, \quad - \cos\big(\frac{\omega s}{\sqrt{1 - \omega^2 R^2}}\big), \quad - \sin\big(\frac{\omega s}{\sqrt{1 - \omega^2 R^2}}\big) \Bigg). \tag{10.D.6b}$$

We then have

$$\frac{d\mathbf{N}}{ds} = \Bigg(0, \quad + \frac{\omega}{\sqrt{1 - \omega^2 R^2}} \sin\big(\frac{\omega s}{\sqrt{1 - \omega^2 R^2}}\big), \quad - \frac{\omega}{\sqrt{1 - \omega^2 R^2}} \cos\big(\frac{\omega s}{\sqrt{1 - \omega^2 R^2}}\big) \Bigg) \tag{10.D.7}$$

and

$$\frac{d\mathbf{N}}{ds} - g\,\mathbf{T} = \begin{pmatrix} - \dfrac{R\omega^2}{(1 - \omega^2 R^2)^{3/2}}, \\[2ex] + \dfrac{\omega}{(1 - \omega^2 R^2)^{3/2}} \sin\big(\dfrac{\omega s}{\sqrt{1 - \omega^2 R^2}}\big), \\[2ex] - \dfrac{\omega}{(1 - \omega^2 R^2)^{3/2}} \cos\big(\dfrac{\omega s}{\sqrt{1 - \omega^2 R^2}}\big) \end{pmatrix} \tag{10.D.8a}$$

$$= \frac{\omega}{1 - \omega^2 R^2} \mathbf{M}, \tag{10.D.8b}$$

where

$$\mathbf{M} = \begin{pmatrix} -\dfrac{R\,\omega}{\sqrt{1 - \omega^2 R^2}}, \\ +\dfrac{1}{\sqrt{1 - \omega^2 R^2}} \sin\left(\dfrac{\omega\, s}{\sqrt{1 - \omega^2 R^2}}\right), \\ -\dfrac{1}{\sqrt{1 - \omega^2 R^2}} \cos\left(\dfrac{\omega\, s}{\sqrt{1 - \omega^2 R^2}}\right) \end{pmatrix}. \tag{10.D.9}$$

Thus, we have, for the torsion, τ,

$$\tau = \frac{\omega}{1 - \omega^2 R^2}. \tag{10.D.10}$$

If we try to compare parameters, we find the equations

$$\frac{1}{2}\sqrt{2\,a^2 + b^2} = \frac{R\,\omega^2}{1 - \omega^2 R^2}, \tag{10.D.11a}$$

$$-\frac{b}{2} = \frac{\omega}{1 - \omega^2 R^2} \tag{10.D.11b}$$

or, upon dividing the first equation by the second, we have

$$R\,\omega = -\frac{\sqrt{2\,a^2 + b^2}}{b} = -\sqrt{1 + \frac{2\,a^2}{b^2}} \tag{10.D.12}$$

and

$$1 - \omega^2 R^2 = 1 - \left(1 + \frac{2\,a^2}{b^2}\right) = -\frac{2\,a^2}{b^2}. \tag{10.D.13}$$

This shows that the motion with tangent vector $\mathbf{e}_t$ is **not** circular motion.

E. The Hyperbolic Analogue of Minkowski Circular Motion in Homogeneous Fields

We take the hyperbolic analogue and put for the worldline

$$\left(\frac{1}{\gamma}\sinh\left(\gamma\, s\sqrt{1 + \rho^2}\right), \quad \frac{1}{\gamma}\cosh\left(\gamma\, s\sqrt{1 + \rho^2}\right), \quad s\,\rho\right) \tag{10.E.1}$$

or, for the tangent vector, we have

$$\mathbf{T} = \Big(\sqrt{1+\rho^2}\,\cosh\big(\gamma\, s\sqrt{1+\rho^2}\big)\,, \\ \sqrt{1+\rho^2}\,\sinh\big(\gamma\, s\sqrt{1+\rho^2}\big)\,, \quad \rho \Big)\,. \tag{10.E.2}$$

Thus

$$\frac{d\mathbf{T}}{ds} = g\,\mathbf{N} = \gamma\,(1+\rho^2)\,\mathbf{N}\,, \tag{10.E.3}$$

where

$$\mathbf{N} = \Big(\sinh\big(\gamma\, s\sqrt{1+\rho^2}\big), \quad \cosh\big(\gamma\, s\sqrt{1+\rho^2}\big), \quad 0 \Big)\,. \tag{10.E.4}$$

We find

$$g = \gamma\,(1+\rho^2)\,. \tag{10.E.5}$$

We now, once again, form

$$\frac{d\mathbf{N}}{ds} - g\,\mathbf{T} \tag{10.E.6}$$

from

$$\frac{d\mathbf{N}}{ds} = \gamma\sqrt{1+\rho^2}\begin{pmatrix} \cosh\big(\gamma\, s\sqrt{1+\rho^2}\big)\,, \\ \sinh\big(\gamma\, s\sqrt{1+\rho^2}\big)\,, \\ 0 \end{pmatrix} \tag{10.E.7}$$

and

$$g\,\mathbf{T} = \gamma\sqrt{1+\rho^2}\begin{pmatrix} (1+\rho^2)\,\cosh\big(\gamma\, s\sqrt{1+\rho^2}\big)\,, \\ (1+\rho^2)\,\sinh\big(\gamma\, s\sqrt{1+\rho^2}\big)\,, \\ \rho\sqrt{1+\rho^2} \end{pmatrix} \tag{10.E.8}$$

with the result of the subtraction being

$$\frac{d\mathbf{N}}{ds} - g\,\mathbf{T} = -\gamma\,\rho\sqrt{1+\rho^2}\begin{pmatrix} \rho\,\cosh\big(\gamma\, s\sqrt{1+\rho^2}\big)\,, \\ \rho\,\sinh\big(\gamma\, s\sqrt{1+\rho^2}\big)\,, \\ \sqrt{1+\rho^2} \end{pmatrix}\,, \tag{10.E.9}$$

which gives the torsion the value

$$\tau = -\gamma\,\rho\sqrt{1+\rho^2}\,. \tag{10.E.10}$$

Comparison of the parameters gives

$$\sqrt{\frac{a^2}{2} + \frac{b^2}{4}} = \gamma\,(1 + \rho^2)\,, \qquad \frac{b}{2} = \gamma\,\rho\,\sqrt{1 + \rho^2}\,, \tag{10.E.11}$$

or, by some manipulation, we can extract ρ

$$\sqrt{1 + \frac{2\,a^2}{b^2}} = \sqrt{1 + \frac{1}{\rho^2}} \quad \longrightarrow \quad \rho = \frac{b}{a\,\sqrt{2}}\,. \tag{10.E.12}$$

That is, we have arrived at

$$\frac{b}{2\,\rho} = \frac{a}{\sqrt{2}} \tag{10.E.13}$$

and we can now determine

$$\begin{aligned} \gamma &= \sqrt{\frac{a^2}{2} + \frac{b^2}{4}} \Big/ \left(1 + \frac{b^2}{2\,a^2}\right) \\ &= \frac{a}{\sqrt{2}} \Big/ \sqrt{1 + \frac{b^2}{2\,a^2}} \\ &= a^2 \Big/ \sqrt{2\,a^2 + b^2}\,. \end{aligned} \tag{10.E.14}$$

Thus, the motion is described in terms of the position vector

$$\left(\frac{\sqrt{2\,a^2 + b^2}}{a^2}\,\sinh\!\left(\frac{a\,s}{\sqrt{2}}\right),\ \frac{\sqrt{2\,a^2 + b^2}}{a^2}\,\cosh\!\left(\frac{a\,s}{\sqrt{2}}\right),\ \frac{s\,b}{a\,\sqrt{2}}\right). \tag{10.E.15}$$

For $b = 0$ the torsion vanishes.

In the general case, the motion is that of a charged particle in a covariantly constant electromagnetic field for which $\mathbf{E} \cdot \mathbf{B} = 0$. For details of that result, see [Honig, 1974].

F. Geodesic Motion in Homogeneous Fields

We shall study, briefly, the geodesic motion in a homogeneous field. A geodesic with tangent vector $\mathbf{U}$ is given by the equation

$$\boldsymbol{\nabla}_{\mathbf{U}}\,\mathbf{U} = dU^a\,(\mathbf{U})\,\mathbf{e}_a + U^a\,\boldsymbol{\nabla}_{\mathbf{U}}\,(\mathbf{e}_a) = 0 \tag{10.F.1}$$

where $\mathbf{U}$ is a unit vector for timelike geodesics

$$\mathbf{U}\cdot\mathbf{U} = -1 \tag{10.F.2}$$

with $\mathbf{U}$ expressed as

$$\mathbf{U} = U^a\,\mathbf{e}_a\,. \tag{10.F.3}$$

For null geodesics we use a distinguished parameter. Derivatives with respect to arc-length s or distinguished parameter will be indicated by a dot. Since

$$\boldsymbol{\nabla}_{\mathbf{U}}\,\mathbf{e}_a = \mathbf{e}_b\,\boldsymbol{\omega}^b{}_a\,(\mathbf{U})\,. \tag{10.F.4}$$

We can write (10.F.1) as

$$\dot{U}^a + \boldsymbol{\omega}^a{}_b\,(\mathbf{U})\,U^b = 0\,. \tag{10.F.5}$$

The connection coefficients from (10.C.10) and (10.C.12) are

$$\boldsymbol{\omega}_{xt} = \frac{a}{\sqrt{2}}\,(\boldsymbol{\omega}^x - \boldsymbol{\omega}^t)\,, \tag{10.F.6a}$$

$$\boldsymbol{\omega}_{yt} = -\frac{b}{2}\,(\boldsymbol{\omega}^x - \boldsymbol{\omega}^t)\,, \tag{10.F.6b}$$

$$\boldsymbol{\omega}_{xy} = -\frac{b}{2}\,(\boldsymbol{\omega}^x - \boldsymbol{\omega}^t)\,. \tag{10.F.6c}$$

We take first the case of vanishing b and put $U^y = 0$. One then gets

$$\dot{U}^t + \boldsymbol{\omega}^t{}_x\,(\mathbf{U})\,U^x = \dot{U}^t + \frac{a}{\sqrt{2}}\,(U^x - U^t)\,U^x = 0\,, \tag{10.F.7a}$$

$$\dot{U}^x + \boldsymbol{\omega}^x{}_t\,(\mathbf{U})\,U^t = \dot{U}^x + \frac{a}{\sqrt{2}}\,(U^x - U^t)\,U^t = 0\,. \tag{10.F.7b}$$

Normalization gives

$$\left(U^t\right)^2 = \left(U^x\right)^2 + 1 \tag{10.F.8}$$

which leads to the relation

$$U^t\,\dot{U}^t = U^x\,\dot{U}^x \tag{10.F.9}$$

which is evidently fulfilled by the equations. We put

$$U^x = \frac{\beta}{\sqrt{1-\beta^2}}, \qquad U^t = \frac{1}{\sqrt{1-\beta^2}} \tag{10.F.10}$$

and have from (10.F.7b)

$$\left(\frac{\beta}{\sqrt{1-\beta^2}}\right)^{\bullet} = \frac{a}{\sqrt{2}}\,\frac{1-\beta}{1-\beta^2} = \frac{a}{\sqrt{2}}\,\frac{1}{1+\beta}. \tag{10.F.11}$$

We have from (10.F.7a,b)

$$\begin{aligned}(\dot{U}^x)^2 - (\dot{U}^t)^2 &= \frac{a^2}{2}\,(U^x - U^t)^2 \\ &= \frac{a^2}{2}\,\frac{(1-\beta)^2}{1-\beta^2} \\ &= \frac{a^2}{2}\,\frac{1-\beta}{1+\beta}.\end{aligned} \tag{10.F.12}$$

That means one would ascribe to the freely falling particle in the homogeneous field in the instantaneous rest frame the acceleration

$$\gamma = \frac{a}{\sqrt{2}}\sqrt{\frac{1-\beta}{1+\beta}}. \tag{10.F.13}$$

This coincides for $\beta = 0$ with minus the acceleration of the frame.

CHAPTER 11

EUCLIDEAN THREE-DIMENSIONAL SPACES

A. Definition of Homogeneous Fields in Euclidean Three-Space

The metric of Euclidean 3-space is locally expressed by three differential one-forms $\boldsymbol{\omega}_i$ as

$$ds^2 = \boldsymbol{\omega}_i \, \boldsymbol{\omega}_i \, . \qquad (11.A.1)$$

Lower case Latin indices run from one to three and repeated indices will be summed. The connection one-forms $\boldsymbol{\omega}_{ij}$ are defined by

$$d\boldsymbol{\omega}_i = \boldsymbol{\omega}_{ij} \wedge \boldsymbol{\omega}_j \, , \qquad (11.A.2a)$$

and satisfy the antisymmetry property

$$\boldsymbol{\omega}_{ij} + \boldsymbol{\omega}_{ji} = 0 \, . \qquad (11.A.2b)$$

We have

$$\begin{aligned} 0 &= dd\boldsymbol{\omega}_i \\ &= d\boldsymbol{\omega}_{ij} \wedge \boldsymbol{\omega}_j - \boldsymbol{\omega}_{ik} \wedge d\boldsymbol{\omega}_k \\ &= (d\boldsymbol{\omega}_{ij} - \boldsymbol{\omega}_{ik} \wedge \boldsymbol{\omega}_{kj}) \wedge \boldsymbol{\omega}_j \, . \qquad (11.A.3) \end{aligned}$$

This puts the torsion equal to zero. Defining the curvature form $\boldsymbol{\Omega}_{ij}$ by

$$\boxed{\boldsymbol{\Omega}_{ij} \equiv d\boldsymbol{\omega}_{ij} - \boldsymbol{\omega}_{ik} \wedge \boldsymbol{\omega}_{kj} = -\boldsymbol{\Omega}_{ji}} \, , \qquad (11.A.4)$$

we can write (11.A.3) as

$$\boldsymbol{\Omega}_{ij} \wedge \boldsymbol{\omega}_j = 0 \, . \qquad (11.A.5)$$

Differentiating the curvature form $\boldsymbol{\Omega}_{ij}$ gives the Bianchi identities

$$\begin{aligned} d\boldsymbol{\Omega}_{ij} &= -d\boldsymbol{\omega}_{ik} \wedge \boldsymbol{\omega}_{kj} + \boldsymbol{\omega}_{ik} \wedge d\boldsymbol{\omega}_{kj} \\ &= -\boldsymbol{\Omega}_{ik} \wedge \boldsymbol{\omega}_{kj} + \boldsymbol{\omega}_{ik} \wedge \boldsymbol{\Omega}_{kj} \, . \qquad (11.A.6) \end{aligned}$$

A *homogeneous field* is defined by connection forms ω_{ij} with

$$\boxed{\omega_{ij} = g_{ijk}\,\omega_k} \tag{11.A.7}$$

with constant components g_{ijk}.

We relate these components to the structure constants of a group as justification for the adoption of the term "homogeneous" to describe these fields. One has

$$g_{ijk} = -\,g_{jik}\,. \tag{11.A.8}$$

Insertion of (11.A.7) into (11.A.2a) gives

$$\begin{aligned} d\omega_i &= g_{ijk}\,\omega_k \wedge \omega_j \\ &= -\,g_{ijk}\,\omega_j \wedge \omega_k \\ &= -\frac{1}{2}\,c_{ijk}\,\omega_j \wedge \omega_k \end{aligned} \tag{11.A.9}$$

where the group's structure constants c_{ijk} have been defined by

$$\boxed{c_{ijk} \equiv g_{ijk} - g_{ikj}} \tag{11.A.10}$$

which have the symmetry

$$c_{ijk} = -\,c_{ikj}\,. \tag{11.A.11}$$

B. Curvature Form

We now compute the curvature form for a homogeneous field. We have from (11.A.7) and (11.A.9)

$$\begin{aligned} d\omega_{ij} &= g_{ijk}\,d\omega_k \\ &= -\frac{1}{2}\,g_{ijk}\,c_{klm}\,\omega_l \wedge \omega_m \end{aligned} \tag{11.B.1}$$

and from (11.A.7) only

$$\omega_{ik} \wedge \omega_{kj} = g_{ikl}\,g_{kjm}\,\omega_l \wedge \omega_m\,. \tag{11.B.2}$$

Thus from (11.A.4), the curvature form can be written as

$$\boldsymbol{\Omega}_{ij} = -\left[\frac{1}{2}\,g_{ijk}\,c_{klm} + g_{ikl}\,g_{kjm}\right]\omega_l \wedge \omega_m\,. \tag{11.B.3}$$

The condition (11.A.5) gives

$$0 = \boldsymbol{\Omega}_{ij} \wedge \boldsymbol{\omega}_j = -\left[\frac{1}{2} g_{ijk}\, c_{klm} + g_{ikl}\, g_{kjm}\right] \boldsymbol{\omega}_l \wedge \boldsymbol{\omega}_m \wedge \boldsymbol{\omega}_j . \quad (11.B.4)$$

Since we have from (11.A.9)

$$g_{kjm}\, \boldsymbol{\omega}_m \wedge \boldsymbol{\omega}_j = \frac{1}{2} c_{kjm}\, \boldsymbol{\omega}_m \wedge \boldsymbol{\omega}_j , \quad (11.B.5)$$

we can write (11.B.4) as

$$0 = -\frac{1}{2} \left(g_{ijk}\, c_{klm} + g_{ikl}\, c_{kjm} \right) \boldsymbol{\omega}_l \wedge \boldsymbol{\omega}_m \wedge \boldsymbol{\omega}_j . \quad (11.B.6)$$

We thus have

$$\boxed{\left(g_{ijk}\, c_{klm} + g_{ikl}\, c_{kjm} \right) + cyclic(l, m, j) = 0} . \quad (11.B.7)$$

C. Jacobi Identity

We want to show that the structure constants fulfill the Jacobi identity. For this purpose we express the g_{ijk} by the structure constants. We have from (11.A.10)

$$c_{ijk} = g_{ijk} - g_{ikj} , \quad (11.C.1a)$$

$$c_{jki} = g_{jki} - g_{jik} , \quad (11.C.1b)$$

$$c_{kij} = g_{kij} - g_{kji} . \quad (11.C.1c)$$

Because of the skew-symmetry of the Ricci coefficients, that is, $g_{ijk} = -g_{jik}$, adding the first two equalities and subtracting the third gives

$$\boxed{2\, g_{ijk} = c_{ijk} + c_{jki} - c_{kij}} . \quad (11.C.2)$$

We then construct expressions

$$\begin{aligned} 2 \left(g_{ijk}\, c_{klm} + g_{ikl}\, c_{kjm} \right) = {} & \left(c_{ijk} + c_{jki} - c_{kij} \right) c_{klm} \\ & + \left(c_{ikl} + c_{kli} - c_{lik} \right) c_{kjm} \end{aligned} \quad (11.C.3)$$

which can be assembled into (11.B.7) and yield

$$\begin{aligned} 0 = &+ 2\, c_{ijk}\, c_{klm} + 2\, c_{ilk}\, c_{kmj} + 2\, c_{imk}\, c_{kjl} \\ &+ c_{jki}\, c_{klm} + c_{lki}\, c_{kmj} + c_{mki}\, c_{kjl} \\ &- c_{kij}\, c_{klm} - c_{kil}\, c_{kmj} - c_{kim}\, c_{kjl} \\ &- c_{ikl}\, c_{kmj} - c_{ikm}\, c_{kjl} - c_{ikj}\, c_{klm} \\ &- c_{kli}\, c_{kmj} - c_{kmi}\, c_{kjl} - c_{kji}\, c_{klm} \\ &+ c_{lik}\, c_{kmj} + c_{mik}\, c_{kjl} + c_{jik}\, c_{klm}\, . \end{aligned} \tag{11.C.4}$$

We obtain thus the Jacobi identity for the structure constants

$$\boxed{c_{ijk}\, c_{klm} + c_{ilk}\, c_{kmj} + c_{imk}\, c_{kjl} = 0}\, . \tag{11.C.5}$$

This result is quite general and holds for all n-dimensional Riemann spaces that allow a homogeneous field.

D. Classification of Homogeneous Fields

We have defined homogeneous fields in terms of constant Ricci rotation coefficients which, in turn, can be written as a sum of the structure constants of a group (11.C.2), if one is assumed to be present. We can, therefore undertake the classification of homogeneous fields by first classifying their Lie groups. The homogeneous field components can be rewritten in terms of quantities derived from the structure constants of our presumed group. From these efforts, we will then classify the Riemann curvature tensor for homogeneous fields.

A Lie algebra is given by the set of commutation relations

$$[\, X_a,\, X_b\,] = c^f{}_{ab}\, X_f \tag{11.D.1}$$

whose structure constants $c^f{}_{ab}$ satisfy

$$c^f{}_{ab} = -\, c^f{}_{ba}\, , \tag{11.D.2}$$

which can also be seen from their definition (11.A.10) in terms of the Ricci rotation coefficients. By the tensor operation of contraction, we may obtain a vector from the structure constants

$$\boxed{c_b = c^a{}_{ab}}\, . \tag{11.D.3}$$

At the end of **Section 7.A**, we exhibited the Lie algebra of the groups described therein.

E. Euclidean Three-Space

We now specialize the theory to the case of Euclidean space with three dimensions. The Jacobi identity is then equivalent to

$$c_{ijk}\,\epsilon_{jlm}\,c_{klm} = 0\,. \tag{11.E.1}$$

We define a second rank tensor, the symmetric part of c_{klm}, by

$$\boxed{b_{jk} \equiv \frac{1}{2}\,\epsilon_{jlm}\,c_{klm}} \tag{11.E.2}$$

and get

$$\begin{aligned} \epsilon_{rjk}\,b_{jk} &= -\frac{1}{2}\,(\,\delta_{rl}\,\delta_{km} - \delta_{rm}\,\delta_{kl}\,)\,c_{klm} \\ &= -\frac{1}{2}\,(\,c_{krk} - c_{kkr}\,) \\ &= c_{kkr}\,. \end{aligned} \tag{11.E.3}$$

The Jacobi identity becomes

$$c_{ijk}\,b_{jk} = 0\,. \tag{11.E.4}$$

This equation involves only the skew-symmetric part of b_{jk} which can be expressed by the vector c_{kkr}. We now adopt the abbreviations

$$\boxed{c_r \equiv c_{kkr}}\,, \qquad \boxed{c^2 \equiv c_r\,c_r}\,. \tag{11.E.5a,b}$$

We have from (11.E.3)

$$\begin{aligned} \epsilon_{rpq}\,\epsilon_{rjk}\,b_{jk} &= (\,\delta_{pj}\,\delta_{qk} - \delta_{pk}\,\delta_{qj}\,)\,b_{jk} \\ &= b_{pq} - b_{qp} \\ &= \epsilon_{rpq}\,c_r\,. \end{aligned} \tag{11.E.6}$$

The Jacobi identity becomes, therefore,

$$\frac{1}{2}\,c_{ijk}\,\epsilon_{rjk}\,c_r = 0\,. \tag{11.E.7}$$

Because of (11.E.2), we have

$$c_r \, b_{ri} \; = \; 0 \, . \tag{11.E.8}$$

By splitting b_{ri} into its symmetric and skew-symmetric components,

$$\boxed{d_{ri} \; \equiv \; \frac{1}{2} \left(\, b_{ri} \; + \; b_{ir} \, \right)} \, , \tag{11.E.9a}$$

we have, from (11.E.6),

$$b_{ri} \; = \; d_{ri} \; + \; \frac{1}{2} \, \epsilon_{ris} \, c_s \, . \tag{11.E.9b}$$

That is, the d_{ri} are the symmetric parts of the b_{ri} which, in turn, were defined as the symmetric parts of the structure constants. We obtain, from (11.E.8),

$$d_{ir} \, c_r \; = \; 0 \, . \tag{11.E.10}$$

We have reached a suitable form of the Jacobi identity and express the structure constants in terms of d_{ir} and c_j. By multiplying (11.E.2) with ϵ_{jpq}, we obtain

$$\epsilon_{jpq} \, b_{jk} \; = \; \frac{1}{2} \left(\, \delta_{pl} \, \delta_{qm} \; - \; \delta_{pm} \, \delta_{ql} \, \right) \; = \; c_{kpq} \, . \tag{11.E.11}$$

Inserting (11.E.9a,b) in this equation gives

$$\begin{aligned} c_{ipq} \; &= \; \epsilon_{rpq} \, b_{ri} \\ &= \; \epsilon_{rpq} \, d_{ri} \; + \; \frac{1}{2} \, \epsilon_{rpq} \, \epsilon_{ris} \, c_s \\ &= \; \epsilon_{rpq} d_{ri} \; + \; \frac{1}{2} \left(\, \delta_{pi} \, c_q \; - \; \delta_{qi} \, c_p \, \right) \, . \end{aligned} \tag{11.E.12}$$

We now use (11.C.2) to express the field components in terms of the d_{ri} and c_s. We obtain

$$\begin{aligned} 2 \, g_{ipq} \; &= \; c_{ipq} \; + \; c_{pqi} \; - \; c_{qip} \\ &= \; + \; d_{ri} \, \epsilon_{rpq} \; + \; \frac{1}{2} \left(\, \delta_{ip} \, c_q \; - \; \delta_{iq} \, c_p \, \right) \\ &\quad \; + \; d_{rp} \, \epsilon_{rqi} \; + \; \frac{1}{2} \left(\, \delta_{pq} \, c_i \; - \; \delta_{pi} \, c_q \, \right) \\ &\quad \; - \; d_{rq} \, \epsilon_{rip} \; - \; \frac{1}{2} \left(\, \delta_{qi} \, c_p \; - \; \delta_{qp} \, c_i \, \right) \end{aligned} \tag{11.E.13}$$

or, upon combining terms,

$$2\, g_{ipq} \;=\; d_{ri}\, \epsilon_{rpq} \;+\; d_{rp}\, \epsilon_{rqi} \;-\; d_{rq}\, \epsilon_{rip} \;+\; \delta_{pq}\, c_i \;-\; \delta_{iq}\, c_p\,. \tag{11.E.14}$$

F. Curvature Form Calculation

The next step is the calculation of the curvature form. We have a simpler calculation if we introduce

$$\boxed{\boldsymbol{\Omega}_w \;\equiv\; \frac{1}{2}\, \epsilon_{wij}\, \boldsymbol{\Omega}_{ij}} \tag{11.F.1}$$

and

$$\boxed{\boldsymbol{\omega}_l \wedge \boldsymbol{\omega}_m \;\equiv\; \epsilon_{lmp}\, \boldsymbol{\chi}_p}\,. \tag{11.F.2}$$

The curvature form (11.B.3) can then be written

$$\boxed{\boldsymbol{\Omega}_w \;=\; \Lambda_{wp}\, \boldsymbol{\chi}_p}\,, \tag{11.F.3a}$$

$$\boxed{\Lambda_{wp} \;\equiv\; -\left[\frac{1}{4}\, \epsilon_{wij}\, g_{ijk}\, c_{klm}\, \epsilon_{lmp} \;+\; \frac{1}{2}\, \epsilon_{wij}\, g_{ikl}\, g_{kjm}\, \epsilon_{lmp}\right]}\,. \tag{11.F.3b}$$

We have from (11.E.9a,b) and (11.E.11)

$$\frac{1}{2}\, c_{klm}\, \epsilon_{lmp} \;=\; \frac{1}{2}\, b_{sk}\, \epsilon_{slm}\, \epsilon_{lmp} \;=\; b_{pk} \;=\; d_{pk} \;+\; \frac{1}{2}\, \epsilon_{pks}\, c_s \tag{11.F.4}$$

and from (11.E.14)

$$\frac{1}{2}\, \epsilon_{wij}\, g_{ijk} \;=\; \frac{1}{2}\, (\, d \cdot \delta_{wk} \;-\; \epsilon_{wkj}\, c_j \;-\; 2\, d_{wk}\,.)\;, \tag{11.F.5}$$

where we introduce here the abbreviations

$$\boxed{d \equiv d_{ii}}\,, \qquad \boxed{d^2 \;\equiv\; d_{ii}\, d_{jj}}\,, \tag{11.F.6a,b}$$

but just use the first for the moment. Multiplying (11.F.4) with (11.F.5), we obtain

$$\frac{1}{4}\, \epsilon_{wij}\, g_{ijk}\, c_{klm}\, \epsilon_{lmp} \;=\; \frac{1}{2}\left(d_{pk} \;+\; \frac{1}{2}\, \epsilon_{pks}\, c_s\right) \times (\, d \cdot \delta_{wk} \;-\; \epsilon_{wkj}\, c_j \;-\; 2\, d_{wk}\,) \tag{11.F.7a}$$

and define

$$\boxed{M_{wp} = \frac{1}{2}\left(d_{pk} + \frac{1}{2}\,\epsilon_{pks}\,c_s \right) \times (\, d\cdot\delta_{wk} - \epsilon_{wkj}\,c_j - 2\,d_{wk}\,)}\,. \tag{11.F.7b}$$

Calculation of M_{wp}

For the expression M_{wp}, we obtain

$$\begin{aligned} M_{wp} &= \frac{1}{2}\left(d_{wp}\cdot d + \frac{d}{2}\,\epsilon_{pws}\,c_s - d_{pk}\,\epsilon_{wkj}\,c_j \right) \\ &\quad - \frac{1}{4}\,(\,\delta_{pw}\,\delta_{sj} - \delta_{pj}\,\delta_{sw}\,)\,c_s\,c_j \\ &\quad - \frac{1}{2}\,(\,-2\,d_{pk}\,d_{wk} - d_{wk}\,\epsilon_{pks}\,c_s\,) \end{aligned} \tag{11.F.8}$$

or

$$\begin{aligned} M_{wp} &= \frac{1}{2}\,(\,-2\,d_{pk}\,d_{wk} - d_{wk}\,\epsilon_{pkj}\,c_j\,) \\ &\quad + \frac{1}{2}\left(d_{wp}\cdot d + \frac{d}{2}\,\epsilon_{pws}\,c_s - d_{pk}\,\epsilon_{wkj}\,c_j \right) \\ &\quad + \frac{1}{2}\left(-\frac{1}{2}\,\delta_{wp}\,c^2 + \frac{1}{2}\,c_w\,c_p \right) \end{aligned} \tag{11.F.9}$$

where we have used the abbreviation (11.E.5b). From (11.F.3a,b) and (11.F.7), we now define

$$\boxed{N_{wp} \equiv \frac{1}{2}\,\epsilon_{wij}\,g_{ikl}\,g_{kjm}\,\epsilon_{lmp}}\,, \tag{11.F.10}$$

and also write

$$\boxed{\Lambda_{wp} \equiv -\,(\,M_{wp} + N_{wp}\,)}\,. \tag{11.F.11}$$

Calculation of N_{wp}

We want to compute N_{wp}. From (11.E.14), as an intermediate stage, we have

$$\begin{aligned} \epsilon_{ijw}\,g_{ikl} &= +\frac{1}{2}\,\{\,+d\cdot\delta_{jk}\,\delta_{wl} - d_{wk}\,\delta_{jl} + d_{jl}\,\delta_{wk}\,\} \\ &\quad + \frac{1}{2}\,\{\,-d\cdot\delta_{jl}\,\delta_{wk} - d_{jk}\,\delta_{wl} - d_{wl}\,\delta_{jk}\,\} \\ &\quad + \frac{1}{2}\,\{\,+d_{jk}\,\delta_{wl} - d_{wk}\,\delta_{jl} - d_{wl}\,\delta_{jk}\,\} \\ &\quad + \frac{1}{2}\,\{\,+d_{jl}\,\delta_{wk} + \epsilon_{ijw}\,c_i\,\delta_{kl} - \epsilon_{ljw}\,c_k\,\}\,. \end{aligned} \tag{11.F.12}$$

or, upon collecting terms,

$$\begin{aligned}\epsilon_{ijw}\, g_{ikl} \;=\; & \frac{1}{2}\{\, d \cdot (\, \delta_{jk}\, \delta_{wl} \;-\; \delta_{jl}\, \delta_{wk}\,)\,\} \\ & + \frac{1}{2}\{\, \epsilon_{jwi}\, c_i\, \delta_{kl} \;-\; \epsilon_{jwl}\, c_k\,\} \\ & + \frac{1}{2}\{\, 2\, d_{jl}\, \delta_{wk} \;-\; 2\, d_{wl}\, \delta_{jk}\,\}\,. \end{aligned} \tag{11.F.13}$$

By multiplication with g_{kjm} comes

$$\begin{aligned}\epsilon_{ijw}\, g_{ikl}\, g_{kjm} \;=\; & \frac{1}{2}\{ -\, d \cdot g_{wlm} \;+\; \epsilon_{jwi}\, c_i\, g_{ljm}\,\} \\ & + \frac{1}{2}\{ -\, \epsilon_{jwl}\, c_k\, g_{kjm} \;+\; 2\, d_{jl}\, g_{wjm}\,\} \end{aligned} \tag{11.F.14}$$

and by further multiplication with ϵ_{lmp} we have

$$\begin{aligned} N_{wp} \;=\; & \frac{1}{4}\{ -\, d \cdot g_{wlm}\, \epsilon_{lmp} \;+\; \epsilon_{jwi}\, c_i\, g_{ljm}\, \epsilon_{lmp}\,\} \\ & + \frac{1}{4}\{ -\, \epsilon_{jwl}\, \epsilon_{lmp}\, c_k\, g_{kjm} \;+\; 2\, d_{jl}\, g_{wjm}\, \epsilon_{lmp}\,\}\,. \end{aligned} \tag{11.F.15}$$

We compute the four terms separately. From (11.F.13), by contracting with respect to the indices w and l, we obtain

$$\epsilon_{ijl}\, g_{ikl} \;=\; \frac{1}{2}\{\, 2\, d_{jk} \;+\; \epsilon_{jki}\, c_i\,\}\,. \tag{11.F.16}$$

For the first term in (11.F.15) with (11.F.16), we now have

$$-\, d \cdot g_{wlm}\, \epsilon_{lmp} \;=\; -\,\frac{d}{2}\,(\, 2\, d_{wp} \;+\; \epsilon_{pwi}\, c_i\,)\,. \tag{11.F.17}$$

For the second term in (11.F.15), with (11.F.16), we get

$$\epsilon_{jwi}\, c_i\, g_{ljm}\, \epsilon_{lmp} \;=\; c_i\, \epsilon_{iwj}\, d_{pj} \;+\; \frac{1}{2}\left(c^2\, \delta_{wp} \;-\; c_w\, c_p \right)\,. \tag{11.F.18}$$

For the third term, with (11.F.13), we obtain

$$\begin{aligned} -\, \epsilon_{jwl}\, \epsilon_{lmp}\, c_k\, g_{kjm} \;=\; & -\,\frac{1}{2}\,(\, d \cdot \epsilon_{kwp}\, c_k \;+\; c_w\, c_p\,) \\ & -\,\frac{1}{2}\left(c^2\, \delta_{wp} \;-\; 2\, d_{wm}\, \epsilon_{kmp}\, c_k \right)\,. \end{aligned} \tag{11.F.19}$$

The fourth term needs (11.E.14) as

$$g_{wjm} = \frac{1}{2}\,(\,d_{rw}\,\epsilon_{rjm} + d_{rj}\,\epsilon_{rmw} - d_{rm}\,\epsilon_{rwj}\,) + \frac{1}{2}\,(\,\delta_{jm}\,c_w - \delta_{wm}\,c_j\,)\,. \tag{11.F.20}$$

We now compute the fourth term as

$$g_{wjm}\,\epsilon_{lmp} = d_{pw}\,\delta_{jl} - d_{lw}\,\delta_{jp} + d_{lj}\,\delta_{wp} - d_{pj}\,\delta_{wl} + \frac{1}{2}\,d\cdot(\,\delta_{wl}\,\delta_{jp} - \delta_{wp}\,\delta_{jl}\,) + \frac{1}{2}\,[\,\epsilon_{ljp}\,c_w - \epsilon_{lwp}\,c_j\,]\,. \tag{11.F.21}$$

For the fourth term, we then obtain

$$2\,d_{jl}\,g_{wjm}\,\epsilon_{lmp} = 3\,d\cdot d_{pw} - d^2\,\delta_{pw} - 4\,d_{pl}\,d_{lw} + 2\,d_{jl}\,d_{lj}\,\delta_{wp} - d_{jl}\,\epsilon_{lwp}\,c_j\,. \tag{11.F.22}$$

By collecting the four expressions, we then get

$$\begin{aligned} 4\,N_{wp} = &-\frac{d}{2}\,(\,2\,d_{wp} + \epsilon_{pwi}\,c_i\,) \\ &+ c_i\,\epsilon_{iwj}\,d_{pj} + \frac{1}{2}\,\left(c^2\,\delta_{wp} - c_w\,c_p\right) \\ &- \frac{1}{2}\,(\,d\cdot\epsilon_{kwp}\,c_k + c_w\,c_p\,) \\ &- \frac{1}{2}\,\left(c^2\,\delta_{wp} - 2\,d_{wm}\,\epsilon_{kmp}\,c_k\right) \\ &+ 3\,d\cdot d_{pw} - d^2\,\delta_{pw} - 4\,d_{pl}\,d_{lw} \\ &+ 2\,d_{jl}\,d_{lj}\,\delta_{wp} - d_{jl}\,\epsilon_{lwp}\,c_j\,. \end{aligned} \tag{11.F.23}$$

This gives

$$N_{wp} = \frac{1}{4}\,\{\,2\,d\cdot d_{pw} - c_w\,c_p + c_i\,(\,\epsilon_{iwj}\,d_{pj} + \epsilon_{ijp}\,d_{wj}\,)\,\} + \frac{1}{4}\{\,-c_j\,d_{jl}\,\epsilon_{lwp} + (\,2\,d_{jl}\,d_{lj} - d^2\,)\,\delta_{pw} - 4\,d_{pl}\,d_{lw}\,\}\,. \tag{11.F.24}$$

Calculation of Λ_{wp}

With (11.F.9), we then find from (11.F.10) and (11.F.11)

$$
\begin{aligned}
-\Lambda_{wp} &= M_{wp} + N_{wp} \\
&= \frac{1}{4}\{-8\, d_{pl}\, d_{lw} - 2\, c_i\, (\epsilon_{iwj}\, d_{pj} + \epsilon_{ipj}\, d_{wj})\} \\
&\quad + \frac{1}{4}\{(2\, d_{jl}\, d_{lj} - d^2 - c^2)\, \delta_{pw} + 4\, d \cdot d_{pw}\} \\
&\quad + \frac{1}{4}\{d \cdot \epsilon_{pwi}\, c_i - d_{jl}\, \epsilon_{lwp}\, c_j\} \\
&\quad + \frac{1}{4}\{c_i\, (\epsilon_{iwj}\, d_{pj} + \epsilon_{ijp}\, d_{wj})\}\,. \qquad (11.F.25)
\end{aligned}
$$

The last three terms in this expression are skew-symmetric in indices w and p. Their vanishing is a consequence of the Jacobi identity (11.C.5), which takes the form (11.E.10). We check this by computing the skew-symmetric part of Λ_{wp} by

$$
-\frac{1}{2}\, \Lambda_{wp}\, \epsilon_{wpq} = -\frac{1}{2}\, d_{iq}\, c_i = 0\,. \qquad (11.F.26)
$$

This result was expected because of the symmetry of the Riemann tensor under interchange of its first and last pairs of indices. With $d_{iq}\, c_i = 0$, we then have

$$
\begin{aligned}
\Lambda_{wp} &= 2\, d_{wl}\, d_{lp} - d \cdot d_{wp} + \frac{1}{4}\, (d^2 + c^2 - 2\, d_{ij}\, d_{ji})\, \delta_{wp} \\
&\quad + \frac{1}{2}\, c_i\, (\epsilon_{iwj}\, d_{pj} + \epsilon_{ipj}\, d_{wj})\,. \qquad (11.F.27)
\end{aligned}
$$

For the trace, we obtain

$$
\Lambda_{pp} = \frac{1}{2}\, d_{ij}\, d_{ji} - \frac{1}{4}\, d^2 + \frac{3}{4}\, c^2\,. \qquad (11.F.28)
$$

G. Riemann Tensor

So far we have not made any assumptions about the Riemann tensor apart from postulating homogeneity for the three-space. We want to find the relations between Λ_{wp} and the Riemann tensor. This will also give us the Ricci tensor and Ricci scalar. Following Pauli [Pauli, 1981], we define the Riemann tensor by

$$\boxed{\boldsymbol{\Omega}_{ij} = \frac{1}{2} R_{ijlm}\, \boldsymbol{\omega}_l \wedge \boldsymbol{\omega}_m} \,. \tag{11.G.1}$$

With (11.F.1) and (11.F.2), we obtain from equations (11.F.3a,b)

$$\boldsymbol{\Omega}_w = \Lambda_{wp}\, \boldsymbol{\chi}_p = \frac{1}{4} \epsilon_{wij}\, R_{ijlm}\, \epsilon_{lmp}\, \boldsymbol{\chi}_p \,, \tag{11.G.2}$$

which defines

$$\boxed{\Lambda_{wp} \equiv \frac{1}{4} \epsilon_{wij}\, R_{ijlm}\, \epsilon_{lmp}} \,. \tag{11.G.3}$$

We can then express the Riemann tensor in the three-dimensional case by means of the Λ_{wp} by putting

$$\boxed{R_{uvqr} \equiv \epsilon_{uvw}\, \Lambda_{wp}\, \epsilon_{pqr}} \,. \tag{11.G.4}$$

H. Ricci Tensor

Defining the Ricci tensor (again following Pauli) as

$$\boxed{R_{vr} \equiv R_{qvqr}} \,, \tag{11.H.1}$$

we obtain

$$\begin{aligned} R_{vr} &= \epsilon_{qvw}\, \Lambda_{wp}\, \epsilon_{pqr} \\ &= \epsilon_{qvw}\, \epsilon_{pqr}\, \Lambda_{wp} \\ &= (\, \delta_{vr}\, \delta_{wp} - \delta_{vp}\, \delta_{wr}\,)\, \Lambda_{wp} \\ &= \delta_{vr}\, \Lambda_{pp} - \Lambda_{rv} \,. \end{aligned} \tag{11.H.2}$$

Further contraction gives

$$\boxed{R = 2\, \Lambda_{pp}} \,. \tag{11.H.3}$$

From (11.F.27) and (11.F.28) we obtain the Ricci tensor as

$$\begin{aligned} R_{vr} &= \left[\frac{1}{2} d_{ij}\, d_{ji} - \frac{1}{4}\, d^2 + \frac{3}{4}\, c^2 \right] \delta_{vr} \\ &\quad - 2\, d_{vl}\, d_{lv} + d \cdot d_{vr} \\ &\quad - \frac{1}{4} \left[d^2 + c^2 - 2\, d_{ij}\, d_{ji} \right] \\ &\quad - \frac{1}{2}\, c_i \left[\epsilon_{ivj}\, d_{rj} + \epsilon_{irj}\, d_{vj} \right] \\ &= \left[d_{ij}\, d_{ji} - \frac{1}{2}\, d^2 + \frac{1}{2}\, c^2 \right] \delta_{vr} \\ &\quad + d \cdot d_{vr} - 2\, d_{vl}\, d_{lv} \\ &\quad - \frac{1}{2} c_i \left[\epsilon_{ivj}\, d_{rj} + \epsilon_{irj}\, d_{vj} \right], \end{aligned} \tag{11.H.4}$$

where we have used (11.E.5) and the additional abbreviations (11.F.6) to compress the contraction of (11.E.9b).

I. Ricci Tensor Eigenvectors ($c_r \neq 0$)

If the vector c_r is different from zero, we obtain from the Ricci tensor by multiplication with c_r

$$R_{vr}\, c_r = \left[d_{ij}\, d_{ji} - \frac{1}{2}\, d^2 + \frac{1}{2}\, c^2 \right] c_v \tag{11.I.1}$$

which says that c_r is an eigenvector of the Ricci tensor belonging to the eigenvalue

$$d_{ij}\, d_{ji} - \frac{1}{2}\, d^2 + \frac{1}{2}\, c^2 . \tag{11.I.2}$$

Choosing a frame such that

$$\boxed{c_r = c\, \delta_{r3}, \qquad c > 0} . \tag{11.I.3}$$

We have d_{ij} in the form

$$d_{ij} = \begin{bmatrix} d_{11} & d_{12} & 0 \\ d_{21} & d_{22} & 0 \\ 0 & 0 & 0 \end{bmatrix} \tag{11.I.4}$$

since

$$c\, d_{i3} = 0\,. \tag{11.I.5}$$

A rotation about the 3-axis diagonalizes d_{ij} to

$$d_{ij} = \begin{bmatrix} a & 0 & 0 \\ 0 & b & 0 \\ 0 & 0 & 0 \end{bmatrix}. \tag{11.I.6}$$

We then get

$$d = a + b, \qquad d_{ij}\, d_{ji} = a^2 + b^2 \tag{11.I.7}$$

and we obtain for the eigenvalue (11.I.2) of the Ricci tensor

$$\begin{aligned} d_{ij}\, d_{ji} - \frac{1}{2} d^2 + \frac{1}{2} c^2 &= a^2 + b^2 - \frac{1}{2}(a + b)^2 + \frac{1}{2} c^2 \\ &= \frac{1}{2}\left[(a - b)^2 + c^2\right]. \end{aligned} \tag{11.I.8}$$

The components R_{13} and R_{23} of the Ricci tensor vanish. We obtain from (11.H.4) the expressions

$$\begin{aligned} R_{11} &= \frac{1}{2}\left[(a - b)^2 + c^2\right] + (a + b)\, a - 2\, a^2 \\ &= \frac{1}{2}\left[(a - b)^2 + c^2\right] + a\,(b - a) \\ &= \frac{1}{2}\left[b^2 + c^2 - a^2\right], \end{aligned} \tag{11.I.9a}$$

$$\begin{aligned} R_{22} &= \frac{1}{2}\left[(a - b)^2 + c^2\right] + (a + b)\, b - 2\, b^2 \\ &= \frac{1}{2}\left[(a - b)^2 + c^2\right] + b\,(a - b) \\ &= \frac{1}{2}\left[a^2 + c^2 - b^2\right], \end{aligned} \tag{11.I.9b}$$

$$\begin{aligned} R_{12} &= -\frac{1}{2} c\,(\epsilon_{312}\, b + \epsilon_{321}\, a) \\ &= \frac{1}{2} c\,(a - b)\,. \end{aligned} \tag{11.I.9c}$$

We have thus far for the Ricci tensor ($c > 0$)

$$R_{vr} = \begin{bmatrix} [b^2 + c^2 - a^2]/2 & c(a - b)/2 & 0 \\ c(a - b)/2 & [a^2 + c^2 - b^2]/2 & 0 \\ 0 & 0 & [(a - b)^2 + c^2]/2 \end{bmatrix} \tag{11.I.10}$$

The eigenvalues of the Ricci tensor ($c > 0$) are thus given by

$$\frac{1}{2}\left[c^2 \pm (a-b)\sqrt{(a+b)^2 + c^2} \right], \tag{11.I.11a}$$

$$\frac{1}{2}\left[(a-b)^2 + c^2 \right]. \tag{11.I.11b}$$

J. Ricci Tensor Eigenvectors ($c_r = 0$)

Turning now to to the case where the vector c_r vanishes, we observe that the Jacobi identities are fulfilled and the matrix d_{ij} can be assumed to be in diagonal form

$$d_{ij} = \begin{bmatrix} a' & 0 & 0 \\ 0 & b' & 0 \\ 0 & 0 & c' \end{bmatrix}. \tag{11.J.1}$$

We then see, from (11.H.4), that R_{vr} will also be in diagonal form. We have

$$d_{ij}\, d_{ji} - \frac{1}{2} d^2 = a'^2 + b'^2 + c'^2 - \frac{1}{2}(a' + b' + c')^2 \tag{11.J.2}$$

and

$$d \cdot d_{vr} - 2\, d_{vl}\, d_{lr} = \tag{11.J.3}$$

$$\begin{bmatrix} (b' + c' - a')\, a' & 0 & 0 \\ 0 & (a' + c' - b')\, b' & 0 \\ 0 & 0 & (a' + b' - c')\, c' \end{bmatrix}.$$

We thus obtain for the Ricci tensor ($c = 0$)

$$R_{vr} = \frac{1}{2}\begin{bmatrix} (b' - c')^2 - a'^2 & 0 & 0 \\ 0 & (a' - c')^2 - b'^2 & 0 \\ 0 & 0 & (a' - b')^2 - c'^2 \end{bmatrix}. \tag{11.J.4}$$

All this is quite general.

K. Homogeneous Field Components

We now want to express the components of the homogeneous fields, the g_{ijk}, in terms of the parameters a, b, c or a', b', c'. It is clear that the Jacobi identity will also result in an algebraic restriction for the g_{ijk}. We restate (11.C.2)

$$\boxed{2\, g_{ijk} \;=\; c_{ijk} \,+\, c_{jki} \,-\, c_{kij}}\,, \tag{11.K.1}$$

and (11.A.10)

$$\boxed{c_{ijk} \;\equiv\; g_{ijk} \,-\, g_{ikj}}\,. \tag{11.K.2}$$

The Jacobi identity (11.C.5) gives for $n = 3$

$$(\, g_{ijk} \,-\, g_{ikj}\,)\, \epsilon_{jlm}\, (\, g_{klm} \,-\, g_{kml}\,) \;=\; 0\,. \tag{11.K.3}$$

We define a tensor h_{pk}

$$\boxed{h_{pk} \;\equiv\; \frac{1}{2}\epsilon_{pij}\, g_{ijk}}\,. \tag{11.K.4}$$

This gives

$$\boxed{g_{uvk} \;=\; \epsilon_{uvp}\, h_{pk}}\,. \tag{11.K.5}$$

We split h_{pk} into its symmetric and skew-symmetric parts and call the symmetric part s_{pk}

$$h_{pk} \;=\; \frac{1}{2}\,(\, h_{pk} \,+\, h_{kp}\,) \,+\, \frac{1}{2}\,(\, h_{pk} \,-\, h_{kp}\,)\,, \tag{11.K.6a}$$

$$\boxed{s_{pk} \;\equiv\; \frac{1}{2}\,(\, h_{pk} \,+\, h_{kp}\,)}\,. \tag{11.K.6b}$$

Contracting (11.K.5) with respect to v and k gives

$$\boxed{\epsilon_{uvp}\, h_{pv} \;=\; g_{uvv} \;\equiv\; g_u}\,. \tag{11.K.7}$$

Multiplying by $\frac{1}{2}\epsilon_{uqr}$ then gives

$$\frac{1}{2}\, g_u\, \epsilon_{uqr} \;=\; \frac{1}{2}\, \epsilon_{uqr}\, \epsilon_{uvp}\, h_{pv} \;=\; \frac{1}{2}\,(\, h_{rq} \,-\, h_{qr}\,)\,. \tag{11.K.8}$$

We then obtain from (11.K.5)

$$\begin{aligned} g_{uvk} \;&=\; \epsilon_{uvp}\, s_{pk} \,+\, \frac{1}{2}\, \epsilon_{uvp}\, g_t\, \epsilon_{tkp} \\ &=\; \epsilon_{uvp}\, s_{pk} \,+\, \frac{1}{2}\,(\, \delta_{ut}\, \delta_{vk} \,-\, \delta_{uk}\, \delta_{vt}\,)\, g_t \\ &=\; \epsilon_{uvp}\, s_{pk} \,+\, \frac{1}{2}\,(\, g_u\, \delta_{vk} \,-\, g_v\, \delta_{uk}\,)\,. \end{aligned} \tag{11.K.9}$$

This gives

$$g_{ijk} - g_{ikj} = + \epsilon_{ijp}\, s_{pk} + \frac{1}{2}\,(\, g_i\, \delta_{jk} - g_j\, \delta_{ik}\,)$$
$$- \epsilon_{ikp}\, s_{pj} - \frac{1}{2}\,(\, g_i\, \delta_{kj} - g_k\, \delta_{ij}\,)$$
$$= \epsilon_{ijp}\, s_{pk} - \epsilon_{ikp}\, s_{pj}$$
$$+ \frac{1}{2}\,(\, g_k\, \delta_{ij} - g_j\, \delta_{ik}\,)\,. \tag{11.K.10}$$

Multiplication with ϵ_{jlm} gives

$$(\, g_{ijk} - g_{ikj}\,)\, \epsilon_{jlm} = (\, \delta_{lp}\, \delta_{mi} - \delta_{li}\, \delta_{mp}\,)\, s_{pk}$$
$$- \epsilon_{ikp}\, s_{pj}\, \epsilon_{jlm}$$
$$+ \frac{1}{2}\,(\, g_k\, \epsilon_{ilm} - g_j\, \epsilon_{jlm}\, \delta_{ik}\,)\,. \tag{11.K.11}$$

We have

$$(\, g_{klm} - g_{kml}\,) = \epsilon_{klq}\, s_{qm} - \epsilon_{kmq}\, s_{ql} + \frac{1}{2}\,(\, g_m\, \delta_{kl} - g_l\, \delta_{km}\,)\,. \tag{11.K.12}$$

After multiplication with (11.K.11) and defining

$$\boxed{s \equiv s_{jj}}\,, \tag{11.K.13}$$

a calculation gives

$$(\, g_{ijk} - g_{ikj}\,)\, \epsilon_{jlm}\, (\, g_{klm} - g_{kml}\,) = 2\,(\, s \cdot g_i - g_l s_{li})\,. \tag{11.K.14}$$

The Jacobi identity (11.C.5) becomes

$$s_{il}\, g_l - s \cdot g_i = 0 \tag{11.K.15}$$

or

$$\boxed{(\, s_{il} - s \cdot \delta_{il}\,)\, g_l = 0}\,. \tag{11.K.16}$$

It remains to establish the relationship between s_{il} and g_i on the one hand and with d_{rs} and c_r on the other. Equation (11.E.5a) defines c_r as

$$\boxed{c_r \equiv c_{kkr}}\,. \tag{11.K.17}$$

We then have, from (11.A.8) and (11.A.10), and with (11.K.7)

$$c_r = c_{kkr} = -g_{iri} = g_{rii} \equiv g_r \, . \tag{11.K.18}$$

Thus we have determined that

$$\boxed{c_r = g_r} \tag{11.K.19}$$

(and, incidentally, that $c^2 = g^2$). We have, from (11.A.10), (11.E.12), and (11.K.12) with (11.K.19),

$$\begin{aligned} c_{klm} &= \epsilon_{klq}\, s_{qm} - \epsilon_{kmq}\, s_{ql} + \frac{1}{2}\,(c_m\, \delta_{kl} - c_l\, \delta_{km}) \\ &= \epsilon_{rlm}\, d_{kr} + \frac{1}{2}\,(c_m\, \delta_{kl} - c_l\, \delta_{km}) \end{aligned} \tag{11.K.20}$$

or

$$\epsilon_{rlm}\, d_{rk} = \epsilon_{klq}\, s_{qm} - \epsilon_{kmq}\, s_{ql} \, . \tag{11.K.21}$$

Multiplication with $\frac{1}{2}\epsilon_{plm}$ gives

$$\begin{aligned} d_{pk} &= \frac{1}{2}\, \epsilon_{plm}\, \epsilon_{rlm}\, d_{rk} \\ &= \frac{1}{2}\,[\, (\delta_{pk}\, \delta_{mq} - \delta_{pq}\, \delta_{mk})\; s_{qm} - (\delta_{pq}\, \delta_{lk} - \delta_{pk}\, \delta_{lq})\; s_{ql}\,] \\ &= \frac{1}{2}\,[\, \epsilon_{plm}\, \epsilon_{klq}\, s_{qm} - \epsilon_{plm}\, \epsilon_{kmq}\, s_{ql}\,] \\ &= \frac{1}{2}\,[\, s \cdot \delta_{pk} - s_{pk} - s_{pk} + s \cdot \delta_{pk}\,] \\ &= s \cdot \delta_{pk} - s_{pk} \, . \end{aligned} \tag{11.K.22}$$

We have thus

$$d_{pk} = s \cdot \delta_{pk} - s_{pk} \, , \tag{11.K.23}$$

which, upon contracting, gives

$$\boxed{d = 2\, s}\, , \tag{11.K.24}$$

and

$$\boxed{s_{pk} = \frac{1}{2}\, d \cdot \delta_{pk} - d_{pk}}\, . \tag{11.K.25}$$

L. Ricci Tensor in Field Strengths

We now express the Ricci tensor in terms of the field strengths. We have from (11.H.4), (11.K.19), (11.K.23), and (11.K.24)

$$\begin{aligned} d_{ij}d_{ji} - \frac{1}{2}d^2 &= (s \cdot \delta_{ij} - s_{ij})(s \cdot \delta_{ji} - s_{ji})\, 2\, s^2 \\ &= s_{ij}\, s_{ji} - s^2 \end{aligned} \qquad (11.L.1)$$

and

$$\begin{aligned} R_{vr} = &\left(s_{ij}\, s_{ji} - s^2 + \frac{1}{2}c^2 \right) \delta_{vr} \\ &- 2\, s_{vl}\, s_{lr} + 2\, s \cdot s_{vr} \\ &+ \frac{1}{2} g_i \left(\epsilon_{ivj}\, s_{rj} + \epsilon_{irj}\, s_{vj} \right) . \end{aligned} \qquad (11.L.2)$$

Field Strengths for c = 0

We first consider the case $c = 0$. We can then assume

$$s_{pk} = \begin{bmatrix} \alpha & 0 & 0 \\ 0 & \beta & 0 \\ 0 & 0 & \gamma \end{bmatrix} . \qquad (11.L.3)$$

This gives, for $c = 0$,

$$R_{jk} = -\, 2 \begin{bmatrix} \beta\,\gamma & 0 & 0 \\ 0 & \alpha\,\gamma & 0 \\ 0 & 0 & \alpha\,\beta \end{bmatrix} , \qquad (11.L.4)$$

a result of remarkable simplicity. It says that the eigenvalues of $-\frac{1}{2}$ times the Ricci tensor are obtained from the eigenvalues of the field strength by Poncelet's Cremona transformation, that is,

$$R_{11} : R_{22} : R_{33} = \frac{1}{\alpha} : \frac{1}{\beta} : \frac{1}{\gamma} . \qquad (11.L.5)$$

By interpreting α, β, and γ as projective coordinates in a plane, we have an essentially one-to-one map from field strengths to curvatures. (See **Appendix G** or any book on algebraic geometry, for example [Griffiths, 1978], for further details on the Cremona transformation.)

The result above shows that the components of the field strength are, in fact, the variables of choice. This leads one to suspect that one might use the s_{pk} right from the beginning to derive the Ricci tensor (and not compute it) by expressing the field strengths in terms of the structure constants. Such a calculation will also provide a check for (11.L.2). This is done in the **Section O** of this chapter to avoid a digression.

M. Ricci Tensor (continued)

We now return to the discussion of the Ricci tensor. We now assume $c > 0$. Since s is an eigenvalue of s_{jk} belonging to the eigenvector g_j, the other two eigenvalues of s_{jk} must be opposite, say ρ and $-\rho$. We then obtain in a frame in which s_{jk} is diagonal from (11.L.2) or (11.O.19)

$$\left(s_{ij}\, s_{ji} \;-\; s^2 \;+\; \frac{1}{2}\, c^2 \right) \delta_{vr} \;-\; 2\, s_{vl}\, s_{lr} \;+\; 2\, s \cdot s_{vr}$$

$$= \begin{bmatrix} \frac{1}{2}\, c^2 \;+\; 2\, \rho\, s & 0 & 0 \\ 0 & \frac{1}{2}\, c^2 \;-\; 2\, \rho\, s & 0 \\ 0 & 0 & \frac{1}{2}\, c^2 \;+\; 2\, \rho^2 \end{bmatrix} . \tag{11.M.1}$$

We further have

$$g_i\, \epsilon_{i1j}\, s_{2j} \;=\; -\, c\, \rho \;=\; g_i\, \epsilon_{i2j}\, s_{1j}\,, \tag{11.M.2}$$

all other indices v, r vanish for this expression in the Ricci tensor. From (11.L.2) or (11.O.19), we get

$$R_{jk} \;=\; \begin{bmatrix} \frac{1}{2}\, c^2 \;+\; 2\, \rho\, s & -\, c\, \rho & 0 \\ -\, c\, \rho & \frac{1}{2}\, c^2 \;-\; 2\, \rho\, s & 0 \\ 0 & 0 & \frac{1}{2}\, c^2 \;+\; 2\, \rho^2 \end{bmatrix} . \tag{11.M.3}$$

For the eigenvalues, λ, of R_{vr}, we obtain

$$\left(\frac{1}{2}\, c^2 \;-\; \lambda_\pm \right)^2 \;-\; 4\, \rho^2\, s^2 \;=\; c^2\, \rho^2\,, \tag{11.M.4a}$$

$$\lambda_0 \;=\; \frac{1}{2}\, c^2 \;+\; 2\, \rho^2\,, \tag{11.M.4b}$$

or

$$\lambda_\pm \;=\; \frac{1}{2}\, c^2 \;\pm\; \rho\, \sqrt{c^2 \;+\; 4\, s^2}\,, \tag{11.M.5a}$$

$$\lambda_0 \;=\; \frac{1}{2}\, c^2 \;+\; 2\, \rho^2\,. \tag{11.M.5b}$$

N. Homogeneous Gravitational Fields in Euclidean $\mathbb{R}^3$

We now turn to a discussion of the results and can answer a question that arose at the end of **Section 7.A**, where we discussed homogeneous fields in the Euclidean $\mathbb{R}^3$.

The Ricci tensor in the Euclidean $\mathbb{R}^3$ vanishes. We thus have, from (11.M.4b), that the constant c has to vanish and therefore the vector g_j also must vanish as a consequence of (11.K.19). It then follows, from (11.L.4), that two of the eigenvalues of s_{pk} have to vanish, which leaves us with (11.L.3) reduced to

$$s_{pk} = \gamma\, \delta^3{}_p\, \delta^3{}_k\,. \tag{11.N.1}$$

We then have, from (11.O.5),

$$g_{ijk} = \gamma\, \epsilon_{ij3}\, \delta^3{}_k \tag{11.N.2}$$

which agrees with equations (7.A.55) if we put $\gamma = 1/\beta$ there.

O. ADDENDUM: Ricci Tensor Calculation from Field Strengths

We now try to derive the Ricci tensor directly from the field strengths. We have

$$\boldsymbol{\omega}_{ij} = g_{ijk}\, \boldsymbol{\omega}_k\,, \tag{11.A.7}$$

and also

$$\begin{aligned} d\boldsymbol{\omega}_{ij} &= g_{ijk}\, d\boldsymbol{\omega}_k \\ &= g_{ijk}\, \boldsymbol{\omega}_{km} \wedge \boldsymbol{\omega}_m \\ &= g_{ijk}\, g_{kml}\, \boldsymbol{\omega}_l \wedge \boldsymbol{\omega}_m\,, \end{aligned} \tag{11.O.1}$$

and

$$\boldsymbol{\omega}_{ik} \wedge \boldsymbol{\omega}_{kj} = g_{ikl}\, g_{kjm}\, \boldsymbol{\omega}_l \wedge \boldsymbol{\omega}_m\,, \tag{11.B.2}$$

Thus (11.A.4) becomes

$$\boxed{\boldsymbol{\Omega}_{ij} = (\, g_{ijk}\, g_{kml} - g_{ikl}\, g_{kjm}\,)\, \boldsymbol{\omega}_l \wedge \boldsymbol{\omega}_m}\,. \tag{11.O.2}$$

This gives

$$\boldsymbol{\Omega}_w = \frac{1}{2}\epsilon_{wij}\, \boldsymbol{\Omega}_{ij} = \frac{1}{2}\epsilon_{wij}\,(\, g_{ijk}\, g_{kml} - g_{ikl}\, g_{kjm}\,)\, \epsilon_{lmp}\, \boldsymbol{\chi}_p\,, \tag{11.O.3}$$

where we reiterate (11.F.2) and (11.F.3a)

$$\boldsymbol{\chi}_p \, \epsilon_{lmp} \equiv \boldsymbol{\omega}_l \wedge \boldsymbol{\omega}_m \,, \tag{11.F.2}$$

and

$$\boxed{\boldsymbol{\Omega}_w \equiv \Lambda_{wp} \, \boldsymbol{\chi}_p} \,. \tag{11.F.3a}$$

We thus have

$$\Lambda_{wp} = \frac{1}{2} \, \epsilon_{wij} \left(g_{ijk} \, g_{kml} - g_{ikl} \, g_{kjm} \right) \epsilon_{lmp} \,. \tag{11.O.4}$$

Defining

$$\boxed{g_{ijk} \equiv \epsilon_{ijp} \, s_{pk} + \frac{1}{2} \left(g_i \, \delta_{jk} - g_j \, \delta_{ik} \right)} \,, \tag{11.O.5}$$

where

$$s_{pk} = s_{kp} \,, \tag{11.O.6}$$

we get

$$\begin{aligned} \frac{1}{2} \, \epsilon_{wij} \, g_{ijk} &= s_{wk} + \frac{1}{4} \left(g_i \, \epsilon_{wik} - g_j \, \epsilon_{wkj} \right) \\ &= s_{wk} + \frac{1}{2} \, g_i \, \epsilon_{wik} \,, \end{aligned} \tag{11.O.7}$$

and

$$\begin{aligned} g_{kml} \, \epsilon_{lmp} &= \left[\epsilon_{kmq} \, s_{ql} + \frac{1}{2} \left(g_k \, \delta_{ml} - g_m \, \delta_{kl} \right) \right] \epsilon_{lmp} \\ &= \left(\delta_{kl} \, \delta_{qp} - \delta_{kp} \, \delta_{ql} \right) s_{ql} - \frac{1}{2} \, g_m \, \epsilon_{kmp} \\ &= s_{pk} - s \cdot \delta_{kp} - \frac{1}{2} \, g_m \, \epsilon_{kmp} \,. \end{aligned} \tag{11.O.8}$$

For the first term in Λ_{wp}, we thus have

$$\begin{aligned} \frac{1}{2} \, \epsilon_{wij} \, g_{ijk} \, g_{kml} \, \epsilon_{lmp} = \; & s_{wk} \, s_{pk} - s \cdot s_{wp} \\ & + \frac{1}{4} \, g^2 \, \delta_{wp} - \frac{1}{4} \, g_w \, g_p \\ & + \frac{1}{2} \, g_i \left(\epsilon_{wik} \, s_{pk} + \epsilon_{pik} \, s_{wk} \right) \\ & - \frac{1}{2} \, s \cdot g_i \, \epsilon_{ipw} \,. \end{aligned} \tag{11.O.9}$$

For the determination of the second term in Λ_{wp}, we first calculate

$$\begin{aligned} g_{kil}\, g_{kjm} &= \delta_{ij}\, s_{rl}\, s_{rm} - s_{jl}\, s_{im} \\ &+ \frac{1}{2}\, g_k \left(\epsilon_{kiq}\, s_{ql}\, \delta_{jm} + \epsilon_{kjq}\, s_{qm}\, \delta_{il} \right) \\ &- \frac{1}{2} \left(g_j\, \epsilon_{miq}\, s_{ql} + g_i\, \epsilon_{ljq}\, s_{qm} \right) \\ &+ \frac{1}{4} \left(g^2\, \delta_{il}\, \delta_{jm} + g_i\, g_j\, \delta_{lm} \right) \\ &- \frac{1}{4} \left(\delta_{il}\, g_j\, g_m + \delta_{jm}\, g_i\, g_l \right). \end{aligned} \tag{11.O.10}$$

Multiplication from the right with ϵ_{lmp} gives

$$\begin{aligned} g_{kil}\, g_{kjm}\, \epsilon_{lmp} &= - s_{jl}\, s_{im}\, \epsilon_{lmp} \\ &+ \frac{1}{2}\, g_k \left(\epsilon_{kiq}\, \epsilon_{ljp}\, s_{ql} + \epsilon_{kjq}\, \epsilon_{ilp}\, s_{ql} \right) \\ &+ \frac{1}{2}\, g_j \left(s_{pi} - s \cdot \delta_{ip} \right) - \frac{1}{2}\, g_i \left(s_{pj} - s \cdot \delta_{jp} \right) \\ &+ \frac{1}{4}\, g^2\, \epsilon_{ijp} - \frac{1}{4}\, g_l \left(g_i\, \epsilon_{ljp} - g_j\, \epsilon_{lip} \right). \end{aligned} \tag{11.O.11}$$

Multiplication from the left with $\frac{1}{2}\epsilon_{wij}$ gives

$$\begin{aligned} \frac{1}{2}\, \epsilon_{wij}\, g_{kil}\, g_{kjm}\, \epsilon_{lmp} &= - \frac{1}{2}\, s_{jl}\, s_{im}\, \epsilon_{wij}\, \epsilon_{lmp} \\ &+ \frac{1}{2}\, g_k\, \epsilon_{wij}\, \epsilon_{kiq}\, \epsilon_{ljp}\, s_{ql} \\ &+ \frac{1}{2}\, g_j \left(s_{pi} - s \cdot \delta_{ip} \right) \epsilon_{wij} \\ &+ \frac{1}{4}\, g^2\, \delta_{wp} - \frac{1}{4}\, g_l\, g_i\, \epsilon_{wij}\, \epsilon_{ljp}. \end{aligned} \tag{11.O.12}$$

We thus, after some intermediate work, obtain

$$\begin{aligned}\frac{1}{2}\,\epsilon_{wij}\,g_{kil}\,g_{kjm}\,\epsilon_{lmp} = &-\frac{1}{2}\,\{\,s\cdot s_{pw} - s_{mw}\,s_{pm}\,\}\\ &-\frac{1}{2}\,\{\,s\cdot s_{pw} - s_{pi}\,s_{iw}\,\}\\ &-\frac{1}{2}\,\{\,\delta_{wp}\,s_{jl}\,s_{lj} - s^2\,\delta_{wp}\,\}\\ &+\frac{1}{2}\,g_j\,s_{pi}\,\epsilon_{wij} - \frac{1}{2}\,s\cdot\,g_j\,\epsilon_{wpj}\\ &+\frac{1}{2}\,g_w\,\epsilon_{lqp}\,s_{ql} - \frac{1}{2}\,g_j\,\epsilon_{ljp}\,s_{wl}\\ &+\frac{1}{4}\,g_w\,g_p\,.\end{aligned} \tag{11.O.13}$$

From (11.O.4) and (11.O.9), we now have

$$\begin{aligned}\Lambda_{wp} = &\; s_{wk}\,s_{kp} - s\cdot s_{wp} + \frac{1}{4}\,g^2\,\delta_{wp} - \frac{1}{4}\,g_w\,g_p\\ &-\frac{1}{2}\,s\cdot g_i\,\epsilon_{ipw} + \frac{1}{2}\,g_i\,(\,\epsilon_{wik}\,s_{pk} + \epsilon_{pik}\,s_{wk}\,)\\ &- s\cdot s_{wp} + s_{wk}\,s_{kp} + \frac{1}{2}\,\delta_{wp}\,(\,s^2 - s_{jl}\,s_{lj}\,)\\ &+\frac{1}{2}\,g_j\,(\,\epsilon_{wlj}\,s_{pl} - \epsilon_{plj}\,s_{wl}\,)\\ &-\frac{1}{2}\,s\cdot g_j\,\epsilon_{wpj} + \frac{1}{4}\,g_w\,g_p\,.\end{aligned} \tag{11.O.14}$$

This gives

$$\begin{aligned}\Lambda_{wp} = &\; 2\,s_{wk}\,s_{kp} - 2\,s\cdot s_{wp}\\ &+\delta_{wp}\left(\frac{1}{2}\,s^2 - \frac{1}{2}\,s_{jl}\,s_{lj} + \frac{1}{4}\,g^2\right)\\ &+\frac{1}{2}\,g_i\,(\,\epsilon_{wik}\,s_{pk} + \epsilon_{pik}\,s_{wk}\,)\\ &+\frac{1}{2}\,g_j\,(\,\epsilon_{wlj}\,s_{pl} - \epsilon_{plj}\,s_{wl}\,)\,.\end{aligned} \tag{11.O.15}$$

Only the last term is skew-symmetric. Its vanishing is equivalent to

$$\begin{aligned}0 &= g_j\,\epsilon_{wlj}\,\epsilon_{wpt}\,s_{pl}\\ &= g_j\,(\,\delta_{lp}\,\delta_{jt} - \delta_{lt}\,\delta_{jp}\,)\,s_{pl}\\ &= g_j\,(\,s\cdot\delta_{jt} - s_{jt}\,)\,,\end{aligned} \tag{11.O.16}$$

which agrees with (11.K.16). We then have

$$\begin{aligned}\Lambda_{wp} &= 2\, s_{wk}\, s_{kp} \;-\; 2\, s \cdot s_{wp} \\ &\quad + \delta_{wp} \left(\frac{1}{2}\, s^2 \;-\; \frac{1}{2}\, s_{jl}\, s_{jl} \;+\; \frac{1}{4}\, g^2 \right) \\ &\quad + \frac{1}{2}\, g_i \left(\epsilon_{wik}\, s_{pk} \;+\; \epsilon_{pik}\, s_{wk} \right) \end{aligned} \tag{11.O.17}$$

and

$$\begin{aligned}\Lambda_{pp} &= 2\, s_{jl}\, s_{jl} \;-\; 2\, s^2 \;+\; \frac{3}{2}\, s^2 \;-\; \frac{3}{2}\, s_{jl}\, s_{jl} \;+\; \frac{3}{4}\, g^2 \\ &= \frac{1}{2} \left(s_{jl}\, s_{jl} \;-\; s^2 \;+\; \frac{3}{2}\, g^2 \right) . \end{aligned} \tag{11.O.18}$$

We obtain the Ricci tensor from

$$\begin{aligned}R_{wp} &= \Lambda_{jj}\, \delta_{wp} \;-\; \Lambda_{wp} \\ &= \delta_{wp} \left(s_{jl}\, s_{jl} \;-\; s^2 \;+\; \frac{1}{2}\, g^2 \right) \\ &\quad + 2\, s \cdot s_{wp} \;-\; 2\, s_{wk}\, s_{kp} \\ &\quad + \frac{1}{2}\, g_i \left(\epsilon_{iwk}\, s_{pk} \;+\; \epsilon_{ipk}\, s_{wk} \right) . \end{aligned} \tag{11.O.19}$$

This again agrees with (11.L.2).

P. ADDENDUM: Ricci Tensor and Ricci Coefficients - The Vector Free Case

We calculate the following expression for the vector-free case

$$\epsilon_{ipq}\,\epsilon_{jmn}\,R_{pm}\,R_{qn}$$

$$= \epsilon_{ipq}\,\epsilon_{jmn}\,\epsilon_{pab}\,\epsilon_{mcd}\,\gamma_{ac}\,\gamma_{bd}\,\epsilon_{qef}\,\epsilon_{ngh}\,\gamma_{eg}\,\gamma_{fh}$$

$$= \epsilon_{ipq}\,\epsilon_{jmn}\,\epsilon_{pab}\,\epsilon_{mcd}\,\epsilon_{qef}\,\epsilon_{ngh}\,\gamma_{ac}\,\gamma_{bd}\,\gamma_{eg}\,\gamma_{fh}$$

$$= \left(\epsilon_{ipq}\,\epsilon_{pab}\,\epsilon_{qef}\right)\,\left(\epsilon_{jmn}\,\epsilon_{mcd}\,\epsilon_{ngh}\right)\,\gamma_{ac}\,\gamma_{bd}\,\gamma_{eg}\,\gamma_{fh}$$

$$= \left[\left(\delta_{qa}\,\delta_{ib} - \delta_{qb}\,\delta_{ia}\right)\epsilon_{qef}\right]\,\left[\left(\delta_{nc}\,\delta_{jd} - \delta_{nd}\,\delta_{jc}\right)\epsilon_{ngh}\right] \times \gamma_{ac}\,\gamma_{bd}\,\gamma_{eg}\,\gamma_{fh}$$

$$= \left[\epsilon_{aef}\,\epsilon_{cgh}\,\delta_{ib}\,\delta_{jd} + \epsilon_{bef}\,\epsilon_{dgh}\,\delta_{ia}\,\delta_{jc} - \epsilon_{bef}\,\epsilon_{cgh}\,\delta_{ia}\,\delta_{jd} + \epsilon_{aef}\,\epsilon_{dgh}\,\delta_{ib}\,\delta_{jc}\right] \times \gamma_{ac}\,\gamma_{bd}\,\gamma_{eg}\,\gamma_{fh}$$

$$= 2\left[\gamma_{ij}\left(\epsilon_{bef}\,\epsilon_{dgh}\,\gamma_{bd}\,\gamma_{eg}\,\gamma_{fh}\right) - \gamma_{ic}\,\gamma_{bj}\,\epsilon_{bef}\,\epsilon_{cgh}\,\gamma_{eg}\,\gamma_{fh}\right]$$

$$= 2\left(6\,\gamma_{ij}\,\det[\gamma_{ij}] - \gamma_{ic}\,\gamma_{jb}\,R_{bc}\right), \tag{11.P.1}$$

that is,

$$6\,\gamma_{ij}\,\det[\gamma_{ij}] - \gamma_{ic}\,\gamma_{jb}\,R_{bc} - \frac{1}{2}\,\epsilon_{ipq}\,\epsilon_{jmn}\,R_{pm}\,R_{qn} = 0\,. \tag{11.P.2}$$

From the equation

$$\gamma_{ic}\,R_{cb} = -\,2\,\delta_{ib}\,\det[\gamma_{ij}]\,, \tag{11.P.3}$$

we get

$$\gamma_{ij} = -\,\frac{1}{16\,\det[\gamma_{ij}]}\,\epsilon_{ipq}\,\epsilon_{jmn}\,R_{pm}\,R_{qn}\,. \tag{11.P.4}$$

CHAPTER 12

HOMOGENEOUS FIELDS IN ARBITRARY DIMENSION

A. Discussion of Homogeneous Gravitational Fields

The interest in homogeneous fields derives from the fact that they may provide a local second order approximation to arbitrary Riemannian manifolds of less than six dimensions. If P is an arbitrary point of an n-dimensional pseudo-Riemannian manifold, we can introduce Riemannian normal coordinates, centered on P. For the metric tensor in a neighborhood of P, we obtain

$$ds^2 = \eta_{ab}\, dx^a\, dx^b + \frac{1}{3} R_{abce}(x^f)\, x^a\, x^c\, dx^b\, dx^e\,, \tag{12.A.1}$$

where $a, b, c, e, f, \cdots \in \{1, \ldots, n\}$. We obtain, therefore, a second order approximation to the metric by representing the functions $R_{abce}(x^f)$, which are assumed to have the symmetries of the Riemann tensor, by their values at the point P, that is, $x^f = 0$.

A homogeneous field is characterized by the constants g_{abc} which are defined by the equations

$$\omega_{ab} = g_{abc}\, \omega^c\,, \qquad g_{abc} = -\, g_{bac}\,, \tag{12.A.2}$$

for the connection forms ω_{ab}. The metric has a second order approximation at P in terms of a homogeneous field if the equations

$$R_{abce}(0) = g_{abf}\left[g^f{}_{ce} - g^f{}_{ec}\right] + g_{fae}\, g^f{}_{bc} - g_{fac}\, g^f{}_{be}\,, \tag{12.A.3}$$

for a given left hand side, have a solution in terms of constants g_{abc}.

One will expect that solutions of (12.A.3) will in general no longer exist for arbitrary left hand sides (with the symmetries of the Riemann tensor) since we then have more equations than unknowns. The number of independent equations is

$$N = \frac{n^2\,(n^2 - 1)}{12}\,, \tag{12.A.4}$$

while the number of unknowns, V, is

$$V = \frac{n^2\,(n - 1)}{2}\,. \tag{12.A.5}$$

Thus, we have

$$N - V = \frac{n^2\,(n - 1)}{12}(n + 1 - 6)\,. \tag{12.A.6}$$

We define structure constants $c^f{}_{ab}$ by

$$\boxed{c^f{}_{ab} \equiv g^f{}_{ab} - g^f{}_{ba}}\,. \tag{12.A.7}$$

If the constants g_{abc} solve (12.A.3), it follows from the well known symmetry of the Riemann tensor

$$\boxed{R_{abcd} + R_{acdb} + R_{adbc} = 0} \tag{12.A.8}$$

that the Jacobi identity is fulfilled. Thus, the structure constants define a locally transitive Lie group of motions and the homogeneous field is a homogeneous space.

The Riemann tensor is quadratic in the field strengths and, roughly, one can say that the field strengths are the square roots of the curvature. A study of the field strengths will result in a better understanding of curvature since the field strengths are—from a mathematical and physical point of view—the simpler quantities. One can compare them with the external curvatures that result from an imbedding of the manifold into a higher dimensional flat space. But unless this imbedding is rigid the external curvatures are largely arbitrary.

To view the field strengths in lower dimensions as intrinsic, and therefore, physical quantities one has to discuss the two problems of existence and uniqueness for the field strengths. For the existence problem we have already mentioned that one has to restrict oneself to the dimensions $n < 6$ if one is seeking a general solution. For $n = 2$ and definite metric one has the exceptional case of the 2–sphere which cannot carry a homogeneous field. While postponing the discussion of the uniqueness problem, we now survey the existence problem for $n = 3$ based on the developments in this chapter. In this case, the Riemann tensor is equivalent to the Ricci tensor and field strengths can be found by a Cremona transformation of the eigenvalues of the Ricci tensor. [**See Appendix G.**] There is also a second solution for a set of "twisted" field strengths where the vector g_i does not vanish. This brings us to the question of uniqueness.

For Minkowski space it was found that non-vanishing field strengths exist besides the standard solution of zero field strength. This non-uniqueness of field strength in the Minkowski case is interesting from the physicist's point of view. It points to the existence of homogeneous gauge transformations by which source-free—so called "fictitious" gravitational fields—can be transformed away. One has here a coordinate independent formulation for homogeneous gravitational fields which entered into the early formulation of Einstein's theory of gravitation. A further study of such fields appears useful for several reasons.

First, one needs to know about the transformations (homogeneous gravitational gauge transformations) for an understanding of the uniqueness problem.

Second, traditionally one studies in Newtonian mechanics motion in rotating frames and discusses Coriolis and centrifugal forces. The homogeneous gauge transformations allow us to extend these treatments to special relativistic mechanics. That can be useful for a systematic discussion of quantum effects in fictitious gravitational fields.

Third, and this is the case we find most interesting, one may learn something for the interpretation of gauge theories of the Kaluza type from a study of the $n = 5$ case in a five-dimensional Minkowski space. The expectation is that one will be able to show that constant electric and magnetic fields can be transformed away in a similar way as homogeneous gravitational fields in the four-dimensional case. The analog of the Einstein elevator here becomes the Kaluza elevator that is built from matter of the same e/m–ratio as that of the particles one studies in it.

B. Homogeneous Fields in N Dimensions

Curvature Forms

The line element for an n-dimensional Riemannian or pseudo-Riemannian space is given by

$$ds^2 = \eta_{jk}\, \omega^j\, \omega^k\,, \qquad \eta_{jk} = \eta_{kj} = constant\,. \tag{12.B.1}$$

We assume that the symmetric constant matrix, η_{jk} is non-singular, that is, it satisfies

$$\det \lfloor \eta_{jk} \rfloor \neq 0 \,. \tag{12.B.2}$$

Lower case Latin indices run from 1 to n.

We define the connection forms, $\boldsymbol{\omega}^j{}_k$, for vanishing torsion by

$$d\boldsymbol{\omega}^j = \boldsymbol{\omega}^j{}_k \wedge \boldsymbol{\omega}^k \tag{12.B.3}$$

with the anti-symmetry condition

$$\boldsymbol{\omega}_{jk} + \boldsymbol{\omega}_{kj} = 0 \,, \qquad \text{and} \qquad \boldsymbol{\omega}_{kl} \equiv \eta_{lj}\, \boldsymbol{\omega}^j{}_k \,, \tag{12.B.4}$$

where our symmetric constant matrix will be used to raise and lower indices.

A *homogeneous field* is defined by expressing the connection forms in terms of the basis

$$\boldsymbol{\omega}_{jk} \equiv g_{jkl}\, \boldsymbol{\omega}^l \,, \qquad \text{where} \qquad g_{jkl} = -\, g_{kjl} \,, \tag{12.B.5}$$

and requiring $g_{jkl} = \textit{constant}$. We have then, using (12.B.3),

$$\begin{aligned} d\boldsymbol{\omega}_{jk} &= g_{jkp}\, d\boldsymbol{\omega}^p \\ &= g_{jkp}\, \boldsymbol{\omega}^p{}_m \wedge \boldsymbol{\omega}^m \\ &= g_{jkp}\, g^p{}_{ml}\, \boldsymbol{\omega}^l \wedge \boldsymbol{\omega}^m \,, \end{aligned} \tag{12.B.6}$$

and

$$\boldsymbol{\omega}_{jp} \wedge \boldsymbol{\omega}^p{}_k = g_{jpl}\, g^p{}_{km}\, \boldsymbol{\omega}^l \wedge \boldsymbol{\omega}^m \,. \tag{12.B.7}$$

We define the curvature forms

$$\begin{aligned} \boldsymbol{\Omega}_{jk} &\equiv d\boldsymbol{\omega}_{jk} - \boldsymbol{\omega}_{jp} \wedge \boldsymbol{\omega}^p{}_k \\ &= (\, g_{jkp}\, g^p{}_{ml} - g_{jpl}\, g^p{}_{km}\,)\, \boldsymbol{\omega}^l \wedge \boldsymbol{\omega}^m \,. \end{aligned} \tag{12.B.8}$$

Riemann Tensor and the Jacobi Identity

The Riemann tensor is defined by

$$\boldsymbol{\Omega}_{jk} = \frac{1}{2}\, R_{jklm}\, \boldsymbol{\omega}^l \wedge \boldsymbol{\omega}^m \,. \tag{12.B.9}$$

This gives, in terms of the field strengths,

$$R_{jklm} = g_{jkp}\, (\, g^p{}_{ml} - g^p{}_{lm}\,) - g_{jpl}\, g^p{}_{km} + g_{jpm}\, g^p{}_{kl} \,. \tag{12.B.10}$$

By defining a new set of constants from the field strengths

$$c^p{}_{ml} \equiv g^p{}_{ml} - g^p{}_{lm}\,, \tag{12.B.11}$$

we can write the Riemann tensor in three cyclic, but entirely equivalent, expressions:

$$R_{jklm} = -g_{jkp}\,c^p{}_{lm} + g^p{}_{kl}\,g_{jpm} - g_{jpl}\,g^p{}_{km}\,, \tag{12.B.12}$$

$$R_{jlmk} = -g_{jlp}\,c^p{}_{mk} + g^p{}_{lm}\,g_{jpk} - g_{jpm}\,g^p{}_{kl}\,, \tag{12.B.13}$$

$$R_{jmkl} = -g_{jmp}\,c^p{}_{kl} + g^p{}_{mk}\,g_{jpl} - g_{jpk}\,g^p{}_{ml}\,. \tag{12.B.14}$$

Because of the cyclic symmetry of the Riemann tensor (12.A.8), addition gives

$$\begin{aligned} 0 = &- g_{jkp}\,c^p{}_{lm} + g_{jpk}\,c^p{}_{lm} \\ &- g_{jlp}\,c^p{}_{mk} + g_{jpl}\,c^p{}_{mk} \\ &- g_{jmp}\,c^p{}_{kl} + g_{jpm}\,c^p{}_{kl}\,. \end{aligned} \tag{12.B.15}$$

We can write this equation as

$$\boxed{0 = c^j{}_{pk}\,c^p{}_{lm} + c^j{}_{pl}\,c^p{}_{mk} + c^j{}_{pm}\,c^p{}_{kl}}\,. \tag{12.B.16}$$

This is the Jacobi identity for the structure constants. A homogeneous field that solves equations (12.B.10) then gives rise to a locally transitive Lie group which has the ω^j as left-invariant differential forms. This follows from the fact that

$$d\omega^j = \omega^j{}_k \wedge \omega^k = g^j{}_{kl}\,\omega^l \wedge \omega^k = -\frac{1}{2}\,c^j{}_{kl}\,\omega^k \wedge \omega^l\,. \tag{12.B.17}$$

These are the Maurer-Cartan equations. Homogeneous fields always give homogeneous metrics and vice versa.

The $n^2(n-1)/2$ components g_{ijk} of a homogeneous field are thus always subject to the $n^2(n-1)(n-2)/2$ equations (12.B.16).

"Straight" and "Twisted Fields"

By contraction, we derive, from the Jacobi identity (12.B.16),

$$0 = c^j{}_{pj}\, c^p{}_{lm} + c^j{}_{pl}\, c^p{}_{mj} + c^j{}_{pm}\, c^p{}_{jl}\,. \tag{12.B.18}$$

From the antisymmetry of the structure constants, $c^i{}_{jk}$, the sum of the last two terms vanishes and with the definition

$$\boxed{c_p \equiv c^j{}_{jp}} \tag{12.B.19}$$

we have that (12.B.18) now becomes

$$c_p\, c^p{}_{lm} = 0\,. \tag{12.B.20}$$

We now express these conditions in terms of the field strengths. We have

$$c_p = g^j{}_{jp} - g^j{}_{pj} = -g^j{}_{pj} = -g_{jp}{}^j = g_{pj}{}^j \equiv g_p\,, \tag{12.B.21}$$

where we have defined the vector

$$\boxed{g_p \equiv g_{pj}{}^j}\,. \tag{12.B.22}$$

The equation (12.B.20) can now be written as

$$g_p\,(g^p{}_{lm} - g^p{}_{ml}) = 0\,. \tag{12.B.23}$$

We will call fields with vanishing vector g_p "straight" and those with non-zero g_p "twisted". The vector g_p will be called the *twist vector*.

Ricci Tensor

We now define the Ricci tensor by contracting the Riemann tensor

$$R_{km} \equiv R^j{}_{kjm} = g^j{}_{kp}\,(g^p{}_{mj} - g^p{}_{jm}) - g^j{}_{pj}\, g^p{}_{km} + g^j{}_{pm}\, g^p{}_{kj}\,. \tag{12.B.24}$$

The second and fourth terms on the right hand side of the equation cancel each other and we have

$$R_{km} = g^j{}_{kp}\, g^p{}_{mj} + g_p\, g^p{}_{km}\,. \tag{12.B.25}$$

We thus obtain in the straight case ($g_p = 0$)

$$R_{km} = g_k{}^j{}_p \, g_m{}^p{}_j \, . \tag{12.B.26}$$

We now define the tensor h_{jkl} which has the same symmetry properties as g_{jkl} but is traceless. We put

$$\boxed{h_{jkl} \equiv g_{jkl} - \frac{1}{n-1} \left(g_j \, \eta_{kl} - g_k \, \eta_{jl} \right)} . \tag{12.B.27}$$

We then have

$$g^j \, h_{jkl} = g^j \, g_{jkl} - \frac{1}{n-1} \left(g_j \, g^j \, \delta_{kl} - g_k \, g_l \right) . \tag{12.B.28}$$

This gives

$$g^j \left(h_{jkl} - h_{jlk} \right) = g^j \left(g_{jkl} - g_{jlk} \right) = 0 \tag{12.B.29}$$

from the contracted Jacobi identity. We obtain for the Ricci tensor

$$\begin{aligned}
R_{km} &= \left[h^j{}_{kp} + \frac{1}{n-1} \left(g^j \, \eta_{kp} - \delta^j{}_p \, g_k \right) \right] \\
&\quad \times \left[h^p{}_{mj} + \frac{1}{n-1} \left(g^p \, \eta_{mj} - \delta^p{}_j \, g_m \right) \right] \\
&\quad + g_p \, h^p{}_{km} + \frac{1}{n-1} \left(g_j \, g^j \eta_{km} - g_k \, g_m \right) \\
&= h^j{}_{kp} \, h^p{}_{mj} + \frac{1}{n-1} \left(g^j \, h_{kmj} + g^p \, h_{mkp} \right) + g_p \, h^p{}_{km} \\
&\quad + \frac{1}{n-1} \left(g_j \, g^j \, \eta_{km} - g_k \, g_m \right) \\
&\quad + \left(\frac{1}{n-1} \right)^2 \left[g_k \, g_m + n \, g_k \, g_m - g_k \, g_m - g_k \, g_m \right] \\
&= h^j{}_{kp} \, h^p{}_{mj} + g_p \, h^p{}_{km} + \frac{1}{n-1} \, g_j \, g^j \, \eta_{km} \, .
\end{aligned} \tag{12.B.30}$$

We interrupt the general development now and study special cases which will revisit some of the work in previous chapters in a more general context.

C. Homogeneous Fields in Two Dimensions

First we will look at $n - 2$ again. We have

$$h_{12}{}^{1} = g_{12}{}^{1} - \left(g_1 \, \delta_2{}^{1} - g_2 \, \delta_1{}^{1} \right) = 0 \,,$$
$$h_{12}{}^{2} = g_{12}{}^{2} - \left(g_1 \, \delta_2{}^{2} - g_2 \, \delta_1{}^{2} \right) = 0 \,. \tag{12.C.1}$$

Thus

$$R_{km} = g_j \, g^j \, \delta_{km} \,. \tag{12.C.2}$$

The spaces are of constant curvature K. The Jacobi identities give no condition. We then have

$$K = - g_j \, g^j \,. \tag{12.C.3}$$

This equation always has solutions except for $K > 0$ and positive definite metrics. Of particular interest are the solutions for $K = 0$ and the Minkowski metric which exist for arbitrary null vectors g^j and give rise to homogeneous gravitational fields. We postpone a detailed discussion until later.

D. Homogeneous Fields in Three Dimensions

Now turn to the case of $n = 3$. We put

$$h_{jkp} = \epsilon_{jkp} \, s^p{}_l \,, \qquad \text{with} \;\; s_{pq} = s_{qp} \tag{12.D.1}$$

and where we have for a totally skew-symmetric ϵ_{jkp}

$$\epsilon_{123} = +1 \tag{12.D.2}$$

and

$$\epsilon_{jkl} \, \epsilon^{pqr} = \det \left[\eta_{ab} \right] \cdot \det \begin{bmatrix} \delta^p{}_j & \delta^q{}_j & \delta^r{}_j \\ \delta^p{}_k & \delta^q{}_k & \delta^r{}_k \\ \delta^p{}_l & \delta^q{}_l & \delta^r{}_l \end{bmatrix} , \tag{12.D.3}$$

which defines ϵ^{pqr}. Calling

$$\boxed{\eta \equiv \det \left[\eta_{ab} \right]} , \tag{12.D.4}$$

we obtain by contraction

$$\epsilon_{jkl} \, \epsilon^{pkl} = 2 \, \eta \, \delta_j{}^p \,, \tag{12.D.5a}$$
$$\epsilon_{jkl} \, \epsilon^{pql} = \eta \, (\delta_j{}^p \, \delta_k{}^q - \delta_k{}^p \, \delta_j{}^q) \,. \tag{12.D.5b}$$

For $n = 3$ the contracted Jacobi identity is equivalent to the uncontracted. Using (12.D.1), we obtain from (12.B.29) by multiplying with $(\eta/2)\epsilon^{klm}$

$$
\begin{aligned}
0 &= \frac{\eta}{2} g^j \epsilon^{klm} \left(\epsilon_{jkp}\, s^p{}_l - \epsilon_{jlp}\, s^p{}_k \right) \\
&= \frac{1}{2} g^j \left\{ \left(\delta^l{}_p \delta^m{}_j - \delta^l{}_j \delta^m{}_p \right) s^p{}_l - \left(\delta^k{}_j \delta^m{}_p - \delta^k{}_p \delta^m{}_j \right) s^p{}_k \right\} \\
&= \frac{1}{2} g^j \left\{ \delta_j{}^m s^l{}_l - s^m{}_j - s^m{}_j + \delta_j{}^m s^l{}_l \right\} \\
&= g^j \left(s^l{}_l \delta^m{}_j - s^m{}_j \right) .
\end{aligned}
\tag{12.D.6}
$$

For straight fields the Jacobi identities give no condition. For twisted fields we find that the twist vector g^j is an eigenvector of the tensor $s^m{}_j$ with the trace of $s^m{}_j$ as its eigenvalue.

For $n = 3$ the Ricci tensor is equivalent to the Riemann tensor. We find for the first term in (12.B.30) with (12.D.1)

$$
\begin{aligned}
h^{jk}{}_p h^p{}_{mj} &= \epsilon^{jkq}\, s_q{}^p\, \epsilon_{pmr}\, s^r{}_j \\
&= \eta\, s_q{}^p s^r{}_j \Big\{ \delta^j{}_p \delta^k{}_m \delta^q{}_r + \delta^j{}_m \delta^k{}_r \delta^q{}_p \\
&\qquad + \delta^j{}_r \delta^k{}_p \delta^q{}_m - \delta^j{}_p \delta^k{}_r \delta^q{}_m \\
&\qquad - \delta^j{}_m \delta^k{}_p \delta^q{}_r - \delta^j{}_r \delta^k{}_m \delta^q{}_p \Big\} \\
&= \eta \Big\{ s^j{}_r s^r{}_j \delta^k{}_m + s^p{}_p s^k{}_m + s^p{}_p s^k{}_m \\
&\qquad - s^j{}_m s^k{}_j - s^k{}_r s^r{}_m - \delta^k{}_m s^r{}_r s^p{}_p \Big\} .
\end{aligned}
\tag{12.D.7}
$$

We put

$$
\boxed{s \equiv s^k{}_k}
\tag{12.D.8}
$$

and now have

$$
h^{jk}{}_p h^p{}_{mj} = \eta \left\{ \delta^k{}_m \left(s^j{}_r s^r{}_j - s^2 \right) - 2\, s^k{}_j s^j{}_m + s\, s^k{}_m \right\} .
\tag{12.D.9}
$$

This gives for the Ricci tensor (12.B.30)

$$
\begin{aligned}
R^k{}_m &= \left[\eta \left(s^j{}_r s^r{}_j - s^2 \right) + \frac{1}{2} g_j g^j \right] \delta^k{}_m \\
&\qquad + 2\eta \left(s\, s^k{}_m - s^k{}_j s^j{}_m \right) + g_p \epsilon^{pkq} s_{qm} ,
\end{aligned}
\tag{12.D.10}
$$

with the eigenvalue problem

$$0 = g^j \left(s\, \delta^m{}_j - s^m{}_j \right) . \tag{12.D.11}$$

By contraction of (12.D.10) we obtain

$$R = \frac{3}{2}\, g_j\, g^j + \eta \left(s^j{}_r\, s^r{}_j - s^2 \right) . \tag{12.D.12}$$

By multiplying (12.D.10) with g_k we have

$$g_k\, R^k{}_m = g_m \left[\eta \left(s^j{}_r\, s^r{}_j - s^2 \right) + \frac{1}{2}\, g_j\, g^j \right] . \tag{12.D.13}$$

This equation states that for twisted fields, g_k is an eigenvector of the Ricci tensor belonging to the eigenvalue ρ_3

$$\rho_3 = \eta \left(s^j{}_r\, s^r{}_j - s^2 \right) + \frac{1}{2}\, g_j\, g^j . \tag{12.D.14}$$

It then follows from (12.D.12) that

$$\rho_1 + \rho_2 = g_j\, g^j \tag{12.D.15}$$

for the sum of the other two eigenvalues ρ_1 and ρ_2.

E. Vanishing Ricci Tensor in Three Dimensions

We now discuss briefly the case of a vanishing Ricci tensor for $n = 3$.

Definite Metrics ($g_j\, g^j > 0$) - Twisted Case

We first deal with the case of a definite metric, that is, Euclidean space, where we assume

$$R_{km} = 0 \tag{12.E.1}$$

and first discuss the "twisted" case

$$g_j\, g^j > 0 \tag{12.E.2}$$

where $g_j \neq 0$. We see from (12.D.15) that this is impossible.

Definite Metrics ($g_j\, g^j = 0$) - Straight Case

Assuming now $g_j = 0$, which is the "straight" case, we see from (12.D.12) that

$$s^j{}_r\, s^r{}_j \;-\; s^2 \;=\; 0 \qquad (12.E.3)$$

and from (12.D.10) that

$$s_{kj}\, s_{jm} \;-\; s \cdot s_{km} \;=\; 0\,. \qquad (12.E.4)$$

The contraction of this equation gives (12.E.3). Let α, β, γ be the eigenvalues of s_{km}. We can bring s_{km} into diagonal form by a orthogonal transformation an then have the equations

$$\alpha\,(\beta + \gamma) = 0\,, \qquad (12.E.5a)$$
$$\beta\,(\alpha + \gamma) = 0\,, \qquad (12.E.5b)$$
$$\gamma\,(\alpha + \beta) = 0\,. \qquad (12.E.5c)$$

If α vanishes, we have

$$\beta\,\gamma = 0\,. \qquad (12.E.6)$$

In this case at least two of the eigenvalues vanish. We now assume

$$\alpha \neq 0\,, \qquad \beta + \gamma = 0\,. \qquad (12.E.7)$$

Addition of the last two equations (12.E.5) then gives

$$\beta\,\gamma = 0\,. \qquad (12.E.8)$$

Again, this means that two of the eigenvalues of s_{km} vanish. We obtain as the general solution

$$s_{km} \;=\; \pm\, a_k\, a_m \qquad (12.E.9)$$

with an arbitrary vector a_k. This gives, with (12.B.27) and (12.D.1),

$$g_{jkl} \;=\; \pm\, \epsilon_{jkp}\, a_p\, a_l\,. \qquad (12.E.10)$$

This is the most general solution for a fictitious homogeneous field in the Euclidean $\mathbb{R}^3$. Putting the vector a_p in the 3-axis and writing α for $\pm\,(a_3)^2$, we obtain the non-vanishing components of the field strengths as

$$g_{123} \;=\; -\,g_{213} \;=\; \alpha \;=\; constant\,. \qquad (12.E.11)$$

Pseudo-Euclidean Metrics ($g_j = 0$) - Straight Case

We turn now to the pseudo-Euclidean case with two space dimensions and one time dimension. We first treat the "straight" case

$$g_j = 0\,, \tag{12.E.12}$$

where the vector necessarily has a zero scalar product. We have $\eta = -1$ and obtain from (12.D.12)

$$s^j{}_r\, s^r{}_j \;-\; s^2 \;=\; 0\,. \tag{12.E.13}$$

From (12.D.10) we then obtain

$$s^k{}_j\, s^j{}_m \;-\; s \cdot s^k{}_m \;=\; 0\,. \tag{12.E.14}$$

The equation (12.E.13) is a consequence of (12.E.14).

We first take the case where $s = 0$ and assume that $s^k{}_j$ does not vanish. Then the only possible Jordan normal form of $s^k{}_j$ is

$$\begin{aligned} S' &\equiv \left[\, s'^k{}_j \,\right] \\ &= \begin{bmatrix} 0 & 1 & 0 \\ 0 & 0 & 0 \\ 0 & 0 & 0 \end{bmatrix} \\ &= A\, S\, A^{-1}\,, \end{aligned} \tag{12.E.15}$$

where $\det\,[\,A\,] \neq 0$. This means that

$$s^k{}_j \;=\; a^k\, b_j \tag{12.E.16}$$

is the product of a contravariant vector with a covariant vector. Pulling down the index "k" with the η_{jk}–metric of (12.B.1) then gives, because of the symmetry,

$$s_{kj} \;=\; a_k\, b_j \;=\; s_{jk} \;=\; a_j\, b_k\,, \tag{12.E.17a}$$

$$a_k\, b_j \;-\; a_j\, b_k \;=\; 0\,, \tag{12.E.17b}$$

which shows that b_k must be a multiple of a_k, for instance

$$b_k \;=\; \lambda\, a_k, \qquad \lambda \neq 0\,. \tag{12.E.18}$$

We then have

$$s^k{}_j = \lambda\, a^k\, a_j\,. \tag{12.E.19}$$

The equation (12.E.14) can only be fulfilled if

$$\lambda^2\, a^k\, a_j\, a^j\, a_m = \left(\lambda^2\, a_j\, a^j\right) s^k{}_m = 0\,. \tag{12.E.20}$$

The vector a^j must, therefore, be a null vector. We thus have

$$s^k{}_j = \pm a^k\, a_j, \qquad a^l\, a_l = 0\,, \tag{12.E.21}$$

where we have re-defined a^k to absorb, up to an overall sign, the factor λ. We also see that the trace condition, $s = 0$, is fulfilled.

Next we consider the case

$$s \neq 0\,. \tag{12.E.22}$$

We first observe that $s^k{}_j$ cannot have an inverse. If it did, we would obtain the equation

$$s^j{}_l - s \cdot \delta^j{}_l = 0 \tag{12.E.23a}$$

and its contraction

$$s - 3\,s = 0 \tag{12.E.23b}$$

from (12.E.14). This would lead to $s = 0$. The determinant of $s^k{}_j$ must, therefore, be zero. Thus, zero is an eigenvalue of $s^k{}_j$. Any vector b^k will do which does not vanish and has been obtained by

$$b^k = s^k{}_j\, c^j \tag{12.E.24}$$

with the some vector c^j. Since $s^k{}_j$ does not vanish, some such c^j must exist. If β is the third eigenvalue of $s^k{}_j$, we have

$$s = \mathrm{trace}\left(s^k{}_j\right) = 0 + \beta + s \tag{12.E.25}$$

which demonstrates that the eigenvalue 0 is doubly degenerate. The possible Jordan normal forms for $s^k{}_j$ are then

$$\left(s'^k{}_j\right) = \begin{bmatrix} 0 & 1 & 0 \\ 0 & 0 & 0 \\ 0 & 0 & s \end{bmatrix} \quad \text{or} \quad \left(s'^k{}_j\right) = \begin{bmatrix} 0 & 0 & 0 \\ 0 & 0 & 0 \\ 0 & 0 & s \end{bmatrix}. \tag{12.E.26}$$

The first of these does not fulfill (12.E.14) for $s'^k{}_j$. It follows then, as before in the case $s = 0$, that

$$s^k{}_j = \pm a^k\, a_j\,, \tag{12.E.27}$$

where the vector a^k can be spacelike or timelike. The null case is excluded since it would lead to a vanishing trace. Summing up this part of the discussion, we find that

$$s'^k{}_j = \pm a^k a_j , \tag{12.E.28}$$

where the vector a^k is arbitrary. (It can also be the zero vector.) We obtain for the field strengths as in (12.E.10)

$$g_{jkl} = \pm \epsilon_{jkp} a^p a_l . \tag{12.E.29}$$

This is the most general solution for the fictitious field (homogeneous) in Minkowski space of three dimensions for vanishing twist. In a previous chapter we have discussed the case of a timelike vector a^k.

Pseudo-Euclidean Metrics ($g_j \neq 0$) - Twisted Case

We now investigate the "twisted" case where $g^j \neq 0$. It follows from equation (12.D.15) that

$$g^j g_j = 0 . \tag{12.E.30}$$

One then obtains from (12.D.12) that

$$s^r{}_j s^j{}_r - s^2 = 0 \tag{12.E.31}$$

and from (12.D.11) that

$$s^m{}_j g^j = s \cdot g^m . \tag{12.E.32}$$

This says that s is an eigenvalue of $s^j{}_m$. The other two eigenvalues of $s^j{}_m$ are, therefore, given by β and $-\beta$ since the sum of the three eigenvalues must be s. We obtain from (12.D.10) the equations

$$0 = 2\eta \left(s \cdot s^k{}_m - s^k{}_j s^j{}_m \right) + g_p \epsilon^{pkq} s_{qm} . \tag{12.E.33}$$

Contraction of these equations leads to (12.E.31). Since the eigenvalues of $s^k{}_j s^j{}_m$ are the squares of the eigenvalues of $s^k{}_m$, we have from (12.E.31) that

$$2\beta^2 + s^2 = s^2 , \tag{12.E.34}$$

or

$$\beta = 0 . \tag{12.E.35}$$

We now introduce a unit vector ξ^j which is defined by the following relations

$$\xi_j \xi^j = 1 , \quad g_j \xi^j = 0 , \quad g_i = \epsilon_{ijk} \xi^j g^k . \tag{12.E.36}$$

These conditions determine ξ^j up to a "gauge" transformation

$$\xi^j \longrightarrow \xi'^j + \lambda\, g^j \,. \tag{12.E.37}$$

We can, however, fix a frame by introducing a third vector h^j by the conditions

$$g_j\, h^j = 2\,, \qquad h_j\, h^j = 0\,, \qquad h_j\, \xi^j = 0\,. \tag{12.E.38}$$

If we put

$$g^j \equiv \mathbf{e}_0{}^j\,, \qquad h^j \equiv \mathbf{e}_1{}^j\,, \qquad \xi^j \equiv \mathbf{e}_2{}^j\,, \tag{12.E.39}$$

we have for their scalar products

$$\eta_{ab} \equiv \mathbf{e}^j{}_a \cdot \mathbf{e}_{jb}\,, \tag{12.E.40}$$

so that in matrix form we have

$$\eta_{ab} = \begin{bmatrix} 0 & 1 & 0 \\ 1 & 0 & 0 \\ 0 & 0 & 1 \end{bmatrix} . \tag{12.E.41}$$

Since the determinant of η_{ab} is different from zero ($\eta = -1$), it follows that the three vectors are linearly independent.

We now assume that s_{jk} has the form

$$\begin{aligned} s_{jk} &= \alpha\, g_j\, g_k + \beta\, h_j\, h_k + \gamma\, \xi_j\, \xi_k \\ &\quad + \delta\, (\, g_j\, h_k + h_j\, g_k\,) + \epsilon\, (\, g_j\, \xi_k + \xi_j\, g_k\,) \\ &\quad + \zeta\, (\, h_j\, \xi_k + \xi_j\, h_k\,)\,. \end{aligned} \tag{12.E.42}$$

Multiplying by g^k, we get from (12.E.32)

$$\begin{aligned} s_{jk}\, g^k &= 2\,\beta\, h_j + 2\,\delta\, g_j + 2\,\zeta\, \xi_j \\ &= (\,\gamma + 4\,\delta\,)\, g_j\,. \end{aligned} \tag{12.E.43}$$

This gives for the trace of s_{jk} the value

$$s = \gamma + 4\,\delta \tag{12.E.44}$$

and for

$$s_{jk}\, s^{jk} = 8\,\alpha\,\beta + \gamma^2 + 8\,\delta^2\,. \tag{12.E.45}$$

This shows that

$$\beta = 0\,, \qquad \zeta = 0\,, \qquad \delta = -2\,\delta\,. \tag{12.E.46}$$

The equation (12.E.31) then gives

$$s_{jk}\, s^{jk} = 12\,\delta^2 = s^2 = 4\delta^2\,. \tag{12.E.47}$$

We have, therefore,

$$s = \gamma = \delta = 0\,. \tag{12.E.48}$$

Collecting results, we obtain

$$s_{jk} = \alpha\, g_j\, g_k + \epsilon\,(\, g_j\, \xi_k + \xi_j\, g_k\,)\,. \tag{12.E.49}$$

We now compute

$$\begin{aligned} s_{jk}\, s^k{}_l &= (\,\alpha\, g_j\, g_k + \epsilon\, g_j\, \xi_k + \epsilon\, \xi_j\, g_k\,) \\ &\quad \times \left(\,\alpha\, g^k\, g_l + \epsilon\, g^k\, \xi_l + \epsilon\, \xi^k\, g_l\,\right) \\ &= \epsilon^2\, g_j\, g_l\,. \end{aligned} \tag{12.E.50}$$

Further, we find

$$\begin{aligned} g^p\, \epsilon_{pkq}\, s^q{}_m &= g^p\, \epsilon_{pkq}\,(\,\alpha\, g^q\, g_m + \epsilon\, g^q\, \xi_m + \epsilon\, \xi^q\, g_m\,) \\ &= \epsilon \cdot \epsilon_{kqp}\, \xi^q\, g^p\, g_m\,. \end{aligned} \tag{12.E.51}$$

This gives with (12.E.36)

$$g^p\, \epsilon_{pkq}\, s^q{}_m = \epsilon\, g_k\, g_m\,. \tag{12.E.52}$$

With (12.E.50) and (12.E.52), the equations (12.E.33) now become

$$\begin{aligned} 0 &= 2\, s_{kj}\, s^j{}_m + g^p\, \epsilon_{pkq}\, s^q{}_m \\ &= \epsilon\,(\,2\,\epsilon + 1\,) \end{aligned} \tag{12.E.53}$$

with the solutions

$$\epsilon = 0\,, \qquad \epsilon = -\frac{1}{2}\,. \tag{12.E.54}$$

The general solutions are given by

$$s_{jk} = \alpha\, g_j\, g_k\,, \tag{12.E.55a}$$

$$s_{jk} = \alpha\, g_j\, g_k - \frac{1}{2}\,(\, g_j\, \xi_k + \xi_j\, g_k\,)\,, \tag{12.E.55b}$$

with arbitrary α and ξ^j subject to conditions (12.E.36). We then obtain from (12.B.27) and (12.D.1)

$$g_{jkl} = \alpha\, \epsilon_{jkp}\, g^p\, g_k + (\, g_j\, \eta_{kl} + g_k\, \eta_{jl}\,) \tag{12.E.56}$$

and

$$g_{jkl} = \alpha\, \epsilon_{jkp}\, g^p\, g_k + \frac{1}{2}\,(\, g_j\, \eta_{kl} - g_k\, \eta_{jl}\,) - \frac{1}{2}\, \epsilon_{jkp}\,(\, g^p\, \xi_k + \xi^p\, g_k\,)\,. \tag{12.E.57}$$

These are the general twisted homogeneous fields for three-dimensional Minkowski space.

CHAPTER 13

SUMMARY

A. Mathematics

The notion of homogeneous fields was introduced by defining them as metrics with constant Ricci coefficients. Such metrics define homogeneous spaces. We described the surprising fact that a flat space like the Euclidean $\mathbb{R}^3$ or Minkowski spaces of all dimensions admit non-trivial homogeneous fields.

Necessary and sufficient for the existence of a local homogeneous field is the existence of a Lie group of n dimensions which is a subgroup of the isometry group of the space and acts locally transitively. The task of finding and classifying all possible types of such subgroups is a difficult problem in the theory of Lie algebras. A solution for the Poincaré group was sketched in **Chapter 10**. However, a proof that there are no other homogeneous fields is still missing.

Thus, from a purely mathematical point of view, much still has to be done in terms of classifications, existence theorems, and uniqueness theorems for the Lie algebras involved. In particular, the revised Bianchi classification by MacCallum [MacCallum, 1979] has been partially carried out for Bianchi A–types [Schücking, 2003A] and for Bianchi B–types [Schücking, 2003B]. See also [Schücking, 2003] for more on classification methodology.

Due to the mathematical complexity of the problem, we discussed thoroughly only the two-dimensional case to get some idea of the properties of homogeneous fields. We then studied a special type of Lie algebra which leads to homogeneous vector fields. Here we found the general solution of the problem in n dimensions.

We saw also that homogeneous fields could be viewed as sections of the orthonormal frame bundle of a pseudo-Riemannian manifold. Having an orthonormal frame in every point means the existence of n unit vectors in each point, orthogonal to each other. Each of these n vector fields generates a one-dimensional group of transformations of the manifold onto itself.

Acting on a point, these transformations create a line through that point. There is a line through each point of the space. Mathematicians call such a line a "congruence".

Therefore, the frames give rise to n congruences. Each frame vector is a unit tangent vector to one of the n congruences. They always intersect each other orthogonally.

If we use only a discrete set of the congruence lines, we can construct for a vacuum a primitive cubic lattice of lines that turns space or spacetime into an infinite jungle-gym. However, as we saw in **Chapter 7**, three-dimensional Euclidean space is homogeneous in more than one way. Besides the group of translations, there is another one generated by the translations in the x-y–plane and screw motions of fixed pitch about the z–axis. The three congruences are now perpendicular systems of parallel straight lines in the planes $z = constant$ and the system of parallels to the z–axis. The lines in the planes $z = constant$ turn around as we climb up the z–axis. The operations of going along a congruence in the plane $z = constant$ and going up parallel to the z–axis no longer commute. The square loop in the primitive lattice ceases to close. The jungle-gym is broken. The bars miss each other. In crystals this phenomenon is known as a screw dislocation. Now the whole $\mathbb{R}^3$ is suffering from this problem.

The fact that the three congruences all consist of straight lines is accidental. If we go to new frames obtained by a constant rotation, giving rise to a different z–axis, we obtain congruences which are helices. The straight lines are simply limiting cases.

Since each line of a given congruence is transformed into itself and its parallel lines by the homogeneity group, it follows that its curvature and torsions must be constant along each line and must be the same for all the lines of the same congruence. Lines of constant curvature and torsions are general helices.

B. Physics

From a physicist's point of view, frames are realized by idealized clocks. The idealized clocks describe, through their permanence, congruences of timelike worldlines if one imagines clocks being everywhere.

Length is nowadays measured in terms of the speed of light which has been defined to be 299,792,458 m/sec. The rate of such a clock is thought to be independent of its acceleration. This assumption has been confirmed for muons in storage rings up to accelerations of $10^{19}g$.

Since most physics experiments are done in accelerated reference frames, for example, an Earth-bound laboratory, one has to define accelerated reference frames. Assuming the validity of special relativity, one is dealing now with timelike congruences in Minkowski space. The simplest curved timelike worldlines are those for which accelerations and torsions are constant. They can be realized experimentally as histories of charged test particles with a given e/m–ratio in a homogeneous electromagnetic field. This holds even in more general spaces if one defines an electromagnetic field as homogeneous if and only if it is covariantly constant.

Again, the simplest step from a single worldline to a congruence is by means of a simply transitive group which contains a one-dimensional subgroup that moves the single worldline into itself.

The known homogeneous fields in Minkowski space seem to be able to give rise to prescribed accelerations and torsions. [**See Chapter 10.**] They appear thus to be able to simulate homogeneous electromagnetic fields in their action on test particles. That was the reason for calling frames with constant Ricci coefficients "*homogeneous fields*".

The timelike congruences which we consider will, in general not fill all of spacetime. There comes a point where particles would have to move with the speed of light to keep up appearances. That is the horizon phenomenon. Mostly, the congruences will not be hypersurface-orthogonal.

A timelike congruence of a homogeneous field provides us with an extended, homogeneous "*reference body*". This body can be realized by test particles in a homogeneous electromagnetic field.

Such reference bodies, whose particles all share the same acceleration and torsions, are not exactly like a piece of copper. For an acceleration g, its particles would drift apart doubling their distance in a year or so. "Real" solids, like a piece of copper, in a state of acceleration at sea level do not disintegrate in a year's time. The interactions between electrons and nuclei prevent that. However, one could use the idealized reference bodies for a formulation of a relativistic elasticity theory.

We saw that a freely falling body seen in a uniformly accelerated frame will show an acceleration in the opposite direction which depends on velocity to the extent of v/c. This effect might be measurable. A thorough discussion of reference frames in actual use is necessary for an analysis of such experiments.

The recent measurements of dark matter, generating deviations from the gravitational force laws used to model galaxy and galactic cluster dynamics as well as gravitational lensing, make it desirable to have a consistent theory of measurements in accelerated frames. A further discussion of homogeneous fields might be of help for this purpose as well as for a formulation of the equivalence principle.

C. Equivalence Principle

Einstein formulated the equivalence principle for the first time as follows [Einstein, 1907]:

> "We consider two reference frames Σ_1 and Σ_2. Let Σ_1 be accelerated in the direction of its X–axis and γ be the amount of the acceleration (constant in time). Let Σ_2 be at rest; but let it be in a homogeneous gravitational field which imparts to all objects the acceleration $-\gamma$ in direction of its X–axis."
>
> "As far as we know, the physical laws with reference to Σ_1 do not differ from those with reference to Σ_2; the reason for this is that all bodies are equally accelerated in a gravitational field. At the present state of our experience we have no reason for the assumption that the systems Σ_1 and Σ_2 are different from each other in any respect, and, therefore, we shall in the following assume the complete physical equivalence of gravitational field and corresponding acceleration of the reference frame."
>
> "This assumption extends the principle of relativity to the case of uniformly accelerated translational motion of the reference frame. The heuristic value of this assumption is that it allows to replace a homogeneous gravitational field by a uniformly accelerated reference frame; this latter case is to a certain degree amenable to theoretical treatment."

We have gone back to the first formulation of the equivalence principle because many later formulations by various authors—and sometimes by the same author at different occasions—do not always appear to be in agreement.

Einstein's treatment following the formulation above was restricted to small accelerations, small velocities, and small time intervals. The theory of homogeneous fields describes uniformly accelerated reference frames for arbitrary constant accelerations and torsions for all velocities and all times. If the principle of equivalence is correct, the homogeneous fields are, in fact, homogeneous gravitational fields, or as we call them, Einstein fields.

APPENDIX A

BASIC CONCEPTS

A. Manifolds

A manifold is a set of objects along with a set of mapping functions, together with their inverses, that enable us to identify individual objects $\mathbf{p}$ with points of $\mathbb{R}^n$. The points of the manifold can be thought of as being labeled with coordinate values

$$\mathbf{p} \Longleftrightarrow x \longleftrightarrow \{x^i\} \longleftrightarrow \left\{ \begin{array}{c} x^0 \\ x^1 \\ \cdot \\ \cdot \\ \cdot \\ x^n \end{array} \right\} \tag{A.A.1}$$

where on the left of the double-headed arrow we have the coordinate-free statement for a point, $\mathbf{p}$, in bold faced notation; on the its right we have its interpretation(s) into successively more specific coordinate language. First we have the general variable, x, and then indexed coordinates, and finally into a column vector of coordinate components.

The points of a manifold are further required to be organized in a model using the terminology derived from cartography. Neighborhoods of manifold points and their associated neighborhoods in $\mathbb{R}^n$ are called "**charts**".

A set of such charts that includes every point of the manifold at least once is called an "**atlas**".

B. Structures on Manifolds

A manifold, M, can have additional mathematical structures or properties attached to it at each point. The manifold serves as a "**base**" for these structures and their properties. Bijection to and from the manifold's coordinate charts allows us to examine the manifold's geometric properties using $\mathbb{R}^n$ as a "**base space**". Familiar concepts from $\mathbb{R}^n$ are used as stepping stones for analysis of manifolds generated by solution's of Einstein's field equations of general relativity.

A system of hypersurfaces $\boldsymbol{\lambda}(\mathbf{p})$ in the manifold provide a coordinate system for the points of a manifold when mapping functions $X_i(\boldsymbol{\lambda})$ and their inverses are provided to locate the mapped point in the $\mathbb{R}^n$. In this way, a set of numbers, the coordinates of each point, can be attached to the point of the manifold.

C. Vectors

If the elements of such an attached structure satisfy the operations that can be performed on a vector space, then we have a vector space at every point of the manifold, that is, we say that it is "attached everywhere". Further, we can study the action of moving from point to point.

We are primarily concerned with two types of vectorial objects: vectors specified at a point, which are denoted $\mathbf{V}(\mathbf{p})$, and linear functions of vectors (or differential forms when their component functions are differentiable), which are denoted $\boldsymbol{\omega}(\mathbf{V}(\mathbf{p}))$. We also concern our selves with linear maps of those linear functions, $\boldsymbol{\vartheta}(\boldsymbol{\omega}(\mathbf{V}(\mathbf{p})))$.

D. The Real Line as Vector Space

The real line can be treated as a vector space. It has only one basis vector:

$$\mathbf{V}(\mathbf{p}) \iff v^1(x)\,\mathbf{e}_1(x)\,, \tag{A.D.1}$$

where the double headed arrow indicates the equivalent expressions for a vector in the abstract vector space on the left and in an explicit basis on the right. For generality, we have allowed the lone base vector, $\mathbf{e}_1(x)$, to depend on position as well as the scaling coefficient, $v^1(x)$.

If we take derivatives, an additional term now appears:

$$\frac{\partial}{\partial x}v^1(x)\,\mathbf{e}_1(x) = \left(\frac{\partial}{\partial x}v^1(x)\right)\mathbf{e}_1(x) + v^1(x)\left(\frac{\partial}{\partial x}\mathbf{e}_1(x)\right), \tag{A.D.2}$$

where we have used the partial derivative in anticipation of generalizing to higher dimensional spaces.

E. Vector Space

In higher dimensional spaces, vectors can be realized as row matrix of functions:

$$\mathbf{V}(\mathbf{p}) \iff \big(e(x^q),\, f(x^q),\, g(x^q),\, h(x^q)\big)$$
$$= \big(e(),\, f(),\, g(),\, h()\big)\,(\,x^q,\, x^q,\, x^q,\, x^q\,)\,, \qquad \text{(A.E.1)}$$

where, for a sense of relevance to our studies, we have written the vector as consisting of four component functions, all of them evaluated at the same manifold point, $\{x^q\}$. In this sense, a vector is a function at a point of a manifold.

Since we may have need of even higher dimensional vectors, we can give the component functions less unique names

$$\mathbf{V}(\mathbf{p}) \iff \big(e^0(\,x^q),\, e^1(\,x^q),\, e^2(\,x^q),\, e^3(\,x^q)\big)$$
$$= \big(\,e^0(\,),\, e^1(\,),\, e^2(\,),\, e^3(\,)\big)\,(\,x^q,\, x^q,\, x^q,\, x^q\,)$$
$$= \big(e^j(x^q)\big)\,. \qquad \text{(A.E.2)}$$

Any element of a vector space can also be expressed in terms of a basis of linearly independent vectors as

$$\mathbf{V}(\mathbf{p}) \iff v^B(x^k)\,\mathbf{e}_B(x^k)\,, \qquad \text{(A.E.3)}$$

where the double headed arrow indicates the equivalent expressions for the same vector in the abstract vector space on the left and in an explicit basis on the right. The vectors $\mathbf{e}_B$ are now basis vectors for the vector space. The Einstein summation convention has been followed: repeated indices are summed over.

F. Frames

When a number of such vectors are attached to a point, they may constitute a "**frame**". We represent a frame by a column matrix of the vectors

$$\mathbf{E}(\mathbf{p}) \iff \{\mathbf{E}_A(x^k)\} = \begin{Bmatrix} \mathbf{E}_0(x^k) \\ \mathbf{E}_1(x^k) \\ \mathbf{E}_2(x^k) \\ \mathbf{E}_3(x^k) \end{Bmatrix}, \qquad \text{(A.F.1)}$$

where the curly braces indicate a column matrix and we have only listed four vectors in the column since generalization to larger spaces is straight forward. When four linearly independent (and possibly orthonormal) vectors are attached at every point, they provide sixteen functions and do constitute a frame if the spacetime is four dimensional.

G. The Covector Space

The "**covector space**" (or "dual space") consists of real valued linear functions of vectors. It too is a vector space. To reduce verbosity, we call these functions "**forms**" or covectors. Other authors sometimes refer to them as "linear functionals". The vector space and the covector space are said to be dual to each other.

We can express any such covector $\boldsymbol{\omega}$ in terms of a basis just like any other kind of vector:

$$\boldsymbol{\omega}(\mathbf{V}(\mathbf{p})) \iff \left[\omega_A(x^k)\,\boldsymbol{\omega}^A(x^k)\right]\left[v^\alpha(x^k)\,\mathbf{e}_\alpha(x^k)\right], \tag{A.G.1}$$

where $\boldsymbol{\omega}^A(x^k)$ are basis forms for the covector space. They have been written as $\boldsymbol{\omega}^A(x^k)$ to emphasize that even the basis elements of the covector space, functions applying to vectors, may themselves be functionally dependent on position. If they are independent of position, then they would be written as $\boldsymbol{\omega}^A(*)$, or as $\boldsymbol{\omega}^A$ when the location independence is readily understood.

We have

$$\begin{aligned}
\boldsymbol{\omega}(\mathbf{V}(\mathbf{p})) &= \omega_A(\mathbf{p})\;\boldsymbol{\omega}^A\Big(\,v^\alpha(\mathbf{p})\;\mathbf{e}_\alpha(\mathbf{p})\Big) && \textit{basis expansion}\\
&\iff \omega_A(x^k)\;\boldsymbol{\omega}^A\Big(v^\alpha(x^k)\,\mathbf{e}_\alpha(x^k)\Big) && \textit{coordinates}\\
&= \omega_A(x^k)\,v^\alpha(x^k)\;\boldsymbol{\omega}^A\Big(\mathbf{e}_\alpha(x^k)\Big) && \textit{linearity}\\
&= \omega_A(x^k)\,v^\alpha(x^k)\;\boldsymbol{\delta}^A{}_\alpha && \textit{orthogonality}\\
&= \omega_j(x^k)\,v^j(x^k)\,. && \textit{sum of scalars}
\end{aligned} \tag{A.G.2}$$

A common computational tactic, irrelevant at the moment, would have been to use distinct indices to denote the coordinates of the very same point in different elements of the expression. We could have used coordinates labeled $\{x^j\}$, $\{x^k\}$, and $\{x^m\}$, different names for the same coordinates, namely those of point $\mathbf{p}$ in a particular coordinate chart of a specific atlas. This is frequently necessary in complex computations to keep track of terms coming from the various pieces of an expression as they are being manipulated. We could have written the otherwise identical expressions

$$\begin{aligned}\boldsymbol{\omega}(\mathbf{V}(\mathbf{p})) \iff & \omega_A(x^p)\, \boldsymbol{\omega}^A\big(v^\alpha(x^q)\, \mathbf{e}_\alpha(x^k)\big) \\ &= \omega_A(x^p)\, v^\alpha(x^q)\, \boldsymbol{\omega}^A\big(\mathbf{e}_\alpha(x^k)\big) \\ &= \omega_A(x^p)\, v^\alpha(x^q)\, \boldsymbol{\delta}^A{}_\alpha \\ &= \omega_j(x^p)\, v^j(x^q) \end{aligned} \tag{A.G.3}$$

instead of those in the right-hand side of (A.G.2).

H. Differentiable Structures

When a vector can be chosen from the structure attached at each point of the manifold in a manner that allows definitions of continuity, we can also examine the change in the vectors from point to point. The change in a vector as we move along a curve gives us the directional derivative

$$D_{\boldsymbol{\lambda}} \mathbf{V}(\mathbf{p}), \tag{A.H.1}$$

where $\mathbf{p}$ is a point of the base manifold, $\boldsymbol{\lambda}$ is a curve in the base manifold, and D indicates the process of measuring the difference in the vector $\mathbf{V}$ along $\boldsymbol{\lambda}$.

The Exterior Derivative

The exterior derivative operation, "d", can be applied to vectors by applying it to each element of the matrix:

$$\begin{aligned} d\mathbf{e}(\mathbf{p}) \iff & d\big(f(x^p),\, g(x^q),\, h(x^r)\big) \\ &= \big(df(x^p),\, dg(x^q),\, dh(x^r)\big) \\ &= \left(\frac{\partial f}{\partial x^p}\, dx^p,\, \frac{\partial g}{\partial x^q}\, dx^q,\, \frac{\partial h}{\partial x^r}\, dx^r\right). \end{aligned} \tag{A.H.2}$$

I. Scalars and Coscalars

The coefficients used to express a vector in terms of a basis are called scalars. This also applies to covectors, though a term such as "coscalars" might be more appropriate. For our purposes, scalars and coscalars are taken from the real number field, $\mathbb{R}$.

J. Forms and Scalar Invariance

A zero-form is an ordinary function taking a point of the base space as its argument. A one-form is a real (scalar) valued linear function of vectors

$$\boldsymbol{\omega}(\mathbf{V}) \equiv \omega_i \, \boldsymbol{\omega}^i (v^k \, \mathbf{e}_k) \,. \tag{A.J.1}$$

Differential one-forms applied to vectors yield a scalar value

$$\boldsymbol{\omega}(\mathbf{V}) = \omega_k \, v^k \,. \tag{A.J.2}$$

Since the basis differential one-forms $\boldsymbol{\omega}_i$ are also vectors, they may be rotated by subjecting them to a linear transformation to rotate the frame $\boldsymbol{\omega}_k$ into one adapted to the task at hand. We have

$$\boldsymbol{\omega}'_j = \mathbf{A}_j{}^k \, \boldsymbol{\omega}_k \,, \qquad \boldsymbol{\omega}_k = [\mathbf{A}^{-1}]_k{}^j \, \boldsymbol{\omega}'_j \tag{A.J.3}$$

where the matrix elements, $A_p{}^q$, representing the linear transformation and its inverse are constants. The line element, when expressed in either of these bases, has the same form

$$ds^2 = \boldsymbol{\omega}_i \, \boldsymbol{\omega}^i = \boldsymbol{\omega}'_j \, \boldsymbol{\omega}'^j \tag{A.J.4}$$

in each such basis of one-forms. The inverse of the transformation matrix satisfies

$$\mathbf{A}_j{}^i \left(\eta_{ik} \, [\mathbf{A}^{-1}]_l{}^k \right) = \eta_{jl} \tag{A.J.5}$$

so that it is an orthogonal matrix. We can now use the matrix $\mathbf{A}$ to obtain other quantities of interest transformed into the new frame.

K. Differential Forms

The differentials of the basis one-forms $\boldsymbol{\omega}_i$ and $\boldsymbol{\omega}'_j$ can themselves be subjected to the same linear transformation and also be expanded in terms of the connection forms and the basis forms:

$$d\boldsymbol{\omega}_i = [\mathbf{A}^{-1}]_i{}^j \, d\boldsymbol{\omega}'_j = \boldsymbol{\omega}_{ik} \wedge \boldsymbol{\omega}^k \,, \tag{A.K.1a}$$

$$d\boldsymbol{\omega}'_j = [\mathbf{A}]_j{}^q \, d\boldsymbol{\omega}_q = \boldsymbol{\omega}'_{jm} \wedge \boldsymbol{\omega}'^m \,. \tag{A.K.1b}$$

The connection forms are also one-forms and are transformed similarly, with the modification that two matrix multiplications must be applied as the connection one forms are multi-indexed,

$$\boldsymbol{\omega}'_{pq} = \mathbf{A}_p{}^j \, \mathbf{A}_q{}^k \, \boldsymbol{\omega}_{jk} \,. \tag{A.K.2}$$

The connection coefficients, which are 0-forms, can now be transformed from frame to frame as well

$$\begin{aligned} g'_{pmq} &= \mathbf{A}_p{}^j\,\mathbf{A}_m{}^k\,\mathbf{A}_q{}^l\,g_{jkl} \\ &= \mathbf{A}_p{}^j\,\mathbf{A}_m{}^k\,\mathbf{A}_q{}^l\,[\,g_j\,\delta_{kl} - g_k\,\delta_{jl}\,] \\ &= \mathbf{A}_p{}^j\,g_j\,\left(\mathbf{A}_m{}^k\,\mathbf{A}_q{}^l\,\delta_{kl}\right) - \mathbf{A}_p{}^j\,g_j\,\left(\mathbf{A}_m{}^k\,\mathbf{A}_q{}^l\,\delta_{kl}\right) \\ &= g'_p\,\delta_{mq} - g'_m\,\delta_{pq}\,, \end{aligned} \tag{A.K.3}$$

where the second step utilized (6.A.8) for a homogeneous vector field. For a homogeneous vector field in n dimensions, we can now use, (6.A.10),

$$\omega_{jk} = g_j\,\omega_k - g_k\,\omega_j\,, \tag{6.K.4}$$

after a suitable rotation has been performed. Inserting it into the first Cartan structure equation (A.P.2), we have

$$(d\boldsymbol{\omega}_j)\,g^j = g^j\,(\,g_j\,\boldsymbol{\omega}_k - g_k\,\boldsymbol{\omega}_j\,) \wedge \boldsymbol{\omega}^k\,. \tag{A.K.5}$$

L. Tensor Product

We can now define a multilinear product on the vector space. When applied to a pair of vectors, **U** and **V**, it evaluates as

$$\left[\boldsymbol{\omega}^i \otimes \boldsymbol{\omega}^j\right](\mathbf{U},\mathbf{V}) \equiv \boldsymbol{\omega}^i(\mathbf{U})\,\boldsymbol{\omega}^j(\mathbf{V})\,. \tag{A.L.1}$$

where the tensor product's symbolic name is $\otimes$ and we call it a product since it obtains its value by the ordinary product of the two real numbers $\boldsymbol{\omega}^i(\mathbf{U})$ and $\boldsymbol{\omega}^j(\mathbf{V})$.

M. Exterior (or Wedge) Product

We adopt the symbol "$\wedge$" for the completely anti-symmetrizing function. When applied to a pair of one–forms its effect is given by

$$\boldsymbol{\omega}^i \wedge \boldsymbol{\omega}^j \equiv \boldsymbol{\omega}^i \otimes \boldsymbol{\omega}^j - \boldsymbol{\omega}^j \otimes \boldsymbol{\omega}^i \tag{A.M.1}$$

where "$\otimes$" indicates the tensor product. When applied to a pair of vectors, **U** and **V**, (A.M.1) evaluates as

$$\begin{aligned} \left[\boldsymbol{\omega}^i \wedge \boldsymbol{\omega}^j\right](\mathbf{U},\mathbf{V}) &= \left[\boldsymbol{\omega}^i \otimes \boldsymbol{\omega}^j - \boldsymbol{\omega}^j \otimes \boldsymbol{\omega}^i\right](\mathbf{U},\mathbf{V}) \\ &= \boldsymbol{\omega}^i(\mathbf{U})\,\boldsymbol{\omega}^j(\mathbf{V}) - \boldsymbol{\omega}^j(\mathbf{U})\,\boldsymbol{\omega}^i(\mathbf{V})\,. \end{aligned} \tag{A.M.2}$$

N. P-forms

A p-form has the explicit expansion as tensor products of its elements

$$\boldsymbol{\omega}^{(1)} \wedge \boldsymbol{\omega}^{(2)} \wedge \cdots \wedge \boldsymbol{\omega}^{(p)} \equiv \delta^{(1)(2)\ldots(p)}_{q_1\, q_2 \cdots\, q_p}\, \boldsymbol{\omega}^{q_1} \otimes \boldsymbol{\omega}^{q_2} \otimes \cdots \otimes \boldsymbol{\omega}^{q_p} \qquad \text{(A.N.1)}$$

where the general Kronecker delta

$$\delta^{(1)(2)\ldots(p)}_{q_1\, q_2 \cdots\, q_p} \qquad \text{(A.N.2)}$$

is defined as

$$\begin{aligned} 0 \text{ if } q_1, q_2 \ldots, q_p &\text{ is not any permutation of } (1)(2)\ldots(p)\,, \\ +1 \text{ if } q_1, q_2 \ldots, q_p &\text{ is an even permutation of } (1)(2)\ldots(p)\,, \\ -1 \text{ if } q_1, q_2 \ldots, q_p &\text{ is an odd permutation of } (1)(2)\ldots(p)\,. \end{aligned}$$

If $\boldsymbol{\alpha}$ and $\boldsymbol{\beta}$ are p-forms, then the "d" operator on p-forms

$$d(\boldsymbol{\alpha} + \boldsymbol{\beta}) = d\boldsymbol{\alpha} + d\boldsymbol{\beta}\,. \qquad \text{(A.N.3)}$$

If $\boldsymbol{\alpha}$ is a p–form and $\boldsymbol{\beta}$ is a q–form, then the Leibnitz rule becomes

$$d(\boldsymbol{\alpha} \wedge \boldsymbol{\beta}) = d\boldsymbol{\alpha} \wedge \boldsymbol{\beta} + (-1)^p\, \boldsymbol{\alpha} \wedge d\boldsymbol{\beta}\,. \qquad \text{(A.N.4)}$$

We also have

$$d(d\boldsymbol{\alpha}) = 0\,. \qquad \text{(A.N.5)}$$

O. Ricci Rotation Coefficients

The any one-form can be written as a linear combination of the basis one-forms $\boldsymbol{\omega}^k$ with scalar coefficients

$$\boldsymbol{\omega}^i = g^i{}_k\, \boldsymbol{\omega}^k\,. \qquad \text{(A.O.1)}$$

The coefficients of this development, $g^i{}_k$, of the one-form $\boldsymbol{\omega}^k$ are 0-forms; that is, they are ordinary scalar functions of position.

P. Maurer-Cartan Equations

The first Maurer-Cartan structure equation is

$$d\boldsymbol{\omega}^i - \boldsymbol{\omega}^i{}_j \wedge \boldsymbol{\omega}^j = \boldsymbol{T}^i , \tag{A.P.1}$$

where $\boldsymbol{\omega}^i{}_j$ are called the connection one-forms and $\boldsymbol{T}^i$ is the vector-valued two-form known as the torsion. When the vector fields are torsion-free, that is, all components are equal to zero, the first Maurer-Cartan structure equation takes the form

$$d\boldsymbol{\omega}^i - \boldsymbol{\omega}^i{}_j \wedge \boldsymbol{\omega}^j = 0 . \tag{A.P.2}$$

As with any one-form, the connection forms can be written in terms of basis one-forms:

$$\boldsymbol{\omega}^i{}_j = g^i{}_{jk}\, \boldsymbol{\omega}^k , \tag{A.P.3}$$

where the coefficients $g^i{}_{jk}$ are called the Ricci rotation coefficients or, also commonly, the connection coefficients. For homogeneous fields the Ricci coefficients are assumed to be constants.

Similarly, the torsion two-form can be written as

$$\boldsymbol{T}^i = T^i{}_{jk}\, \boldsymbol{\omega}^j \wedge \boldsymbol{\omega}^k . \tag{A.P.4}$$

From (A.P.2) and (A.P.3), we can now calculate the differentials of the one-forms that appear in a metric thus obtaining the resulting two-forms

$$d\boldsymbol{\omega}^i = \frac{1}{2}\left[g^i{}_{jk} - g^i{}_{kj} \right] \boldsymbol{\omega}^k \wedge \boldsymbol{\omega}^j \tag{A.P.5a}$$

$$= -\frac{1}{2}\, c^i{}_{jk}\, \boldsymbol{\omega}^j \wedge \boldsymbol{\omega}^k \tag{A.P.5b}$$

$$= c^i{}_{jk}\, \boldsymbol{\omega}^k \wedge \boldsymbol{\omega}^j , \qquad j < k , \tag{A.P.5c}$$

where equations (A.P.5b,c) are known as the Maurer-Cartan structure equations. In (A.P.5a,b) we have allowed unrestricted summation and inserted a factor of one-half; in (A.P.5c) we have added a restriction on the indices to avoid the double counting in the previous expressions.

Constant Ricci coefficients give, as their skew-symmetric part with respect to the indices j and k, the structure constants of an n-dimensional local Lie group. We have

$$c^i{}_{jk} = g^i{}_{jk} - g^i{}_{kj} , \tag{A.P.6}$$

where the c^i_{jk} are the structure constants of that local Lie group and obey the skew-symmetry condition

$$c^i{}_{jk} = -c^i{}_{kj}\,, \tag{A.P.7}$$

and also the Jacobi identity

$$c^i{}_{jk}\, c^m{}_{il} + c^i{}_{kl}\, c^m{}_{ij} + c^i{}_{lj}\, c^m{}_{ik} = 0\,. \tag{A.P.8}$$

The Jacobi identity follows from

$$dd\boldsymbol{\omega}^i = 0\,. \tag{A.P.9}$$

When the vector fields are torsion-free, so that the first Maurer-Cartan structure equation takes the form (A.P.2), all the components of the torsion tensor,

$$T^i{}_{jk} = g^i{}_{jk} - g^i{}_{kj} - c^i{}_{jk} = 0\,, \tag{A.P.10}$$

are zero. This justifies the Maurer-Cartan equations (A.P.5b).

Equations (A.P.5a,b) for constant coefficients (A.P.6) are the necessary and sufficient conditions for the existence of a local Lie group [Sattinger, 1986]. The relationship between the structure constants of the Lie group and the Ricci rotation coefficients is given by

$$g_{ijk} = \frac{1}{2}\left[\, c_{ijk} + c_{jki} - c_{kij}\,\right]. \tag{A.P.11}$$

Q. Homogeneous Fields and Spaces

Metrics with constant Ricci coefficients are homogeneous spaces. The metric of these spaces is invariant under the operation of a transitive Lie group.

We shall also use the four vector fields $\mathbf{e}_j(x)$ that are the duals of the differential forms $\boldsymbol{\omega}^k$

$$\boldsymbol{\omega}^k(\mathbf{e}_j) = \delta^k{}_j\,. \tag{A.Q.1}$$

These vector fields are orthonormal. Their scalar products are

$$\mathbf{e}_j \cdot \mathbf{e}_k = \eta_{jk}\,. \tag{A.Q.2}$$

When the vector fields are more general, they need not be orthonormal and their scalar products give

$$\mathbf{e}_j \cdot \mathbf{e}_k = g_{jk}\,, \tag{A.Q.3}$$

which are the metric coefficients that appear in the line element.

The signature of the four-dimensional Minkowski spacetime is -2, so that the line element is given by:

$$ds^2 = dt^2 - dx^2 - dy^2 - dz^2 . \tag{A.Q.4}$$

In matrix form this metric is given by

$$\eta_{ij} = \begin{bmatrix} 1 & 0 & 0 & 0 \\ 0 & -1 & 0 & 0 \\ 0 & 0 & -1 & 0 \\ 0 & 0 & 0 & -1 \end{bmatrix} . \tag{A.Q.5}$$

R. Covariant Derivatives in Homogeneous Spaces

We have for the covariant derivative of the vector $\mathbf{e}_k$

$$\nabla_{\mathbf{e}_j}\mathbf{e}_k = -\mathbf{e}_m\, \boldsymbol{\omega}^m{}_k(\mathbf{e}_j) , \tag{A.R.1}$$

which is the equation dual to (A.P.2). According to (A.P.3) and (A.Q.1), this gives

$$\nabla_{\mathbf{e}_j}\mathbf{e}_k = -\mathbf{e}_m\, g^m{}_{kj} . \tag{A.R.2}$$

We now consider a geodesic congruence whose unit tangent vector will be identified with the frame vector $\mathbf{e}_0$. The geodesic condition gives

$$\nabla_{\mathbf{e}_0}\mathbf{e}_0 = 0 . \tag{A.R.3}$$

That means the covariant derivative of the tangent vector in the direction of the tangent vector vanishes. Comparison with (A.R.2) then gives

$$g^m{}_{00} = 0 \tag{A.R.4}$$

as the geodesic condition.

S. Local Inertial Systems in Homogeneous Spaces

If we want to describe a situation such that the geodesic congruence with tangent $\mathbf{e}_0$ gives rise to a local inertial system, we take $\mathbf{e}_0$ timelike and the three spacelike vectors $\mathbf{e}_\alpha$ ($\alpha \in 1, 2, 3$) of the frame parallelly transferred in the $\mathbf{e}_0$–direction. This condition of no rotation means

$$\dot{\mathbf{e}}_\alpha \equiv \nabla_{\mathbf{e}_0}\mathbf{e}_\alpha = 0 , \tag{A.S.1}$$

or, according to (A.R.2),

$$g^m{}_{\alpha 0} = 0 \tag{A.S.2}$$

for the Ricci coefficients.

This Appendix is referenced by **Chapter 3, Section A and Section D** and also **Chapter 9, Section B**.

APPENDIX B

A NON-TRIVIAL GLOBAL FRAME BUNDLE

A. Extreme Sections for Surfaces

The question of preferred frames for curved manifolds began to interest physicists through the interpretation of Einstein's theory of gravitation. The concepts of physics like energy and momentum were based on the Poincaré group. Their application to curved spacetimes—thus no longer invariant under the group—became questionable. However, there was the desire to continue to apply these notions in the case of spatially isolated systems which tend to become flat at spatial infinity.

It seems advantageous here to view this problem by using the concept of the frame bundle. To keep the geometry simple and as a help for visualization, here we shall deal with two-dimensional surfaces.

Suppose we have a two-dimensional orientable manifold, M^2. This manifold can be covered by overlapping coordinate charts $\{x^1_{(\rho)}, x^2_{(\rho)}\}$, where (ρ) belongs to some index set. We further suppose that the manifold is sufficiently differentiable and has a Riemannian metric. The orientability means that there is a cover (that is, a set of charts) such that the Jacobian of any two charts is positive (> 0) in the overlap.

At each point of each chart, one can form, from the local tangent vectors, along the coordinate lines

$$X_{(\rho)1} = \frac{\partial}{\partial x^1_{(\rho)}}\,, \qquad X_{(\rho)2} = \frac{\partial}{\partial x^2_{(\rho)}}\,, \tag{B.A.1}$$

their scalar products

$$g_{(\rho)\alpha\beta} = < X_{(\rho)\alpha}, X_{(\rho)\beta} >\,, \qquad \alpha, \beta \in 1, 2 \tag{B.A.2}$$

which define the metric tensor. We have

$$\det\left[\, g_{(\rho)\alpha\beta}\,\right] > 0\,. \tag{B.A.3}$$

We can now use the coordinate vectors to construct in each point an orthogonal reference frame that is positively oriented like the tangent vectors. We put

$$\mathbf{e}_{(\rho)1} = \frac{1}{\sqrt{g_{(\rho)11}}} X_{(\rho)1}, \tag{B.A.4a}$$

$$\mathbf{e}_{(\rho)2} = \frac{X_{(\rho)2} - \left(g_{(\rho)12} / g_{(\rho)11}\right) X_{(\rho)1}}{\sqrt{g_{(\rho)22} - g^2_{(\rho)12}/g_{(\rho)11}}}. \tag{B.A.4b}$$

These vectors will then be orthonormal and positively oriented.

We are now ready to define the frame bundle of M^2. Let

$$E_{(\rho)1}(x^\alpha_{(\rho)}) + i\, E_{(\rho)2}(x^\alpha_{(\rho)}) = e^{i\phi_{(\rho)}} \left(\mathbf{e}_{(\rho)1}(x^\alpha_{(\rho)}) + i\, \mathbf{e}_{(\rho)2}(x^\alpha_{(\rho)}) \right) \tag{B.A.5}$$

be a positively oriented orthonormal frame at the point $x^\alpha_{(\rho)}$ rotated in the positive sense by the angle $\phi_{(\rho)}$. The manifold of all such frames at all points of M^2 is the frame bundle FM^2. FM^2 is a three-dimensional manifold locally described by the coordinates $\{x^1_{(\rho)}, x^2_{(\rho)}, \phi_{(\rho)}\}$. We write (B.A.5) in the abbreviated form

$$E_{(\rho)} = e^{i\phi_{(\rho)}}\, \mathbf{e}_{(\rho)}. \tag{B.A.6}$$

Suppose we now have, in the overlap of the charts (ρ) and (ρ'), also

$$E_{(\rho')} = e^{i\phi_{(\rho')}}\, \mathbf{e}_{(\rho')} \tag{B.A.7}$$

and

$$\mathbf{e}_{(\rho')} = e^{i\phi_{(\rho'\rho)}}\, \mathbf{e}_{(\rho)} \tag{B.A.8}$$

where $\phi_{\rho'\rho}$ is the angle between the tangent vectors $X_{(\rho)}$ and $X_{(\rho')}$, we get

$$E_{(\rho')} = e^{i(\phi_{(\rho')} + \phi_{(\rho'\rho)})}\, \mathbf{e}_{(\rho)}. \tag{B.A.9}$$

We thus have

$$\phi_{(\rho')} - \phi_{(\rho)} = \phi_{(\rho'\rho)} \quad modulo\, 2\pi. \tag{B.A.10}$$

If we have a third chart (ρ''), we get, in the common overlap of (ρ), (ρ'), and (ρ''),

$$\phi_{(\rho'')} - \phi_{(\rho')} = \phi_{(\rho''\rho')} \quad modulo\, 2\pi, \tag{B.A.11a}$$

$$\phi_{(\rho)} - \phi_{(\rho'')} = \phi_{(\rho\rho'')} \quad modulo\, 2\pi \tag{B.A.11b}$$

and by adding (B.A.10) to the equations (B.A.11)

$$\phi_{(\rho\rho'')} + \phi_{(\rho''\rho')} + \phi_{(\rho'\rho)} = 0 \quad modulo\, 2\pi\,. \tag{B.A.12}$$

This equation is known as the cocycle condition. It assures us that FM^2 is a manifold; to be more precise, it is a "circle bundle". If the manifold M^2 can be covered with a single chart the topology of FM^2 is given by $\mathbb{R}^2 \times \mathbb{S}^1$, the direct product of a plane with a circle. The three-dimensional manifold FM^2 is a fiber space with M^2 as the base manifold and circles as fibers. The angle ϕ is a coordinate along the fiber.

However, the frame bundle $F\mathbb{S}^2$ of the sphere $\mathbb{S}^2$ has a different topology.

B. The Non-trivial Global Frame Bundle of the Two-Sphere

To give a simple case of a non-trivial global frame bundle of a manifold we choose the frame bundle of a two-sphere $\mathbb{S}^2$. We define the $\mathbb{S}^2$ by means of two coordinate charts

$$\left(x^1, x^2\right) \quad \text{and} \quad \left(x'^1, x'^2\right)\,. \tag{B.B.1}$$

It is convenient to introduce complex variables defined by

$$y \equiv x^1 + ix^2 \quad \text{and} \quad y' \equiv x'^1 + ix'^2\,. \tag{B.B.2}$$

The two coordinate charts are then defined by the y and y' planes respectively, which can be considered to be in contact with the North and South poles of the sphere. We postulate that the y-plane minus its origin and the y'-plane minus its origin are related by

$$y' = \frac{1}{y} \quad \text{and} \quad y = \frac{1}{y'} \tag{B.B.3}$$

which defines an atlas for the $\mathbb{S}^2$. We compute the Jacobian of this transformation from

$$\frac{i}{2}\, dy' \wedge d\bar{y}' = \frac{i}{2}\, [y\,\bar{y}]^{-2}\, dy \wedge d\bar{y} \tag{B.B.4a}$$

$$= dx'^1 \wedge dx'^2 \tag{B.B.4b}$$

$$= \left[\left[x^1\right]^2 + \left[x^2\right]^2\right]^{-2} dx^1 \wedge dx^2\,. \tag{B.B.4c}$$

This shows that the $\mathbb{S}^2$ is orientable. A metric is given by

$$ds^2 = \frac{4\,dy\,dy}{[1 + y\,\bar{y}]^2} = \frac{4\,dy'\,d\bar{y}'}{[1 + y'\,\bar{y}']^2} . \tag{B.B.5}$$

We now introduce, according to (B.B.5),

$$\boldsymbol{\omega}(y, \phi) \equiv \frac{2\,dy}{[1 + y\,\bar{y}]}\, e^{i\phi} \tag{B.B.6a}$$

and

$$\boldsymbol{\omega}'(y', \phi') \equiv \frac{2\,dy'}{[1 + y'\,\bar{y}']}\, e^{i\phi'} . \tag{B.B.6b}$$

Since

$$\frac{dy'}{[1 + y'\,\bar{y}']} = -\frac{1}{y^2}\left[1 + \frac{1}{y\,\bar{y}}\right]^{-1} dy \tag{B.B.7}$$

we have from the equation (B.B.6)

$$\boldsymbol{\omega}'(y', \phi') = \boldsymbol{\omega}(y, \phi) \tag{B.B.8}$$

in the overlap

$$\frac{\bar{y}}{y}\, e^{i\phi'} = e^{i\phi} \tag{B.B.9}$$

or

$$\phi' = \phi + 2\,\arg(y) , \qquad \Phi = 2\,\arg(y) . \tag{B.B.10}$$

This is just the relationship that equations (5.B.14) specify for the frame bundle in the overlap.

This Appendix is referenced by **Chapter 5, Section B**.

APPENDIX C

GEODESICS OF THE POINCARÉ HALF-PLANE

A general introduction to the Poincaré half-plane can be found in Stahl's text [Stahl, 2008]. Here we provide a limited introduction adapted to our discussions of homogeneous fields. The metric of the Poincaré half-plane is given by

$$ds^2 = \frac{R^2}{y^2}\left[dx^2 + dy^2\right], \tag{C.1}$$

where $R = constant$ and $y > 0$, according to (5.E.6). The geodesics can be obtained as solutions of the Euler-Lagrange equations of the variational problem

$$\delta \int \mathcal{L}\, ds = 0, \qquad \mathcal{L} = \frac{R^2}{2\,y^2}\left(\dot{x}^2 + \dot{y}^2\right), \tag{C.2}$$

where a dot denotes derivation with respect to s. The equation

$$\frac{\partial \mathcal{L}}{\partial x} - \left(\frac{\partial \mathcal{L}}{\partial \dot{x}}\right)^{\cdot} = 0 \tag{C.3}$$

has the integral

$$\frac{\partial \mathcal{L}}{\partial \dot{x}} = \frac{R^2}{y^2}\dot{x} = A, \qquad A = constant. \tag{C.4}$$

On the other hand, we have from (C.1) that

$$\dot{x}^2 + \dot{y}^2 = \frac{y^2}{R^2}. \tag{C.5}$$

Dealing first with the case $A = 0$, which by (C.4) gives vanishing $\dot{x}$, and

$$x = constant, \qquad y = y_0 \exp\left[\frac{s}{R}\right], \qquad y_0 > 0. \tag{C.6}$$

These geodesics appear in the Poincaré half-plane as Euclidean parallels of the y-axis. However, the factor y^2 in the denominator of (C.1) shows that these geodesics converge towards each other as y goes toward infinity. They are the kind of parallels in the hyperbolic space which all intersect "at infinity." Expressed differently, these geodesics have a common origin at infinity and are constantly diverging until they eventually drift apart toward infinite distances from each other as they approach $y = 0$.

For the other geodesics, we assume now $A \neq 0$, which leads to

$$\dot{x} \neq 0\,, \tag{C.7}$$

and (C.4) then gives

$$\dot{x} = \frac{A}{R^2}\, y^2\,, \qquad A = constant\,. \tag{C.8}$$

Instead of parameterizing the geodesics with s/R, we now use x as the independent variable determining y. This is always possible since $\dot{x}$ is now assumed never to be equal to zero, and equation (C.5) requires that $\dot{x}$ and $\dot{y}$ cannot simultaneously be zero except at $y = 0$. We obtain from (C.1), (C.5) and (C.8)

$$1 + \left[\frac{dy}{dx}\right]^2 = \frac{R^2}{A^2\, y^2}\,. \tag{C.9}$$

This gives

$$x - x_0 = \frac{A}{R}\int \frac{y\, dy}{\sqrt{1 - [A^2/R^2]\, y^2}} = -\frac{R}{A}\sqrt{1 - [A^2/R^2]\, y^2}\,. \tag{C.10}$$

Thus, we obtain equations for the geodesics with $\dot{x} \neq 0$ and x_0 a constant of integration

$$[\,x - x_0\,]^2 + y^2 = \frac{R^2}{A^2}\,. \tag{C.11}$$

Since we are restricted to the upper half plane $y > 0$, the geodesics $\dot{x} \neq 0$ are given by the semicircles with radius $|R/A|$ with center x_0 on the x-axis. **Figures C.1** and **C.2** illustrate this [Adapted from Guggenheimer, 1963].

It would seem that the geodesics given by $x = constant$ are different in nature from those described as semicircles. This is, however, not the case. If we take the point x_0 in **Figure C.1** and move it along the positive y-axis out to infinity, the semicircles through x_0 will become flatter and flatter. In the limit they will become parallels to the y-axis given by $x = constant$. The

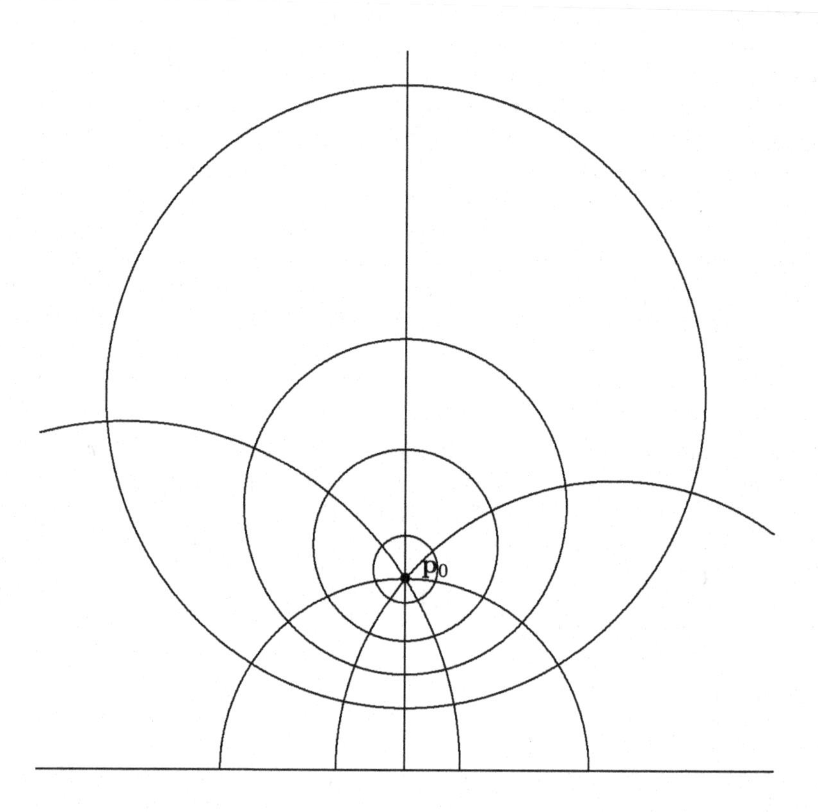

Figure C.1 Geodesics of the Poincaré Plane Three geodesics through the point $\mathbf{p}_0$ on the y-axis appear in the Poincaré plane as a semi-circle and two arcs of semicircles. They intersect the horizontal x-axis orthogonally. The y-axis is also a geodesic. Four geodesic circles about the common center $\mathbf{p}_0$ are also shown. They intersect the geodesics originating from $\mathbf{p}_0$ orthogonally. They appear in the Poincaré half-plane as circles.

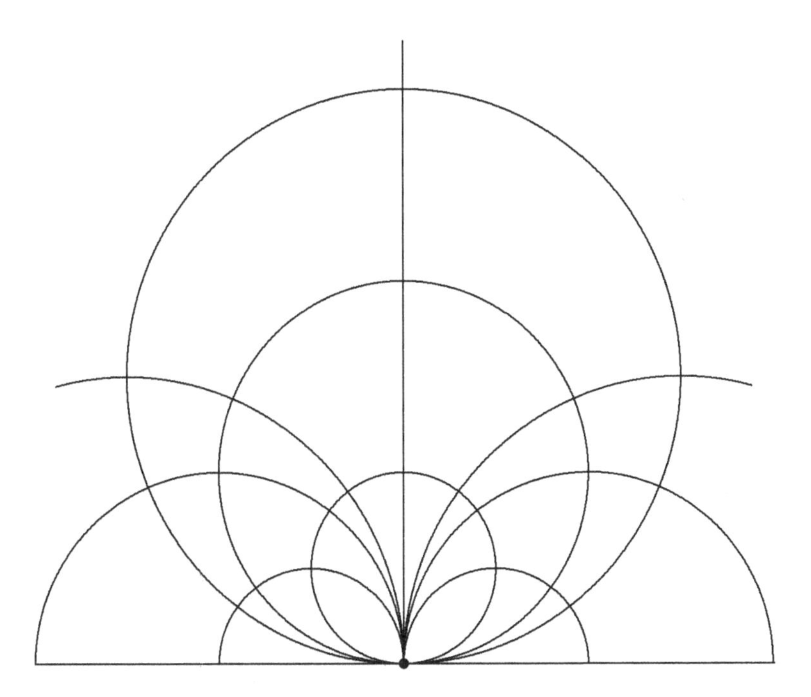

Figure C.2 Pencil of Geodesics in the Poincaré Plane The semicircles through the origin together with the y-axis form a pencil of geodesics which are parallel in the sense that they all intersect in one common point that lies at infinity (which is the origin). The full circles orthogonal to all geodesics of the pencil also go through this common point at infinity. Their centers lie on the y-axis and they all have the x-axis as a common tangent. These circles are not geodesics. They are "circles" with an infinite radius which have no real analog in spherical or Euclidean geometry. They are known as horocycles.

geodesic circles about x_0 become, in the limit, horocycles through the point $x = 0$, $y = +\infty$ and degenerate into the lines $y = constant > 0$. Together with geodesics $x = constant$, they form an orthogonal net. In order to see this one can study the isometries of the Poincaré half-plane. They move geodesics into geodesics. These isometries are obtained by considering the Poincaré half-plane as the upper part of the complex number plane by writing

$$z \equiv x + i\,y\,. \tag{C.12}$$

We can write the metric (C.1) as

$$ds^2 = \frac{R^2}{y^2}\,[\,dz\,d\bar{z}\,] \tag{C.13}$$

where $\bar{z}$ denotes the complex conjugate of z.

The transformations

$$z' = \frac{A\,z + B}{C\,z + D}\,, \qquad A, B, C, D \in \mathbb{R}\,, \qquad A\,D - B\,C = 1 \tag{C.14}$$

form the group $SL(2, \mathbb{R})$. They give

$$y' = \frac{1}{2\,i}\,(z' - \bar{z}') = \frac{1}{2\,i}\,(z - \bar{z})\,\frac{A\,D - B\,C}{|\,C\,z + D\,|^2} \tag{C.15a}$$

$$= \frac{y}{|\,C\,z + D\,|^2}\,. \tag{C.15b}$$

These transformations take the upper half-plane into itself. Coefficients (A, B, C, D) and $(-A, -B, -C, -D)$ give rise to the same transformation. Taking the factor group of $SL(2, \mathbb{R})$ with respect to this identification results in the group $PSL(2, \mathbb{R})$.

We want to show that this group leaves the metric (C.13) unchanged. We have

$$dz' = \frac{A\,dz\,(\,C\,z + D\,) - (\,A\,z + b\,)\,C\,dz}{(\,C\,z + D\,)^2} \tag{C.16a}$$

$$= \frac{dz}{(\,C\,z + D\,)^2} \tag{C.16b}$$

and thus with (C.13) and (C.15)

$$ds^2 = \frac{R^2}{y'^2}\,[\,dz'\,d\bar{z}'\,]\,. \tag{C.17}$$

The transformations are, therefore, isometries. The discrete transformation

$$z' = -\bar{z} \quad \longrightarrow \quad x' = -x\,, \qquad y' = y\,, \tag{C.18}$$

which is a reflection on the y-axis that transforms the upper half-plane into itself, leaves the metric (C.13) unchanged. Together with the $PSL(2,\mathbb{R})$ this reflection generates the full isometry group of the hyperbolic plane.

We are interested here in a subgroup of the $PSL(2,\mathbb{R})$ given by

$$z' = \mu\, z + \sigma\,, \qquad \mu, \sigma \in \mathbb{R}\,, \qquad \mu > 0\,, \tag{C.19a}$$

$$\mu = A^2\,, \qquad \sigma = AB\,, \qquad C = 0\,, \qquad D = \frac{1}{A}\,. \tag{C.19b}$$

This group of dilations by μ and translations along the x-axis acts transitively on the upper half-plane. This is seen by rewriting (C.14) as

$$x' = \mu\, x + \sigma\,, \qquad y' = \mu\, y\,, \qquad y' > 0\,, \qquad y > 0\,. \tag{C.20}$$

These equations give

$$\sigma = x' - \frac{y'}{y}\, x\,, \qquad \mu = \frac{y'}{y}\,. \tag{C.21}$$

This shows that, for arbitrary z' and z in the upper half-plane, there is one and only one transformation with parameters μ and σ which moves z into z' as the group's action.

From (C.20), we see that

$$dx' = \mu\, dx\,, \qquad dy' = \mu\, dy\,. \tag{C.22}$$

It follows then

$$\frac{1}{y'}\, dx' = \frac{1}{y}\, dx\,, \qquad \frac{1}{y'}\, dy' = \frac{1}{y}\, dy\,. \tag{C.23}$$

The complex differential form

$$\boldsymbol{\omega} \equiv \omega_1 + i\, \omega_2 = R\, \frac{dx + i\, dy}{y} = \frac{R}{y}\, dz \tag{C.24}$$

is therefore invariant under the subgroup. In particular, this is true for the connection form $\boldsymbol{\omega}$ of (5.E.4)

$$\boldsymbol{\omega} = \frac{1}{R}\, \omega_1\,. \tag{C.25}$$

The motions (C.19) transform the geodesics $x = constant$ among themselves. To see the equivalence of all geodesics under the motions (C.14), one finds it convenient to transform from the Poincaré half-plane to the Poincaré disc. A transformation which does this is given by

$$z' = \frac{z - i}{1 - i\, z}. \tag{C.26}$$

This transformation maps the upper half of the complex z-plane into the interior of the unit circle of the z'-plane, the Poincaré disc. The real axis of the z-plane (together with $z = \infty$) becomes the unit circle of the z'-plane centered about the origin. Since the map (C.26) is conformal and transforms circles into circles (including straight lines), we now find that the geodesics in a space of negative curvature appear in the Poincaré disc as arcs of circles which intersect the unit circle orthogonally.

The geodesics $x = constant$ of the z-plane appear in the z'-plane, the Poincaré disc, as circles which all go through the same point on the periphery of the unit circle. They have a common tangent there. One of these circles becomes a straight diameter of the Poincaré disc. The horocycles $y = constant$ of the Poincaré half-plane all become circles in the Poincaré disc going through the same point on the periphery, touching the unit circle in only this point from the interior. The system of the orthogonal lines $x = constant$ and $y = constant$ of the Poincaré half-plane appears on the disc as a system of two pencils of circles orthogonal to each other as they are familiar from the Smith chart of electrical engineers [**See Figure C.3**].

While the Poincaré half-plane puts into evidence the homogeneity of the hyperbolic plane, it seems to distinguish the subgroup (C.19). The Poincaré disc, on the other hand, demonstrates that the pencil of geodesics that emanates from the point with coordinates $(0, i)$ on the unit circle is on the same footing with all the other pencils emanating from any other point of the unit circle. One is obtained from the other by a simple rotation about the origin. The subgroup (C.19) is one of a one-parameter set which are similar to each other.

After this excursion into hyperbolic terminology, we can talk now sensibly about pencils of parallel geodesics. They all "intersect" in their common point at infinity if extended infinitely.

One such pencil of geodesics parallel to the y-axis appears as the set of lines $x = constant$ in the Poincaré half-plane. On the Poincaré half-plane, the parallel geodesics to the y-axis appear as Euclidean parallels and their homogeneity becomes obvious. Other parallel geodesics appear as pencils of semi-circles. They look, therefore, quite different. On the Poincaré disc,

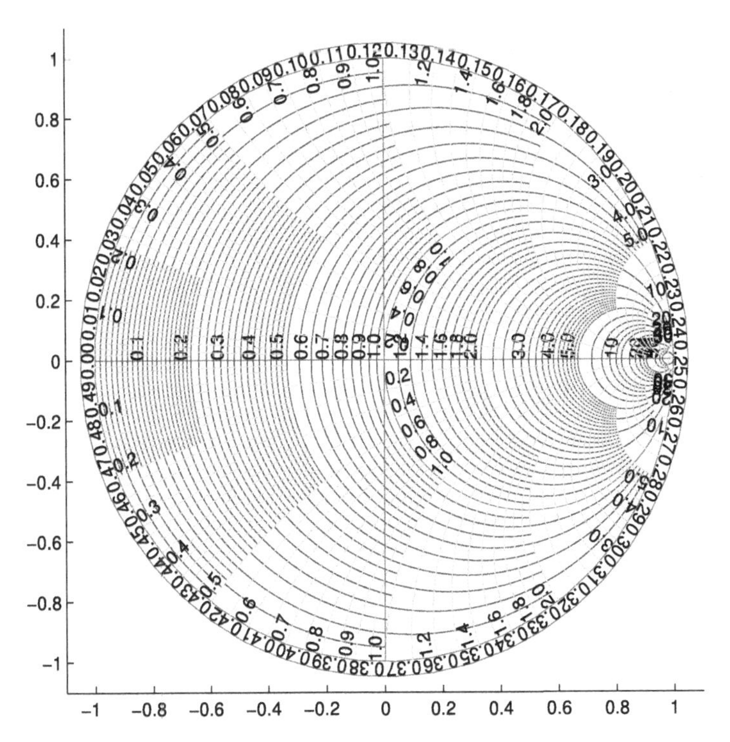

Figure C.3 Smith Chart An example of a Smith chart that is used in electrical engineering to determine the impedance of transmission lines. [Plotted via MATHLAB using Victor Aprea's "plotsmithchart.m" contribution to MATHLAB's file exchange.]

in contrast, all pencils of parallels are obtained from each other by a simple rotation about the origin.

A final remark should be added for completeness. There is another chart of the hyperbolic plane known as the Beltrami-Klein model. This map is sort of an intermediate between the two Poincaré maps. The hyperbolic plane is the interior of a unit circle and its geodesics are the segments of straight lines in its interior.

The word "parallel" is sometimes used more generally to describe straight lines (that is, geodesics) on the surface of negative curvature that do not meet. Our notion of parallelity (meeting at infinity) goes back to Gauss and Lobachevski.. Bolyai called these lines asymptotic.

We close with a description of geodesies in terms of their leg components in a homogeneous field. A homogeneous field is given by

$$\boldsymbol{\omega}^j = \frac{R}{y}\, dx^j\,, \qquad \mathbf{e}_k = \frac{y}{R}\,\frac{\partial}{\partial x^k}\,, \tag{9.27}$$

with

$$\boldsymbol{\omega}^j\,(\mathbf{e}_k) = \delta^j_k\,. \tag{9.28}$$

The unit tangent vector $\mathbf{T}$ of a curve is given by the leg components t^j

$$\mathbf{T} = t^j\,\mathbf{e}_j\,, \qquad \left(t^1\right)^2 + \left(t^2\right)^2 = 1\,. \tag{9.29}$$

A geodesic characterized by the constant A has, according to (9.8) and (9.27), the t^j-components

$$t^1 = \frac{A\,y}{R}\,, \qquad t^2 = \sqrt{1-(t^1)^2}\,. \tag{9.30}$$

The leg-curvature vector $\dot{\mathbf{T}}$ has the components

$$\dot{\mathbf{T}} = (\,\dot{t}^1,\,\dot{t}^2\,) \tag{9.31}$$

with

$$\dot{t}^1 = \frac{A\,\dot{y}}{R}\,, \qquad \dot{t}^2 = -\,\frac{1}{\sqrt{1-(t^1)^2}}\;t^1\,\dot{t}^1 = -\,\frac{t^1\,\dot{t}^1}{t^2}\,. \tag{9.32}$$

We have from (9.5) and (9.8)

$$\dot{y} = \frac{y}{R}\sqrt{1-\frac{A^2\,y^2}{R^2}} = \frac{y}{R}\,t^2\,. \tag{9.33}$$

This gives

$$\dot{t}^1 = \frac{t^1\,t^2}{R}\,, \qquad \dot{t}^2 = -\,\frac{(t^1)^2}{R}\,. \tag{9.34}$$

The leg-curvature of a geodesic is thus

$$C = \sqrt{(\dot{t}^1)^2 + (\dot{t}^2)^2} = \frac{t^1}{R} \tag{9.35}$$

and its radius of leg-curvature ρ is

$$\rho = \frac{1}{C} = \frac{R}{t^1}\,. \tag{9.36}$$

This Appendix is referenced by **Chapter 5, Section E.**

APPENDIX D

DETERMINATION OF HOMOGENEOUS FIELDS IN TWO-DIMENSIONAL RIEMANNIAN SPACES

We have not shown in the text that the homogeneous fields we discussed were the only possible ones. For flat space this question could be settled very easily. We get

$$ds^2 = [\omega_1]^2 + [\omega_2]^2 , \qquad \omega_1 \wedge \omega_2 \neq 0 , \tag{D.1}$$

since the metric is Riemannian. The field strength vanishes for flat space, thus the connection form vanishes, and we have

$$d\omega_1 = 0 , \qquad d\omega_2 = 0 . \tag{D.2}$$

This means that there are locally independent functions x and y such that

$$\omega_1 = dx , \qquad \omega_2 = dy . \tag{D.3}$$

We are dealing simply with a Euclidean space without any further restrictions. The geodesically complete ones give rise to the five known space forms (namely, the plane, cylinder, torus, Klein bottle, and Moebius strip).

For the case of constant negative curvature, we stated that the invariance of the Poincaré disc shows us the equivalence of homogeneous fields. We want to prove now that one can obtain all homogeneous fields from a given one by rotation with a constant angle and reflections. We have for the connection form ω_{12}

$$d\omega_1 = \omega_{12} \wedge \omega_2 , \qquad d\omega_2 = -\,\omega_{12} \wedge \omega_1 . \tag{D.4}$$

Combining ω_1 and ω_2 as

$$\kappa \equiv \omega_1 + i\,\omega_2 , \tag{D.5}$$

we can write equations (D.4) as

$$d\boldsymbol{\kappa} = -i\,\omega_{12} \wedge \boldsymbol{\kappa}\,. \tag{D.6}$$

One sees now that the transformation

$$\boldsymbol{\kappa} \longrightarrow \boldsymbol{\kappa}\, e^{i\phi} = \boldsymbol{\kappa}' \tag{D.7}$$

will not change the metric (D.1),

$$ds^2 = \boldsymbol{\kappa}\,\bar{\boldsymbol{\kappa}} = \boldsymbol{\kappa}'\,\bar{\boldsymbol{\kappa}}'\,, \tag{D.8}$$

and gives, with

$$d\boldsymbol{\kappa}' = -i\,\omega'_{12} \wedge \boldsymbol{\kappa}'\,, \tag{D.9}$$

the new connection form

$$\omega'_{12} = \omega_{12} + d\phi\,. \tag{D.10}$$

Rotations of the frame by a constant angle ϕ, for which $d\phi = 0$, do not change the connection form. Under a reflection,

$$\omega_1 \longrightarrow \omega_2\,, \qquad \omega_2 \longrightarrow \omega_1\,, \tag{D.11}$$

the connection form,

$$\omega_{12} \equiv \omega\,, \tag{D.12}$$

changes sign:

$$\omega' = -\omega\,. \tag{D.13}$$

After these general statements, we consider the hyperbolic plane. By a rotation of a homogeneous field we can arrange that the connection form ω_{12} becomes

$$\omega_{12} = \frac{1}{r}\,\omega_1\,. \tag{D.14}$$

We then have, from (D.4),

$$d\omega_2 = 0 \tag{D.15}$$

and can write with a suitable variable v

$$\omega_2 = \frac{r}{v}\,dv\,. \tag{D.16}$$

Inserting this into the first equation (D.4), we get

$$d\omega_1 - \frac{1}{v}\,\omega_1 \wedge dv = 0\,, \tag{D.17}$$

or, after multiplying with v,

$$d\left[\,\omega_1\, v\,\right] = \left[\,d\omega_1\,\right] v - \omega_1 \wedge dv = 0\,. \tag{D.18}$$

We put, therefore,

$$\omega_1\, v = r\, du\,, \tag{D.19}$$

with a suitable variable u and obtain

$$\omega_1 = \frac{r}{v}\, du\,, \qquad \omega_2 = \frac{r}{v}\, dv\,, \tag{D.20}$$

and, from (D.1), we get

$$ds^2 = \frac{r^2}{v^2}\left[\,du^2 + dv^2\,\right]. \tag{D.21}$$

This shows that a homogeneous field on the hyperbolic plane is given by (5.E.4) and the fields obtained from it by a constant rotation of the frame.

This Appendix is referenced by **Chapter 5, Section E**.

APPENDIX E

SPACE EXPANSION

The concept of "space expansion" in n-dimensional spaces can be given a precise form by using the formula for geodesic deviation [Levi-Civita,1977]. This formula can be derived as follows: Let $u^\mu(x^\lambda)$ be the unit tangent vector of a congruence of geodesics

$$u^\mu\, u_\mu = 1, \qquad u^\mu{}_{;\nu}\, u^\nu = 0, \qquad \mu,\, \lambda \in 1, \ldots, n\,, \tag{E.1}$$

where a semicolon denotes the covariant derivative. Further, let $\eta^\mu(x^\lambda)$ be a vector field orthogonal to the congruence. Then $\epsilon\, \eta^\mu(x^\lambda)$ is a infinitesimal vector connecting a geodesic with a neighboring one. We have then

$$u^\mu\, \eta_\mu = 0\,, \qquad u^\mu{}_{;\nu}\, \eta^\nu - \eta^\mu{}_{;\nu}\, u^\nu = 0\,. \tag{E.2}$$

The last equation states that the Lie derivative of η^μ along u^μ vanishes.

We now write $\dot\eta^\mu$ for the directional derivative of the vector η^μ along the geodesic with tangent vector u^μ and arc length s. Because of (E.2) we have

$$\dot\eta^\mu \equiv \eta^\mu{}_{;\nu}\, u^\nu = u^\mu{}_{;\nu}\, \eta^\nu\,. \tag{E.3}$$

We obtain for the second derivative, $\ddot\eta^\mu$,

$$\ddot\eta^\mu = (\, u^\mu{}_{;\nu}\, \eta^\nu\,)_{;\rho}\, u^\rho = (\, u^\mu{}_{;\nu;\rho}\, \eta^\nu\,)\, u^\rho + u^\mu{}_{;\nu}\, \eta^\nu{}_{;\rho}\, u^\rho\,. \tag{E.4}$$

The second term can be converted because of (E.3) and gives

$$\ddot\eta^\mu = (\, u^\mu{}_{;\nu;\rho}\, \eta^\nu\,)\, u^\rho + u^\mu{}_{;\nu}\, u^\nu{}_{;\rho}\, \eta^\rho\,. \tag{E.5}$$

By interchanging the summation indices in the first term and observing that

$$u^\mu{}_{;\nu}\, u^\nu{}_{;\rho}\, \eta^\rho = -\, u^\mu{}_{;\nu;\rho}\, \eta^\rho\, u^\nu \tag{E.6}$$

because of (E.1), we obtain

$$\begin{aligned} \ddot{\eta}^{\mu} &= (u^{\mu}{}_{;\rho;\nu} \; - \; u^{\mu}{}_{;\nu;\rho}) \, \eta^{\rho} \, u^{\nu} \\ &= R^{\mu}{}_{\lambda\rho\nu} \, u^{\lambda} \, u^{\nu} \, \eta^{\rho} \, . \end{aligned} \tag{E.7}$$

Here $R^{\mu}{}_{\lambda\rho\nu}$ is the Riemann tensor. We define the tensor

$$E^{\mu}{}_{\rho} \equiv - R^{\mu}{}_{\lambda\rho\nu} \, u^{\lambda} \, u^{\nu} \tag{E.8}$$

and now write the equation of geodesic deviation

$$\ddot{\eta}^{\mu} + E^{\mu}{}_{\rho} \, \eta^{\rho} = 0 \, . \tag{E.9}$$

The tensor $E^{\mu}{}_{\rho}$ is symmetric and has u^{μ} as eigenvector with the eigenvalue zero

$$E_{\mu\rho} = E_{\rho\mu} , \qquad u^{\rho} \, E^{\mu}{}_{\rho} = 0 \, . \tag{E.10}$$

For the case $n = 2$ of a two-dimensional surface one obtains Jacobi's equation

$$\frac{d^2\eta}{ds^2} + K(s) \, \eta = 0 \, , \tag{E.11}$$

where $K(s)$ is the Gaussian curvature along a geodesic with arc length s while $\eta(s)$ is the infinitesimal distance to a neighboring geodesic.

For a space of constant curvature, we have

$$R_{\mu\lambda\rho\nu} = K \, (g_{\mu\nu} \, g_{\lambda\rho} \; - \; g_{\mu\rho} \, g_{\lambda\nu}) \tag{E.12}$$

and thus

$$\begin{aligned} E_{\mu\rho} &= - K \, (g_{\mu\nu} \, g_{\lambda\rho} \; - \; g_{\mu\rho} \, g_{\lambda\nu}) \, u^{\lambda} \, u^{\nu} \\ &= K \, (g_{\mu\rho} - u_{\mu} u_{\rho}) \, . \end{aligned} \tag{E.13}$$

This gives for the equation of geodesic deviation

$$\ddot{\eta}^{\mu} + K \, \eta^{\mu} = 0 \, . \tag{E.14}$$

More generally, if the eigenvalues of the tensor $E^{\mu}{}_{\rho}$ are all smaller than, but not equal to, zero, we have, according to (E.8), space creation along u^{μ}. This results because two Lie-parallel geodesics with $\dot{\eta} = 0$ will diverge at this point since $\ddot{\eta} > 0$.

This Appendix is referenced by **Chapter 5, Section E.**

APPENDIX F

THE REISSNER-NORDSTROM ISOTROPIC FIELD

The Reissner-Nordstrom solution to Einstein's gravitational field equations, written in coordinates that reduce to the Schwarzschild solution, is given as

$$
\begin{aligned}
ds^2 = & \left[1 - \frac{2m}{r} + \frac{e^2}{r^2}\right] dx_0^2 - \left[1 - \frac{2m}{r} + \frac{e^2}{r^2}\right]^{-1} dr^2 \\
& - r^2 \left(d\theta^2 + \sin^2\theta \, d\phi^2 \right) . \qquad \text{(F.1)}
\end{aligned}
$$

When we transform to isotropic coordinates, we are finding coordinates that treat all the spatial coordinates on an equal basis. The final result will give the line element in coordinates that resemble rectangular coordinates with an overall factor scaling the axes. The Schwarzschild line element has an isotropic form given as

$$
\begin{aligned}
ds^2 = & \left(1 - \frac{m}{2\rho}\right)^2 \left(1 + \frac{m}{2\rho}\right)^{-2} dx_0^2 \\
& - \left(1 + \frac{m}{2\rho}\right)^4 \left[d\rho^2 + \rho^2 \left(d\theta^2 + \sin^2\theta \, d\phi^2\right)\right] . \qquad \text{(F.1a)}
\end{aligned}
$$

To obtain an isotropic form for the Reissner-Nordstrom line element, we will start by taking the spatial coordinates in Reissner-Nordstrom, (F.1), and converting them so that they all share a common factor. This brings the line element into the form

$$
\begin{aligned}
ds^2 = & \left[1 - \frac{2m}{r} + \frac{e^2}{r^2}\right] dx_0^2 \\
& - \lambda^2(\rho) \left[d\rho^2 + \rho^2 \left(d\theta^2 + \sin^2\theta \, d\phi^2 \right)\right] , \qquad \text{(F.2)}
\end{aligned}
$$

where we have defined

$$r^2 \equiv \lambda^2(\rho)\, \rho^2 \tag{F.3}$$

and assumed that λ will need to depend only on a radial coordinate. This assumption is based on the recognizing that the original line element (F.1) is spherically symmetric. We need to determine λ as it is incompletely specified by (F.3), which gives λ implicitly, in terms of both r and ρ. To eliminate r from the expression for λ, we must first obtain an expression for the radial coordinate r solely as a function of ρ. Since (F.3) contains three mutually dependent variables, we need another relationship to replace $\lambda(\rho)$ by an explicit expression. The invariant ds^2 provides

$$\lambda^2\, d\rho^2 = \left(\frac{r}{\rho}\right)^2 d\rho^2 = \left[1 - \frac{2m}{r} + \frac{e^2}{r^2}\right]^{-1} dr^2 \tag{F.4}$$

by equating $ds^2(t, r, \theta, \phi) = ds^2(t, \rho, \theta, \phi)$ at fixed time and angular position. Integration of the square root of (F.4) provides the required function of $r\,(\rho)$. We have

$$I = \int \frac{d\rho}{\rho} = \pm \int \frac{dr}{\sqrt{r^2 - 2mr + e^2}}\,. \tag{F.5}$$

Our integral is a specific instance of the general form

$$I = \pm \int \frac{dx}{\sqrt{a\,x^2 - b\,x + c}} \tag{F.6}$$

with the values, in our case,

$$a = 1\,, \quad b = -\,2\,m\,, \quad c = e^2\,, \tag{F.7a}$$

$$q^2 = 4\left(e^2 - m^2\right)\,, \tag{F.7b}$$

where $q^2 \equiv 4ac - b^2$. The general form of our indefinte integral is

$$I = \frac{1}{\sqrt{a}} \ln\left[\sqrt{a\,x^2 - b\,x + c} + \sqrt{a}\,x + \frac{b}{2\sqrt{a}}\right] - \ln\left[D\right] \tag{F.8}$$

for $a > 0$ and where we must determine the integration constant D from boundary conditions appropriate to our problem. Substituting our parameter values (F.7a) yields

$$I = \ln\left[\,\rho\,\right] = \pm \ln\left[\sqrt{r^2 - 2mr + e^2} + r - m\right] - \ln\left[D\right]. \tag{F.9}$$

We obtain the value of $\ln[D]$ by observing that for large r ($r \to \infty$) we must have $r \to \rho$ and

$$\ln[\rho D] \approx \pm \ln[2r] \,, \tag{F.10}$$

which gives us $D = 2$, and with the choice of the positive sign, we have

$$2\rho = \sqrt{r^2 - 2mr + e^2} + r - m\,. \tag{F.11}$$

We note that

$$\begin{aligned} e^2 - m^2 &= \left[\sqrt{r^2 - 2mr + e^2} + (r - m)\right] \\ &\quad \times \left[\sqrt{r^2 - 2mr + e^2} - (r - m)\right] , \end{aligned} \tag{F.12}$$

which allows us to solve for $r(\rho)$. By using relation (F.11), we can eliminate the square roots from (F.12); the resulting expression for $r(\rho)$ is

$$\begin{aligned} r &= \rho + m + \frac{m^2 - e^2}{4\rho} = \frac{4\rho^2 + 4m\rho + m^2 - e^2}{4\rho} \\ &= \frac{(2\rho + m)^2 - e^2}{4\rho} \,, \end{aligned} \tag{F.13}$$

which has the proper Schwarzschild isotropic coordinate limit

$$r = \rho\left(1 + \frac{m}{2\rho}\right)^2 \tag{F.14}$$

when $e = 0$. Since $\lambda = r/\rho$, from (F.13), we have

$$\lambda = \frac{1}{4\rho^2}\left[4\rho(\rho + m) + \left(m^2 - e^2\right)\right] . \tag{F.15}$$

With these forms for $r(\rho)$ and $\lambda(\rho)$, we can rewrite (F.2) in the required isotropic form. The coefficient of the squared "time" differential, dx_0^2, is given, in terms of the isotropic coordinate, as

$$\begin{aligned} 1 - \frac{2m}{r} + \frac{e^2}{r^2} &= \frac{r^2 - 2mr + e^2}{r^2} \\ &= \frac{(r - m)^2 - m^2 + e^2}{r^2} \end{aligned}$$

$$= \frac{\left[(\rho+m)+\frac{m^2-e^2}{4\rho}-m\right]^2-\left(m^2-e^2\right)}{\left[(\rho+m)+\frac{m^2-e^2}{4\rho}\right]^2}$$

$$= \frac{\left[\rho+\frac{\left(m^2-e^2\right)}{4\rho}\right]^2-\left(m^2-e^2\right)}{\left[(\rho+m)+\frac{\left(m^2-e^2\right)}{4\rho}\right]^2}$$

$$= \frac{\left[4\rho^2+\left(m^2-e^2\right)\right]^2-(4\rho)^2\left(m^2-e^2\right)}{\left[4\rho\,(\rho+m)+(m^2-e^2)\right]^2}$$

$$= \frac{\left[4\rho^2-\left(m^2-e^2\right)\right]^2}{\left[4\rho\,(\rho+m)+(m^2-e^2)\right]^2}\,. \tag{F.16}$$

To reduce the growth of expressions, we can define

$$A(\rho,m,e) = \left[4\rho^2-\left(m^2-e^2\right)\right], \tag{F.17a}$$
$$B(\rho,m,e) = \left[4\rho\,(\rho+m)+\left(m^2-e^2\right)\right]. \tag{F.17b}$$

In the derived isotropic coordinates, the line element (F.2) becomes:

$$\begin{aligned} ds^2 &= \left[\frac{A(\rho,m,e)}{B(\rho,m,e)}\right]^2 dx_0^2 \\ &- \left[\frac{B(\rho,m,e)}{4\rho^2}\right]^2\left[d\rho^2+\rho^2\left(d\theta^2+\sin^2\theta\,d\phi^2\right)\right]. \end{aligned} \tag{F.18}$$

We now convert to rectilinear coordinates by rewriting

$$\rho^2\left(d\theta^2+\sin^2\theta\,d\phi^2\right) = dx_1^2+dx_2^2+dx_3^2\,, \tag{F.19}$$

where we use rectilinear coordinates to express the complete equivalence of all spatial directions. Factoring the polynomials gives our final result

$$\begin{aligned} ds^2 = &+\left[\frac{(2\rho+m)\,(2\rho-m)+e^2}{(2\rho+m+e)\,(2\rho+m-e)}\right]^2 dx_0^2 \\ &-\left[\left(1+\frac{m}{2\rho}\right)^2-\frac{e^2}{4\rho^2}\right]^2\left[dx_1^2+dx_2^2+dx_3^2\right]. \end{aligned} \tag{F.20}$$

This Appendix is referenced by **Chapter 9, Section H.**

APPENDIX G

THE CREMONA TRANSFORMATION

The Cremona transformation that we referred to in **Section 11.L**, shown in **Figure G.1**, maps the X–plane with projective coordinates x^1, x^2, x^3 into the Y–plane with projective coordinates y^1, y^2, y^3. When none of the coordinates vanish the map is given by

$$x^1 : x^2 : x^3 = \frac{1}{y^1} : \frac{1}{y^2} : \frac{1}{y^3} , \tag{G.1}$$

and

$$y^1 : y^2 : y^3 = \frac{1}{x^1} : \frac{1}{x^2} : \frac{1}{x^3} . \tag{G.2}$$

As the figure shows, the fully extended sides of the coordinate triangle are mapped into the vertices of the other and vice versa.

We have seen in **Section 11.L** that the field strengths in the Bianchi spaces with vanishing vector are related to the eigenvalues of the Ricci tensor by a Cremona transformation. Going to metric coordinates one can normalize the coordinates x^1, x^2, x^3 by putting

$$x^1 + x^2 + x^3 = \frac{\sqrt{3}}{2} , \tag{G.3}$$

where the coordinates x^j measure the distance from the side $x^j = 0$ of an equilateral triangle with a side length of unity. These Viviani coordinates are all positive in the interior of the triangle. We obtain, in this way, a geometric classification, shown in **Figure G.2**, for the degeneracy cases of the Bianchi types of class A [MacCallum, 1979].

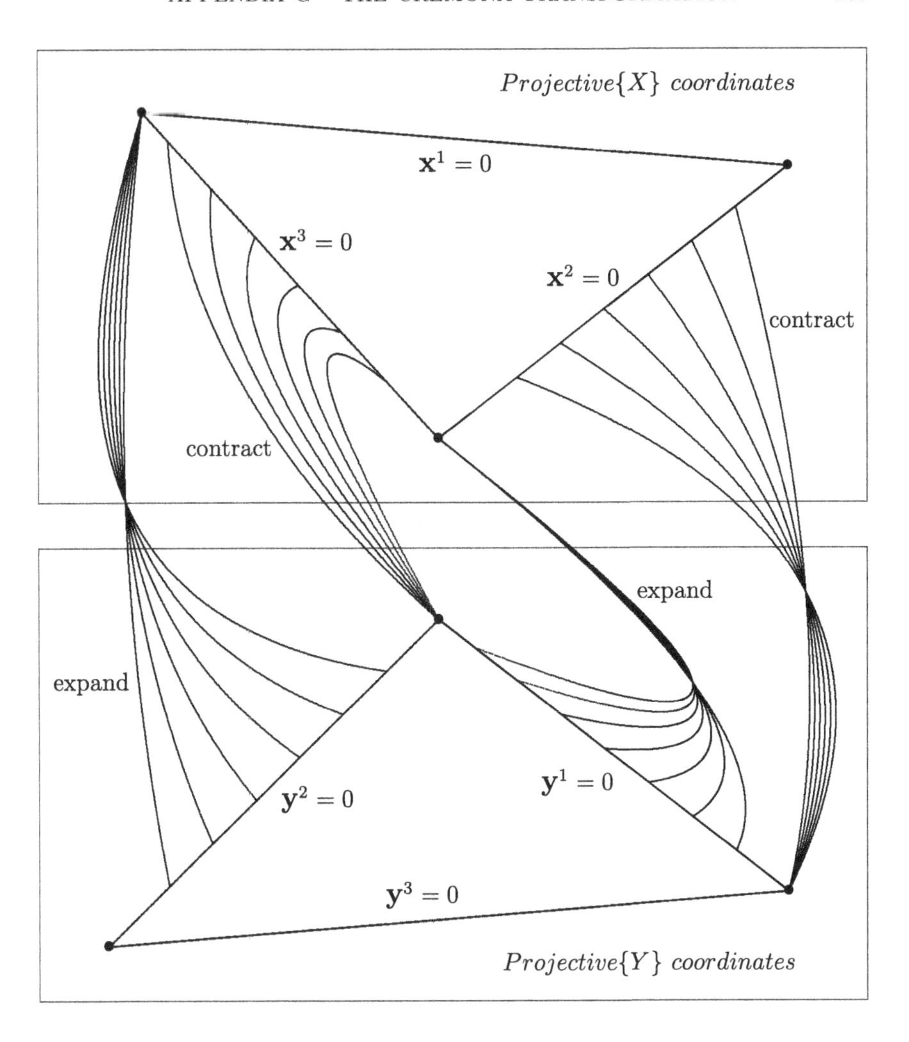

Figure G.1 The Cremona Transformation. The Cremona transformation maps the X–plane of the field strengths α, β, γ (which are essentially the structure constants of the Lie group) into the Y–plane of the eigenvalues of the Ricci tensor times $-1/2$. [Adapted from Mumford, 1976, p. 31]

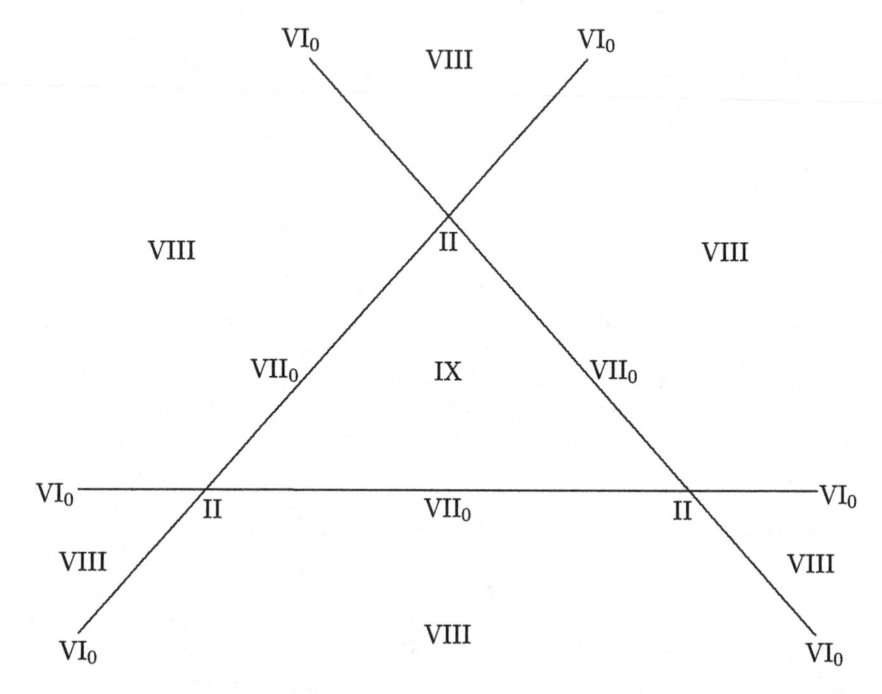

Figure G.2 Bianchi Class A Degeneracy Pattern. This diagram shows the pattern of degeneracy for the Bianchi types of class A.

These types are given in the following table where N_1, N_2, N_3 are MacCallum's equivalent to our α, β, γ of **Section 11.L.**

Class	*Type*	N_1	N_2	N_3
A	*I*	0	0	0
	II	1	0	0
	VI_0	0	1	−1
	VII_0	0	1	1
	VIII	1	1	−1
	IX	1	1	1

This Appendix is referenced by **Chapter 11, Section L** and **Chapter 12, Section A**.

APPENDIX H

HESSENBERG'S "VECTORIAL FOUNDATION OF DIFFERENTIAL GEOMETRY"

When Einstein discovered teleparallelism in 1928 he was apparently not aware of the fact that the teleparallel Ricci Grid had been discovered a dozen years earlier. The man who first recognized this possibility was a professor of mathematics in Breslau, well known in the profession for his work on the foundations of geometry. In a paper finished in June 1916 Gerhard Hessenberg replaced Christoffel's cumbersome calculations by an invariant co-vector method that reached later an even more elegant form in Élie Cartan's papers. Hessenberg discovered the torsion tensor, T_{kjl}, and also the contortion tensor. By introducing auto-parallel curves for a teleparallel Ricci Grid he proved that the torsion tensor vanishes if and only if all auto-parallel curves are geodesics, that is, shortest lines in the Riemann metric. Hessenberg was the first to discover that the geometry of a gravitational field is characterized by the torsion of teleparallelism.

I have nowhere seen his discovery of this special kind of torsion acknowledged.

Hermann Weyl, who missed finding torsion when generalizing the notion of connections, refers to Hessenberg only by crediting him with the proof that the symmetry of the Riemann tensor in its first and second pairs of the indices follows from the cyclic symmetry in the last three indices.

Hessenberg introduces an n-leg, $\underline{p}_j$ $(j, k = 1, \ldots, n)$, into every point of a n-dimensional Riemannian manifold. I simplify his representation by taking these n-legs to be orthonormal. Then all indices can be kept downstairs. I refer to equations in his paper by putting them into square brackets "[..]". This gives his equation [24]

$$\underline{p}_j \cdot \underline{p}_k = \delta_{jk} \,. \tag{H.1}$$

He defines the differential one-form db_{jk} in [39] by (I lower his indices according to the footnote on his page [198])

$$d\underline{p}_j \cdot \underline{p}_k \equiv db_{jk} \tag{H.2}$$

skew-symmetric in indices j and k. His equation [41] gives

$$db_{jk} + db_{kj} = 0\,. \tag{H.3}$$

Hessenberg calls the differential one-form db_{jk} the "Orientation Tensor". Nowadays one would write

$$d\underline{p}_j = \omega_{jl}\,\underline{p}_l\,, \quad d\underline{p}_j \cdot \underline{p}_k = \omega_{jk}\,, \quad \text{where} \quad \omega_{jk} + \omega_{kj} = 0 \tag{H.4}$$

where ω_{jk} is now the connection one-form. We thus have to identify db_{jk} with ω_{jk}. He next introduces in [47] a cogredient differential $d\underline{A}$ of a tensor A with α indices

$$\underline{A} = A_{j_1 \ldots j_\alpha}\,\underline{p}_{j_1} \cdots \underline{p}_{j_\alpha} \tag{H.5}$$

in terms of the covariant differentials $\delta A_{j_1 \ldots j_\alpha}$

$$d\underline{A} = \delta A_{j_1 \ldots j_\alpha}\,\underline{p}_{j_1} \cdots \underline{p}_{j_\alpha} \tag{H.6}$$

where this differential is defined by [48]

$$\delta A_{j_1 \ldots j_\alpha} = dA_{j_1 \ldots j_\alpha} - db_{j_1 k}\,A_{k \ldots j_\alpha} - \ldots - db_{j_\alpha k} A_{j_1 \ldots k}\,. \tag{H.7}$$

In his Section 20, on page [205], he introduces differential one-forms ω_j that give the Riemannian metric

$$ds^2 = \omega_j\,\omega_j\,. \tag{H.8}$$

These differential forms ω_j are his $u^{j\rho}\,dt_\rho$ (confusingly, also denoted as du^j). They are dual to his orthonormal vectors $\underline{p}_k$

$$\omega_j(\underline{p}_k) = \delta_{jk}\,. \tag{H.9}$$

Hessenberg's equation [87] is the necessary and sufficient condition that all straightest lines are geodesics in his more general geometry. In this case the connection form specializes to ω'_{jk} obeying what appears now as the first Cartan structural equation for vanishing torsion

$$0 = -\,d\omega_j + \omega'_{jk} \wedge \omega_k\,. \tag{H.10}$$

His notation "$D_{12}\, u^j$" precisely explained in the first footnote on page [211] with the opposite sign against Cartan's "d"-operator. In this case Hessenberg's one-form db_{jk} specializes into the Levi-Civita connection form ω'_{jk}. On the other hand Hessenberg's equation [94] reads now

$$- d\omega_j = \frac{1}{2} U_{ljm}\, \omega_l \wedge \omega_m \,, \quad U_{ljm} = - U_{mjl} \,. \tag{H.11}$$

This equation according to Cartan's first structural equation defines the right-hand side of (H.10) as the negative of the torsion form Θ_j for a vanishing connection that defines the teleparallelism of Hessenberg's straightest lines

$$\frac{1}{2} U_{ljm}\, \omega_l \wedge \omega_m = - \Theta_j = - d\omega_j + \omega_{jk} \wedge \omega_k \,, \quad \omega_{jk} = 0 \,. \tag{H.12}$$

Comparing now equations (H.9) and (H.10) he needs the development of the Levi-Civita connection form ω'_{jl} in terms of the ω_m. This gives the Ricci rotation coefficients h_{jlm}

$$\omega'_{jl} = h_{jlm} \omega_m \,, \quad h_{jlm} = - h_{ljm} \,. \tag{H.13}$$

This is the meaning of [97] where the Christoffel symbol of the second kind vanishes because of our simplification (H.1). From (H.9) and (H.10) follows now

$$0 = \frac{1}{2} U_{ljm}\, \omega_l \wedge \omega_m + h_{ljm}\, \omega_l \wedge \omega_m \,, \tag{H.14}$$

or simply Hessenberg's equation [98]

$$0 = U_{ljm} + h_{ljm} - h_{mjl} \,. \tag{H.15}$$

By cyclic interchange of the indices one obtains the equations

$$0 = U_{jml} + h_{jml} - h_{lmj} \tag{H.16}$$

and

$$0 = U_{mlj} + h_{mlj} - h_{jlm} \,. \tag{H.17}$$

Then (H.15) + (H.16) - (H.17) gives

$$2\, h_{mjl} = U_{ljm} + U_{jml} - U_{mlj} \,. \tag{H.18}$$

Changing to the indices used in Hessenberg's second footnote on page [211] one obtains (I have replaced the index "i" by "j")

$$2\,h_{jlk} \;=\; U_{ljk} \,+\, U_{klj} \,-\, U_{jkl}\,. \tag{H.19}$$

This does not agree with the expression in Hessenberg's second footnote on page [211]. The reason is that he re-defines U by turning its upper index into its first lower index instead of into its second as he stated as a general rule in the footnote on page [198]. The interchange of the first two indices in U turns (H.19) into U', skew-symmetric in its last two indices,

$$2\,h_{jlk} \;=\; U'_{jlk} \,+\, U'_{lkj} \,-\, U'_{kjl}\,. \tag{H.20}$$

This agrees with Hessenberg. With U'_{jlk} being the negative of the torsion tensor the tensor h_{jlk} becomes now the negative of the contorsion tensor. The contorsion tensor g_{lkj} becomes by cyclic permutation of the indices

$$2\,g_{jlk} \;=\; C_{lkj} \,+\, C_{kjl} \,-\, C_{jlk}\,, \tag{H.21}$$

where the tensor C_{lkj} is skew-symmetric in its first two indices. With

$$C_{lkj} \;=\; -\,U'_{jlk}\,, \tag{H.22}$$

comparing (19) and (20) gives

$$h_{jlk} \;=\; -\,g_{jlk}\,, \tag{H.23}$$

identifying the contortion tensor with the negative Ricci rotation coefficients of Christoffel's covariant derivative (which became called the Levi-Civita connection). In this way Hessenberg discovered a special case of torsion, namely, as we would nowadays say, a case where the symmetric part of the connection coefficients vanishes, a case of teleparallelism. The fact that he has a geometric interpretation for it in terms of auto-parallel curves shows that he is writing about a geometric phenomenon, the discrepancy between the straightest and the shortest curves in a geometry with torsion. This is exactly the example that Cartan used to explain his geometry to Einstein pointing out the distinction between rhumbs and geodesics on the sphere.

This Appendix is referenced by **Chapter 0, Section B.**

APPENDIX K

GRAVITATION IS TORSION

X. Comment

The text of this Appendix was presented by E. L. Schücking as an entry for the 2007 Babson Essays on Gravitation in slightly different form.

A. Abstract

The mantra about gravitation as curvature is a misnomer. The curvature tensor for a standard of rest does not describe acceleration in a gravitational field but the *gradient* of the acceleration (e.g., geodesic deviation). The gravitational field itself [Einstein, 1907] is essentially an accelerated reference system. It is characterized by a field of orthonormal four-legs in a Riemann space with Lorentz metric. By viewing vectors at different events having identical leg-components as parallel (teleparallelism) the geometry in a gravitational field defines torsion. This formulation of Einstein's 1907 principle of equivalence uses the same Riemannian metric and the same 1916 field equations for his theory of gravitation and fulfills his vision of General Relativity.

B. Why Gravity is Not "Manifestation of Space-Time Curvature" [Ciufolini and Wheeler, 1995]

By gravity one understands the force that draws apples toward the Earth. Its field strength near the surface of our planet is characterized by a radial acceleration $g = 9.8m/sec^2$ that it imparts to all matter.

The curvature of the pseudo-Riemannian spacetime manifold is described by its Riemann tensor that generalizes the intrinsic curvature of 2-dimensional surfaces defined by Carl Friedrich Gauss.

To investigate the relation between gravitation and curvature in General Relativity I shall discuss the falling of a mass point (e.g., an idealized apple) near the surface of an idealized spherical non-rotating Earth. If the mass point falls radially, its motion takes place in the spacetime "plane" spanned by a radial coordinate r and a time coordinate t.

Karl Schwarzschild and Johannes Droste discovered the geometry in this plane by finding the exact solution of Einstein's vacuum field equations for a spherically symmetric gravitational field. They wrote for the square of the line element ds

$$ds^2 = \left(1 - \frac{2m}{r}\right)(d(ict))^2 + \left(1 - \frac{2m}{r}\right)^{-1}(dr)^2 . \qquad \text{(K.B.1)}$$

In this formula ds describes the infinitesimal distance between two points with coordinates (t, r) and $(t + dt, r + dr)$. The radial distance r is fixed by stating that the area A of a concentric sphere with radius r is given by Archimedes' formula $A = 4\pi r^2$. The constant m stands for

$$m = G\frac{M}{c^2} \qquad \text{(K.B.2)}$$

where G is Newton's gravitational constant, M the mass of the Earth and c the speed of light in vacuo. Finally, i is short for $\sqrt{-1}$.

I use the structural equations of Élie Cartan [Nomizu, 1963] for calculation of the Gaussian curvature K in the t-r–plane. In a n-dimensional Riemann space Cartan writes the square of the line element ds in terms of n differential one-forms $\boldsymbol{\omega}_j$ with the index j running from 0 to $n - 1$

$$ds^2 = (\boldsymbol{\omega}_0)^2 + ... + (\boldsymbol{\omega}_{n-1})^2 . \qquad \text{(K.B.3)}$$

The first structural equation says

$$\Theta_j = d\boldsymbol{\omega}_j + \boldsymbol{\omega}_{jk} \wedge \boldsymbol{\omega}_k . \qquad \text{(K.B.4)}$$

Here $\boldsymbol{\Theta}_j$ are the torsion two-forms while $\boldsymbol{\omega}_{jk}$ are the connection one-forms. Einstein summation over the repeated index k is implied and the symbol '$\wedge$' denotes the skew-symmetric multiplication of differential forms. For a metric connection that preserves length under parallel transport we have

$$\boldsymbol{\omega}_{jk} = -\boldsymbol{\omega}_{kj} . \qquad \text{(K.B.5)}$$

In our case $n = 2$ and we have, from (K.B.1) and (K.B.3), the two differential forms

$$\boldsymbol{\omega}_0 = \left(1 - \frac{2m}{r}\right)^{1/2} d(ict), \quad \boldsymbol{\omega}_1 = \left(1 - \frac{2m}{r}\right)^{-1/2} dr . \qquad \text{(K.B.6a,b)}$$

If we assume that the torsion forms vanish, that is, $\boldsymbol{\Theta}_j = 0$, then the connection form $\boldsymbol{\omega}_{01}$ $(= -\boldsymbol{\omega}_{10})$ defines the Levi-Civita connection, subject to the two equations from (K.B.4):

$$0 = d\boldsymbol{\omega}_0 + \boldsymbol{\omega}_{01} \wedge \boldsymbol{\omega}_1 \,, \qquad 0 = d\boldsymbol{\omega}_1 - \boldsymbol{\omega}_{01} \wedge \boldsymbol{\omega}_0 \,. \tag{K.B.7a,b}$$

According to (K.B.6) the differential two-form $d\boldsymbol{\omega}_1$ vanishes. The second equation (K.B.7b) tells us that the connection form $\boldsymbol{\omega}_{01}$ must be proportional to the one-form $\boldsymbol{\omega}_0$, say, $\boldsymbol{\omega}_{01} = \lambda\, \boldsymbol{\omega}_0$, with some scalar function λ. Inserting this relation into the first equation (K.B.7) gives, with (K.B.6),

$$0 = \frac{m}{r^2}\left(1 - \frac{2m}{r}\right)^{-1/2} dr \wedge d(ict) + \lambda\, d(ict) \wedge dr \,. \tag{K.B.8}$$

We read off that

$$\lambda = \frac{m}{r^2}\left(1 - \frac{2m}{r}\right)^{-1/2} . \tag{K.B.9}$$

The connection form $\boldsymbol{\omega}_{01}$ becomes

$$\boldsymbol{\omega}_{01} = \lambda\, \boldsymbol{\omega}_0 = \frac{m}{r^2}\left(1 - \frac{2m}{r}\right)^{-1/2} \boldsymbol{\omega}_0 = \frac{m}{r^2}\, d(ict) \,. \tag{K.B.10}$$

The curvature follows from Cartan's second structural equation, giving for $n = 2$, the curvature two-form $\boldsymbol{\Omega}_{01}$

$$\boldsymbol{\Omega}_{01} = R_{0101}\, \boldsymbol{\omega}_0 \wedge \boldsymbol{\omega}_1 = d\boldsymbol{\omega}_{01} = \frac{2m}{r^3}\, \boldsymbol{\omega}_0 \wedge \boldsymbol{\omega}_1 = K\, \boldsymbol{\omega}_0 \wedge \boldsymbol{\omega}_1 \,. \tag{K.B.11}$$

By (K.B.11) the curvature form defines the component R_{0101} of the Riemann tensor and the Gaussian curvature K

$$R_{0101} = K = \frac{2m}{r^3} \,. \tag{K.B.12}$$

For a discussion of the connection form $\boldsymbol{\omega}_{01}$, I introduce the reference frame $\mathbf{e}_k$. It consists of n orthonormal vector fields $\mathbf{e}_k$ tangent to the manifold that are dual to Cartan's differential one-forms $\boldsymbol{\omega}_j$

$$\boldsymbol{\omega}_j(\mathbf{e}_k) = \delta_{jk} \,, \tag{K.B.13}$$

where Kronecker's δ_{jk} is the unit matrix. Katsumi Nomizu's process of covariant differentiation $\nabla_{\mathbf{v}}\mathbf{w}$ of a vector field $\mathbf{w}$ with respect to a tangent vector $\mathbf{v}$ gives, for $\mathbf{w} = \mathbf{e}_k$,

$$\nabla_{\mathbf{v}}\mathbf{e}_k = \mathbf{e}_j\, \boldsymbol{\omega}_{jk}(\mathbf{v}) \,, \tag{K.B.14}$$

defining the connection one-forms $\boldsymbol{\omega}_{jk}$. The covariant derivative of the timelike vector field $\mathbf{e}_0$ with respect to $\mathbf{v} = \mathbf{e}_0$ defines the negative of the geodesic acceleration, $-a\,\mathbf{e}_1$,

$$\nabla_{\mathbf{e}_0}\mathbf{e}_0 \;=\; -\,\mathbf{e}_1\,\omega_{01}(\mathbf{e}_0) \;=\; -\,\lambda\,\mathbf{e}_1 \;=\; -a\,\mathbf{e}_1\,, \tag{K.B.15}$$

where a is the magnitude of the geodesic acceleration. This equation shows that

$$a \;=\; \lambda \;=\left(\frac{m}{r^2}\right)\left(1-\frac{2m}{r}\right)^{-1/2}. \tag{K.B.16}$$

Calculating the radial gradient of the geodesic acceleration da/ds, we obtain

$$\frac{da}{ds} =\left(1-\frac{2m}{r}\right)^{1/2}\frac{da}{dr} \;=\; -\left(\frac{2m}{r^3}\right)\left(1-\frac{3m}{2r}\right)\left(1-\frac{2m}{r}\right)^{-1}. \tag{K.B.17}$$

When the radius r is large compared to the Schwarzschild radius $2m$, the geodesic acceleration a becomes Newton's g/c^2 and da/ds its gradient. We identify thus the geodesic acceleration with the gravitational field strength and the spacetime curvature $2m/r^3$ with its gradient. While the geodesic acceleration has dimension of inverse length, the spacetime curvature has dimension of inverse length squared and is thus not suitable to manifest gravity. Even the square root of the curvature, a sort of "radius of curvature", bears no order of magnitude relation to the geodesic acceleration as one can easily infer by putting numbers into the equations for the surface of the Earth. A drastic proof of this fact is the simple observation that by letting m and r going to infinity in (K.B.16), but keeping the geodesic acceleration a fixed, one can have arbitrary acceleration for vanishing curvature. This shows that there are gravitational fields of arbitrary strength in Minkowski spacetime.

The statement that gravity is a manifestation of spacetime curvature is misleading. And misleading too are the illustrations in physics texts, or models in exhibitions, purporting to explain Einstein's theory of gravitation by having balls orbiting on curved surfaces as fake planets or satellites. Equally flawed are the "proofs" in current texts that the gravitational redshift near the Earth's surface shows that spacetime is curved.

C. The Geometry of the Gravitational Field

In 1928 Einstein discovered distant parallelism. This is the geometry of the gravitational field as confirmed by the Pound-Rebka experiment in 1960. Riemann's geometry can be interpreted in terms of distant parallelism as follows: Above, we had set the torsion forms $\boldsymbol{\Theta}_j$ equal to zero for determining the $\boldsymbol{\omega}_{jk}$ of the Levi-Civita connection. We can write instead

$$\Theta_j = -\boldsymbol{\omega}_{jk} \wedge \boldsymbol{\omega}_k = d\boldsymbol{\omega}_j \tag{K.C.1}$$

for Einstein's distant parallelism that has torsion but vanishing connection forms. Here vectors are now considered to be equal and parallel if they have the same components with respect to the given orthonormal frame. In particular, the frame vector $\mathbf{e}_0$ becomes now tangent to timelike worldlines that provide the standard of rest in the gravitational field. Time is now no longer measured along geodesics.

In a rectangle of the four points (r_1, t_1), (r_1, t_2), (r_2, t_1), (r_2, t_2) its opposite timelike sides show, according to (1), a ratio of length

$$\left[\frac{(1-2m/r_2)}{(1-2m/r_1)}\right]^{1/2} \approx 1 + m\,\frac{r_2 - r_1}{{r_1}^2}\,. \tag{K.C.2}$$

This failure of a metric rectangle to close is precisely the result of the Pound-Rebka experiment. It also demonstrates that the geometry of a gravitational field can be described by distant parallelism (torsion) without appealing to curvature. Curvature terms on the left hand side of (K.C.2) are of higher order in r_1.

The interpretation of Einstein's theory in terms of distant parallelism uses the same metric, variational principle, and field equations, but gives a clear definition of a gravitational field, thus providing a satisfying account of General Relativity, a vision that had shriveled to not much more than "Einstein's Field Equations".

This Appendix is referenced by **Chapter 5, Section F**.

APPENDIX R

REFERENCES

Boldface indicates the character string used as a key to these references in the chapter and section given in the brackets.

Adler, 1975 [9.H]
Adler Ronald; Bazin, Maurice; Schiffer, Menahem
Introduction to General Relativity (2nd.ed.)
McGraw-Hill Book Co.; New York, NY (1975)

Bergmann, 1976 [0.B]
Bergmann, Peter G.
Introduction to the Theory of Relativity
Dover; New York, NY (1976/1942)

Bianchi, 1918 [7.A]
Bianchi, Luigi
Lezioni Sulla Teoria dei Gruppi Continui Finiti di Trasformazioni
Pisa, Italy (1918)

Bondi, 1979 [4.A]
Bondi, Hermann
in: "Relativity, Quanta, and Cosmology" vol. **1** p. 181
editors: M. Pantaleo and F. de Finis
Johnson Reprint Corporation; New York (1979)

Bowler, 1976 [9.L]
Bowler, M. G.
Gravitation and Relativity
Pergamon; Oxford, UK (1976)

Cartan, 1922 [2.A]
Cartan, Élie
"*Sur une généralisation de la notion de courbure de Riemann et les espaces à torsion*",
Comptes Rendus (Paris), **174** (1922), pp. 593–595;
English translation by G. D. Kerlick in:
"Cosmology and Gravitation: Spin, Torsion, Rotation, and Supergravity"
edited by P. G. Bergmann and V. de Sabbata
Plenum Press, New York (1980)

Cartan, 1923 [2.A]
Cartan, Élie
"*Sur les variétés à connexion affine et la théorie de la relativité généralisée (premiére partie)*" [On Manifolds with an Affine Connection and the Theory of General Relativity (first part)]
Ann. École Norm. Sup. **40** pp. 325–412
translated by Anne Magnon and Abhay Ashtekar with a Foreword by Andrzej Trautman; Bibliopolis, Naples, Italy (1986)

Cartan, 1924 [2.A]
Cartan, Élie
"*Sur les varietés à connexion affine et la théorie de la relativité généralisée (suite partie)*" [On Manifolds with an Affine Connection and the Theory of General Relativity (final part)]
Ann. École Norm. Sup. **41** pp. 1–25
translated by Anne Magnon and Abhay Ashtekar with a Foreword by Andrzej Trautman; Bibliopolis, Naples, Italy (1986)

Cartan, 1926 [1.E]
Cartan, Élie; Schouten, J. A.
"*On the geometry of the group manifold of simple and semi-simple groups*"
Proc. Akad. van Wetens., Amsterdam, Proc., vol. 29, pp. 803–815 (1926)

Cartan, 1927 [0.B][9.B]
Cartan, Élie
Riemannian Geometry in an Orthogonal Frame
World Scientific; River Edge, NJ (2002) [See p. 162.]

Cartan, 1930 [4.C]
Cartan, Élie
"*Notice historique sur la notion de parallélisme absolu*"
Math. Annalen vol. 102 pp. 698–706 (1930)

Carter, 1972 [9.G]
Carter, Brandon
Black Hole Equilibrium States
in: DeWitt, C.; DeWitt, B. (eds.)
Les Houches Lectures 1972: Black Holes
Gordon & Breach; New York (1973)

Ciufolini, 1995 [K.B]
Ciufolini, I.; Wheeler, J. A.
Gravitation and Inertia
Princeton University Press (1995) p. 1:
"gravity is a manifestation of **spacetime curvature**".

DeWitt-Morette, 1977 [0.B]
DeWitt-Morette, Cecile; Choquet-Bruhat, Yvonne; Dillard-Bleick, Margaret
Analysis, Manifolds, and Physics (rev. ed.)
Elsevier/North-Holland; New York, NY (1977)

EDM2, 1993 [2.A]
Mathematical Society of Japan
Encyclopedic Dictionary of Mathematics, second edition
The MIT Press, Cambridge, MA (1993); vol. 2, §.417 Tensor Calculus, p. 1571

Einstein, 1907 [0.A][13.C][K.A]
Einstein, Albert
"*Über das Relativitätsprinzip und die aus demselben gezogenen Folgerungen*"
[On the Relativity Principle and the Conclusions drawn from it];
Jahrbuch der Radioaktivität und Elektronik, **4**, pp. 411–462 (1907).
An English translation is available in
"The Collected Papers of Albert Einstein", vol. 2
A. Beck, translator; P. Havas, consultant
Princeton University Press, Princeton, NJ, (1989) pp. 252–311
The quote is on p. 302.

Einstein, 1908 [0.B][1.A]
Einstein, Albert
Jahrbuch der Radioaktivität und Elektronik, **5**, pp. 98–99 (1908)
An English translation is available in "The Collected Papers of Albert Einstein";
A. Beck, translator; P. Havas, consultant
Princeton University Press, Princeton, NJ, (1989);
vol. 2; pp. 316–317; The quote is on p. 317

Einstein, 1911 [0.B][0.E][4.E]
Einstein, Albert
Über den Einfluss der Schwerkraft auf die Ausbreitung des Lichtes
[On the Influence of Gravitation on the Propagation of Light]
Ann. Phys. (Germany); v.**35** pp. 898–908 (1911)
An English translation is available in:
"The Collected Papers of Albert Einstein, vol. 3 ";
Anna Beck, translator; Don Howard, consultant;
Princeton University Press, Princeton, NJ, (1993)

Einstein, 1929 [4.C]
Einstein, Albert
in: Élie Cartan – Albert Einstein; Letters on Absolute Parallelism 1929–1932
translated by Jules Leroy and Jim Ritter
Princeton University Press and Académie Royale de Belgique (1979)
The quote is on p. 11 in the letter dated 10 May 1929.

Eisenhart, 1925 [1.E]
Eisenhart, L. P.
"*Linear connections of a space which are determined by simply transitive groups*"
Nat. Acad. Sci., Proc., vol. 11, pp. 246–250 (1925)

Ellis, 1969 [0.B]
Ellis, G. F. R.; MacCallum, M. A. H.
"*A Class of Homogeneous Cosmological Models*"
Comm. Math. Phys.; v.**12** pp. 108–141 (1969)

Fermi, 1922 [2.C]
Fermi, Enrico
"*Sopra i fenomeni che avvengono in vicinanza di una linea oraria*"
[On the phenomena that occur in the neighborhood of a time-like worldline]
Rendiconti Acc. Lincei; **31**(1), pp. 21–23, pp. 51–52, pp. 101–103 (1922)
Reprinted in Fermi's collected papers which were published by U. Chicago Press;
"*Note e Memorie*", vol. 1 (1962)

Flanders, 1963 [0.B][5.C]
Flanders, Harley
Differential Forms
Academic Press; New York, NY (1963)

Flückinger, 1974 [0.A]
Flückinger, Max
Albert Einstein in Bern
Bern. Paul Haupt. (1974) p. 63

Fock, 1964 [9.D]
Fock, V.
The Theory of Space, Time, and Gravitation (2nd.ed.)
Pergamon Press; New York, NY (1964)

Geroch, 1969 [0.B][9.A]
Geroch, Robert
"*Limits of Spacetimes*"
Commun. Math. Phys.; v.**13** p. 180 (1969)

Gogala, 1980 [4.D]
Gogala, Borut
"*Torsion and Related Concepts: An Introductory Overview*"
Int. J. Theor. Phys. v.**19**, p. 573 (1980)

Griffiths, 1978 [11.L]
Griffiths, Phillip; Harris, Joseph
Principles of Algebraic Geometry
John Wiley; New York, NY (1978)

Guggenheimer, 1963 [C.0]
Guggenheimer, Heinrich W.
Differential Geometry
Dover; New York, NY (1963/1977)
[see Figures 11–5 and 11–5 on p. 277]

Hawking, 1975 [0.B]
Hawking, S. W.
"*Particle Creation by Black Holes*"
Comm. Math. Phys.; v.**43** n. 3 p. 199 (1975)

Hicks, 1965 [4.D]
Hicks, Noel J.
Notes on Differential Geometry
D. Van Nostrand Co., Inc.; Princeton, NJ (1965) p. 59

Honig, 1974 [10.E]
Honig, E.; Schücking, E. L.; Vishveshwara, C. V.
"*Motion of Charged Particles in Homogeneous Electric Fields*"
J. Math. Phys.; v.**15** n. 6 p. 774 (June 1974)

Israel, 1966 [9.D]
Israel, Werner
"*New Interpretation of the Extended Schwarzschild Manifold*"
Phys. Rev.; v. **143** n. 4 p. 1016 (March 1966)

Koszul, 1950 [3.C]
Koszul, Jean-Louis
"*Homologie et cohomologie des algebres de Lie*
Bulletin de la Société Mathématique 78: p. 65127 (1950)

Lee, 1986 [0.B]
Lee, T. D.
"*Are Black Holes Black Bodies?*"
Nuc. Phys. B; v. **234** n. 2/3 p. 437 (17 February 1986)

Levi-Civita, 1977 [E.0]
Levi-Civita, Tullio
The Absolute Differential Calculus
Dover; New York, NY (1977)

Levi-Civita, 1917 [0.B][9.E]
Levi-Civita, Tullio
"ds^2 *Einsteiniani in Campi Newtoniani*"
R. C. Accad. Lincei (5) V.26 P.519 (1917)
in: Opere Matematiche di Tullio Levi-Civita v. **4** (1917–1928) p. 89
Zanichelli, Nicola (ed.)
Accademia Nazionale dei Lincei; Bologna, Italy (1960)

MacCallum, 1985 [9.A]
MacCallum, M. A. H.
"*On Some Einstein–Maxwell Fields of High Symmetry*"
Gen. Rel. Grav.; v. **17** n. 7 p. 659 (July 1985)

MacCallum, 1979 [7.A][G.0]
MacCallum, M. A. H.
"*Anisotropic and Inhomogeneous Relativistic Cosmologies*"
in: Hawking, S. W.; Israel, W. (eds.)
General Relativity: An Einstein Centenary Survey
Cambridge U. Pr.; New York, NY (1979)

Misner, 1973 [4.D][5.C]
Misner, Charles W.; Thorne, Kip S.; Wheeler, John A.
Gravitation
Freeman; New York, NY (1973)

Mumford, 1976 [G.0]
Mumford, David
Algebraic Geometry I: Complex Algebraic Varieties
Springer–Verlag; New York, NY (1976) p. 31

Nester, 1984 [4.D]
Nester, James
"*Gravity, Torsion and Gauge Theory*"
in: ed. by H. C. Lee
"An Introduction to Kaluza-Klein Theories"
World Scientific; Singapore (1984) p. 98

Newton, 1687 [0.A]
Newton, Isaac
"*Principia Philosophiae Naturalis*"
The Principia, A New Translation
translated by: Cohen, I. Bernhard; Whitman, Anne
University of California Press; Berkeley (1999); p. 807

Nomizu, 1963 [K.B]
Kobayashi, S.; Nomizu, K.
Foundations of Differential Geometry
Interscience, New York (1963) p. 121

Okun, 1991 [9.L]
Okun, Lev B.
"*The Concept of Mass*"
Physics Today; v. **42** n. 6 p. 31 (June 1989)

Pauli, 1929 [4.C]
Pauli, Wolfgang
"*Wissenschaftlicher Briefwechsel*"
edited by A. Hermann, K. V. Meyenn, V. F. Weisskopf
Springer-Verlag; New York (1979) p. 527
Also see the letter at CERN archive: einstein_0079.pdf

Pauli, 1981 [11.G][11.H]
Pauli, Wolfgang
Theory of Relativity
Dover; New York, NY (1921:1981)

Plebanski, 1976 [9.A]
Plebanski, Jerzy; Demianski, M.
"*Rotating, Charged, and Uniformly Accelerating Mass in General Relativity*"
Ann. Phys.; v. **98** p. 98 (1976)

Reissner, 1916 [9.F][F.0]
Reissner, H.
"*Über die Eigengravitation des elektrischen Feldes nach der Einsteinschen Theorie*"
Ann. Phys. (Lpz.); v. **50** pp. 106–120 (1916)

Rindler, 1966 [0.B]
Rindler, Wolfgang
"*Kruskal Space and the Uniformly Accelerated Frame*"
Am. J. Phys.; v. **34** n. 12 pp. 1174 (December 1966)

Runge, 1909 [0.B]
Runge, Carl
Göttinger Nachrichten n. 1 pp. 37–41 (1909)

Saletan, 1961 [9.A]
Saletan, Eugene J.
"*Contraction of Lie Groups*"
J. Math. Phys.; v.**2** n. 1 p. 1 (January 1961)

Sattinger, 1986 [A.C]
Sattinger, D. H.; Weaver, O. L.
Lie Groups and Algebras with Applications to Physics, Geometry, and Mechanics
Springer-Verlag; New York, NY (1986)
[see p. 113]

Schücking, 2007 [K.X][K.C]
Schücking, E. L.; Surowitz, E. J.
Einstein's Apple: His First Principle of Equivalence
[gr-qc/0703149]

Schücking, 2003B [13.A]
Schücking, Engelbert L.; Surowitz, Eugene J.; Zhao, J.
"A Diagram for Bianchi–Types B"
Unpublished

Schücking, 2003A [13.A]
Schücking, Engelbert L.; Surowitz, Eugene J.; Zhao, J.
"A Diagram for Bianchi A–Types"
Gen. Rel. Grav.; v. **35** n. 9 p. 1519 (September 2003)

Schücking, 2003 [13.A]
Engelbert Schücking; Behr, Christoph G.; Ellis, George F.R.;
Estabrook, Frank B.; Jantzen, Robert; Krasiński, A.;
Kundt, Wolfgang; Wahlquist, Hugo D.
"The Bianchi classification in the Schücking-Behr approach"
— Based on a seminar by Engelbert Schücking and Notes taken by Wolfgang Kundt
Gen. Rel. Grav.; v. **35** p. 475 (2003)

Schücking, 1985B [9.E]
Schücking, E. L.
"A Uniform Static Magnetic Field in Kaluza-Klein Theory"
in: Bergmann, Peter G.; De Sabbata, Venzo, editors
Topological Properties and Global Structure of Space-Time
NATO Advanced Study Institute; Erice, Italy; May 12–22, !985
Plenum Press; New York, NY (1986)

Schücking, 1985A [0.B]
Schücking, Engelbert L.
"The Homogeneous Gravitational Field"
Found. Phys.; v. **15** n. 5 p. 571 (May 1985)

Schutz, 1980 [3.A]
Schutz, Bernard F.
Geometrical Methods of General Relativity
Cambridge University Press; New York, NY (1980)

Schutz, 2009 [3.A]
Schutz, Bernard F.
A First Course in General Relativity, 2nd.ed.
Cambridge University Press; New York, NY (2009)

Stachel, 1980 [4.E]
Stachel, John
"The Rigidly Rotating Disk as the 'Missing Link' in the History of General Relativity"
in: Held, A.
General Relativity and Gravitation
Plenum Press; New York, NY (1980)
vol. 1 pp. 1–15

Stahl, 2008 [C.A]
Stahl. Saul
A Gateway to Modern Geometry: The Poincaré Half-Plane
Jones & Bartlett Publishers; Sudbury, MA (2008)

Stephani, 1980.1st [9.A]
Herlt, E.; Kramer, D.; MacCallum, M.; Stephani, H.
Exact Solutions of Einstein's Equations ; first edition
Cambridge U. Pr.; New York, NY (1980)
[See section 30.6.1 on p. 341.]

Stephani, 2003.2nd [9.A]
Herlt, Eduard; Hoenselaers, Cornelius; Kramer, Dietrich;
MacCallum, Malcolm; Stephani, Hans
Exact Solutions of Einstein's Equations ; second edition
Cambridge U. Pr.; New York, NY (1980)

Stukeley, 1936 [4.F]
Stukeley, William
Memoirs of Sir Isaac Newton's Life
Ed. by A. Hastings White
London (1936) p. 19

Synge, 1960 [0.B] [0.D]
Synge, John L.
Relativity: The General Theory
Elsevier/North-Holland; New York, NY (1960)
[See Preface pp. ix–x.]

Taub, 1986 [9.A]
Taub, A. H.
Personal communication (January 1986)

Taub, 1980 [9.A]
Taub, A. B
"Space-times with Distribution Valued Curvature Tensors"
J. Math. Phys.; v. **21** n. 6 p. 1423 (June 1980)

Unruh, 1976 [0.B] [0.D]
Unruh, W. G.
"Notes on Black-hole Evaporation"
Phys. Rev. **D14** pp. 870-892 (1976)

Wald, 1994 [0.D]
Wald, Robert M.
Quantum Field Theory in Curved Spacetime
University of Chicago Press; Chicago, IL (1994)

Walker, 1932 [2.C]
Walker, Arthur G.
"Relative Coordinates"
Proc. of the Royal Society Edinburgh, v. **52**, pp. 345–353 (1932)

Wigner, 1939 [8.E]
Wigner, E. P.
"On Unitary Representations of the Inhomogeneous Lorentz Group"
Ann. Math.; v.40 pp. 149–204 (1939)

Wigner, 1953 [9.A]
Wigner, E. P.; Inonu, E.
"On the Contraction of Groups and their Representations"
Proc. Natl. Acad. Sci. (U.S.); v.**39** n. 6 p. 510 (15 June 1953)

APPENDIX X

INDEX

This index is generated from the raw TEX input for the manuscript. Rather than the conventional page number, the index provides the chapter and section numbers in which a word appears with the numbers separated by a period. A zero is used for the section number when there is only one section; this occurs only in the appendices. When an entry contains an asterisk(*) for the section number, this indicates that the word in question appears in title or heading text.

APPENDIX N

NOTATIONS AND CONVENTIONS

A. Notation

The text and mathematical expressions of this work generally adhere to the following conventions but this statement of intention may include extensions to what actually appears in the body of the work.

$\mathbb{R}$ is the real numbers and real line.

$\mathbb{C}$ is the complex numbers and their plane.

V denotes a vector space with no dimension specified.

V^n denotes a vector space of dimension n.

$\mathcal{V}$ denotes a dual vector space with no dimension specified.

$\mathcal{V}^n$ denotes a dual vector space of dimension n.

$\mathbf{V}$ is a vector in vector space V.

$\mathbf{V}(\mathbf{p})$ is a vector in vector space V which is dependent on position $\mathbf{p}$.

$\mathbf{e}_j(\mathbf{p})$ is a basis vector which is dependent on position $\mathbf{p}$.

$\boldsymbol{\omega}$ denotes a form, an element of the covector space.

$\boldsymbol{\omega}(\mathbf{V})$ denotes a form applied to $\mathbf{V}$, an element of the vector space.

$\boldsymbol{\omega}_{(\mathbf{p})}$ when the form is dependent on position $\mathbf{p}$.

$\boldsymbol{\omega}^j_{(\mathbf{p})}$ a basis form dependent on position $\mathbf{p}$.

$\boldsymbol{\omega}^j_{(*)}$ when the basis form, basis or otherwise, is independent of position.

$[g_{ij}]$ is the matrix which has g_{ij} as its elements.

$\det [g_{ij}]$ is the determinant of the matrix which has g_{ij} as its elements.

$|u|$ is the absolute value of the variable u.

$|\mathbf{V}|$ is the magnitude of the vector $\mathbf{V}$.

$\mathbf{S}^1$ is the circle with no specific radius supplied or implied.

$\mathbf{S}^2$ is the 2-sphere, or just "the sphere".

$\mathbf{S}^n$ is the n-sphere.

B. Indices

Each calculation has its own selection of indices, however we follow the general patterns of common usage. The index domain for tensors is the ordered set $\{0, 1, 2, 3, \ldots, n\}$.

Lower case Greek characters denote the indices using this set. Lower case Greek characters also denote the set $\{1, 2, 3, \ldots, n\}$ when the metric is positive definite and does not contain an index associated with time. Upper case Greek letters are not used for indices.

Capitalized Latin characters will denote the set $\{1, 2, 3, \ldots, n-1\}$ when the metric is *not* positive definite and where we have used a complex time coordinate as the n-th coordinate.

C. Einstein Summation Convention

The Einstein summation convention is followed throughout: Repeated indices are summed over unless explicitly stated otherwise. We frequently adopt the use of orthonormal frames to bury the distinction between co-variant and contravariant quantities.

Printed in the United States
By Bookmasters